Earth and Environmental Sciences

Earth and Environmental Sciences

Edited by **John Wayne**

New York

Published by Callisto Reference,
106 Park Avenue, Suite 200,
New York, NY 10016, USA
www.callistoreference.com

Earth and Environmental Sciences
Edited by John Wayne

International Standard Book Number: 978-1-63239-155-1 (Hardback)

Contents

Preface

We have come to face a situation where new environmental troubles arise every day and we need to make some fast and binding decisions. In many cases information regarding geological processes is crucial to find out an appropriate explanation. Earth and environmental sciences is a dynamic field in which new problems arise with time and the old ones often re-emerge. The main aim of this book is to present the readers with a broad overview of Earth and Environmental Sciences by analyzing geology, geochemistry, seismology and hydrology. We are hopeful that this recent research will offer readers with a helpful foundation for discussing and analyzing specific environmental issues and will help in generating new techniques to tackle problems.

After months of intensive research and writing, this book is the end result of all who devoted their time and efforts in the initiation and progress of this book. It will surely be a source of reference in enhancing the required knowledge of the new developments in the area. During the course of developing this book, certain measures such as accuracy, authenticity and research focused analytical studies were given preference in order to produce a comprehensive book in the area of study.

This book would not have been possible without the efforts of the authors and the publisher. I extend my sincere thanks to them. Secondly, I express my gratitude to my family and well-wishers. And most importantly, I thank my students for constantly expressing their willingness and curiosity in enhancing their knowledge in the field, which encourages me to take up further research projects for the advancement of the area.

Editor

Part 1

Hydrogeology

1

Natural Radioactive Isotopes in Water in Relation with Geology and Hydrological Investigations in the Territory of Luxembourg

Zornitza Tosheva, Harald Hofmann and Antoine Kies
[1]University of Luxembourg,
[2]Monash University,
[1]Luxembourg
[2]Australia

1. Introduction

Luxembourg is situated in the middle of Europe and is surrounded by Belgium in the north-west, France in the south-west and south, and Germany in the east (Figure 1). It is one of the smaller countries in Europe with a surface area of 2.600 km² and a population of approximately 500.000 inhabitants. The northern third, known as 'Eislek', has a plateau character with an average elevation of 450 m, whereas the southern two-third are know as the 'Gutland', good land in English. This region has a more varied topography and an average elevation of 240 m.

Groundwater is the main water source for drinking water supply in the Grand Duchy of Luxembourg. About 60% of the country's drinking water comes from so called "Luxemburg Sandstone", which forms the most important aquifer. Most of the groundwater captures are natural springs or shallow wells and represent 80 % of the volume of this aquifer. The majority of the springs are distributed along the main river valleys. The unconfined part of the sandstone is located in the centre of the country and covers about 300 km² with a surface to water table depths ranging from 0 to 100 meters (Colbach, 2006).

Except from some bores in the north of Luxembourg and some in the south, natural springs are mainly used for public drinking purpose.

The radionuclide content of more than 300 springs and bores was determined as part of 2 monitoring campaigns, in a first step to relate the physico-chemical parameters with local geological conditions and in a second to obtain data on groundwater pollution regarding agricultural contaminants, such as pesticides, nitrate etc. The second campaign allows a general quality assessment of the groundwater resources in the country, which is also required by the European Water Framework Directive (EU, 2000).

Radio-nuclides are of interest for two reasons. Firstly, a high activity poses a health risk. An example is dissolved radon ($222Rn$) in groundwater that degasses when the water leaves the aquifer. Radon concentrations increase in the air in buildings where the water is trapped, such as spring trappings, or when used like in households. The inhalation of radon and the deposition of its daughter products in the lungs can cause lung cancer. Secondly, radio-nuclides can be useful tools as tracers in the hydrological / hydro-geological cycle.

Radioactive carbon and tritium are widely known and utilized tracers for groundwater dating and follows the diffusion and migration processes. The radioactive carbon and tritium are cosmogenic. Most of the cosmogenically produced radio-nuclides in the atmosphere are oxidized and become attached to aerosol particles. These particles act as condensation nuclei for the formation of cloud droplets and eventually coagulate to form precipitation. Similar is the mechanism of naturally produced ^{210}Pb, a long-lived decay product of the noble gas radon from the uranium decay series, which reach the atmosphere and are transported with the atmospheric circulations in form of particles until deposition on the surface with the precipitations. Other radioactive elements, besides these isotopes, such those of the uranium and thorium decay series, can be used as groundwater flow and mixing tracers.

Fig. 1. Geological map of the main units in Luxembourg. The area of the Devonian formations in the North forms the 'Eislek', whereas the Southern 'Gutland' is represented by the Triassic and Jurassic units.

The uranium and thorium series isotopes derive from the release out of uranium / thorium bearing minerals in the soils and rocks. These naturally occurring radio-isotopes are supplementary tracers to conventional major ion chemistry and stable isotopes in hydrological applications, e.g. creates a particularly interest such as surface water/groundwater interactions.

There had been some research on the use of radio-nuclides in these fields but still only little was known about their distribution and the mobility over the country. If there presence is ubiquitous, there are physical and chemical aquifer characteristics, such as porosity,

Natural Radioactive Isotopes in Water in
Relation with Geology and Hydrological Investigations in the Territory of Luxembourg

5

hydraulic conductivity or organic matter content that play major roles in their occurrence and migration in the environment.

This study does not aim to solve all questions, which are related to the distribution and mobility of radio-nuclides in the groundwater but it helps to get closer to some solutions.

In this work we present two different kind of investigations performed with the natural radioactive isotopes in relation with geological and hydrological studies.

The aim of the first part is to deliver a high resolved spatial radioactive isotope distribution maps and an analysis for the causes of activity variations. We rely on the major ion chemistry data and physical parameters that have been measured in Luxembourg's drinking water supplies. The relations of the radionuclide concentrations with the major ion chemistry were studied. Comparison of the obtained results versus local geology was followed in this approach.

Besides a dense spatial variability and chemical behavior along the hydro-geological regions, the second part of this work aims analyses of temporal variations of radio-isotopes, especially in the case of radon. These results are crucial for the use of radio-isotope data as groundwater and surface water tracers.

2. Sampling and methodology

The samples were taken in 2 litres, airtight glass bottles directly from the spring. Where possible they were taken right at the emergence of the water from the host rock. Water from wells and bores was pumped with the preinstalled pumps from the water suppliers. Purging was necessary because the sampled bores are production bores that are constantly in use. The bottles were filled completely without air space and stored in insulated coolers to avoid degassing during the transport to the laboratory.

The electrical conductivity and the temperature of the water was measured in-situ using a WTW i340 multi probe. The electrical conductivity was temperature corrected to 20°C.

Major ion chemistry was done at the National Water Laboratory of Luxembourg by ion chromatography for the anions and induced-coupled mass spectrometry for the cation content and trace elements.

The radio-nuclides were measured at the Radio-physics Laboratory of the University of Luxembourg. Radon (^{222}Rn) was measured a few hours after sampling by liquid scintillation counting (LSC) on a Perkin-Elmer "Guardian" Liquid Scintillation Counter, using 10 ml of Perkin-Elmer Ultima Gold F scintillator and 12 ml of sample. For some choosen samples 3H were measured by LSC after purification with Eichrom tritium columns. The other isotopes (^{238}U, ^{235}U, ^{234}U, ^{232}Th, ^{230}Th, ^{228}Th, ^{228}Ra, ^{226}Ra, ^{224}Ra, ^{210}Pb) were separated with sequential extraction (Tosheva et al. 2003, 2009) and ^{210}Po by electro-deposition on silver discs. After separation they were measured on Canberra alpha detector or LSC respectively.

3. Local geology and hydrogeology

3.1 Geology

In the official web page of Luxembourg geological survey (geology.lu) is written that Luxembourg shows on its small area of about 2500 km^2 a dense geological diversity.

As mentioned above, Luxembourg is orographicaly and geologically divided in two major natural regions, which are geologically different. On the one hand, there is the 'Eislek' in the

North, which is built up from Lower Devonian formations in Siegen facies (Figure 1). It belongs to the rhenohercynian block of the Eifel Mountains. The outcropping rocks consist mostly of folded and slated medium grained sediments with actual thickness of several thousand meters, transformed into schist, meta-sandstones, and bluish-grey low-grade methamorphosed shists sediments. Inter-beddings of quartzite sandstone occur in some parts of the formation. The geological substrate is covered by a thin soil layer of 0.5 to 1.5 m thickness.

On the other hand, there is the 'Gutland' in the South, which encompasses Mesozoic units forms the Perm-Triassic boundary to the Middle Jurassic. They belong to the units of the Paris Basin and extend about 150 km from the southern Eifel in Germany, via Luxemburg towards Belgium (Berners, 1983). The sequence starts with the Lower Triassic units of the 'Buntsandstein' with fluvial red bed sand-, siltstones and clays, followed by marls of the Upper Keuper (Rhät). However, the unit, which bears most drinking water springs in the country, is the 'Luxembourg Sandstone'. It is carbonate bound sandstone from the Middle Liassic (li), which can be assigned to a shallow marine shelf environment with wave influencing reworking and re-sedimentation (Berners, 1983). Heterogeneous layering boundaries can be observed within and in between the sandstone formation and the Psilonotas marls. It consists of light yellow, fine to medium grained sandstones with a low to medium carbonate content (15-50 %), which origins from the spartic cement and bioclastic components (Colbach, 2006). Some layers contain very well rounded pebbles (ca. 10 cm).

The Luxembourg sandstone is covered by younger formations in the South-West of the country. They encompass the iron Minette ore formation carbonate units of the Middle Dogger and carbonates of particularly the Upper Dogger. These south-west part ironstones are present in the form of limonitic and pisolithic surface formations.

3.2 Hydrography and hydrogeology

Luxembourg has a temperate climate with an average rainfall of 800 mm per year and an average temperature of 9°C. The land-use distribution is as follows: forested 34%, pastures 26%, agricultural-cultivated 23%, other 17%.

The majority of the samples come from the Luxembourg sandstone (grès de Luxembourg) aquifer, which is the most important aquifer for the drinking water supply in the country. It is generally limited on its bottom and top by the limestone-shale alternations of the Lorraine facies, although interbeddings with these exist (Colbach, 2005). The porosity of the aquifer is closely related to the carbonate content with approximately 5% in well cemented sandy limestones to 35% in poorly cemented sections, whereas water flow occurs in the well developed open fracture network. It is a dual-porosity aquifer and shows very heterogeneous permeabilites across its occurrence. The general flow pattern is dominated by the dip to the south-west of the unit.

The other hydrogeological units, which were encompassed in the campaigns, are Devonian shists and sandstone in Siegen facies, red-bed sandstones (Buntsandstein), Muschelcalc sandstone, dolomitic Keuper limestones, carbonate-bound sandstones from the middle Lias and iron-bearing limestones form the Dogger.

The Devonian schist and sandstones in Siegen and Ems build up the northern part of the country and belong to the rheno-hercynian block of the Rhenish Slate Mountains and the Ardennes Mountains. They are folded and fractured schist and sandstones with a low hydraulic conductivity. Flow and storage happens mostly in the fracture network.

Natural Radioactive Isotopes in Water in
Relation with Geology and Hydrological Investigations in the Territory of Luxembourg

7

The Buntsandstein in Luxembourg shows a sandy facies of fluvial environments with conglomerate and silt/clay interbeddings. Groundwater flow and storage happens in the fracture network as well. It has mixture permeability of fissures and pores. The dolomitic limestones of the Keuper are characterized by a more developed fracture network than the older formations. Carbonate dissolution has created a reasonable specific storage and hydraulic conductivity. The same hold true for the middle Lias carbonate-bound sandstones. The Dogger formations consist of oolithic iron ores from the Minette-type. The hydrogeology of the Minette rocks is complicated due to the mining activities in the past but general characteristics of the aquifer encompass a medium to high hydraulic conductivity. This aquifer is a dual porosity aquifer with a well developed open fracture network. The pore space in the matrix is high in the weathered areas and low in the non-weathered areas.

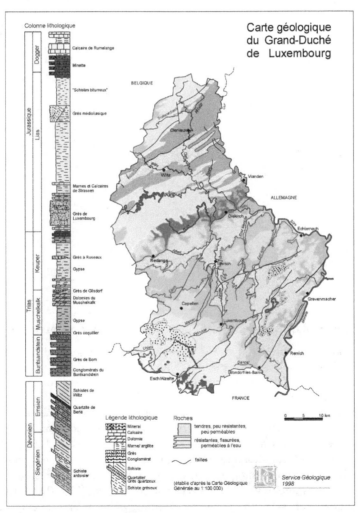

Fig. 2. Geo-morphological map of the main units in Luxembourg.

4. Results and discussion

The major ion chemistry of the water in Luxembourg is dominated by Ca^{2+}, Mg^{2+}, HCO_3^- and SO_4^{2-} and therefore the waters can be assigned to a calcium-bicarbonate type (figure 3). However Na^+, K^+ and Cl^- play a significant role, especially in places with higher permeability and fractures. Another not negligible contribution comes from NO_3^- ions, especially on the plateaus of Luxembourg Sandstones when agriculture affects the water table or organic matter is present in the rocks. The carbonate species are dominant in the carbonate units of the Luxembourg Sandstones and the younger Jurassic lime stones formations and carbonate units if Luxembourg's sandstones with carbonate cements and bioclastic components. Calcium values range from a few mg/l to over 200 mg/l, where values of the main aquifers, such as the Luxemburg sandstone, range from 80 to 200 mg/l. The bicarbonate follows the same trend with concentrations in between 1 to 40 mg/l.

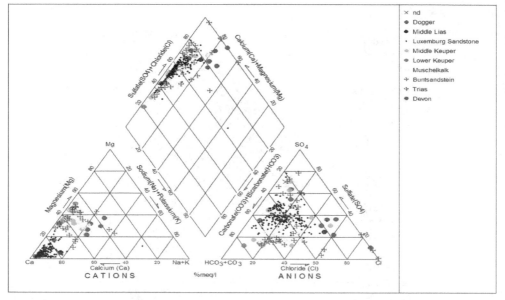

Fig. 3. Piper-diagram of the sampled water in the Spatial Monitoring campaign.

Radionuclide concentrations mainly vary with the local geology and in many cases are supported by different chemical mecanismus. Values for activity concentration of uranium vary from 0.2 to 74 mBq/L with predominant 10–30 mBq/L. They correspond to aquifers with similar mineralogy described by[Porceli et al., 2003].

Radium in natural water is derived mainly by interaction between the water and radium bearing materials, such as rock, soil and deposits due to its good solubility. As it forms sulfates, carbonates, nitrates and chlorides at different degrees and as such is a good indicator of the geochemical situation. In the analyzed samples the concentration varied from 3 to 35 mBq/L. Only five samples show significantly increased results, which can be explained with the different geological material in the aquifers (Devonian and Trias formations). Another data set with higher than median values was from a spring where the waters have a quite different chemistry and high salinity [Oliveira et al, 2006].

Natural Radioactive Isotopes in Water in
Relation with Geology and Hydrological Investigations in the Territory of Luxembourg

9

Lead mobility in water depends on the pH, hardness, sulfate and carbonate ion concentration and the presence of organic complexion agents [Oliveira et al, 2006]. In our study the spatial distribution of ^{210}Pb follows the same pattern as for ^{238}U and ^{230}Th (figure 4). These data prove that the investigated sources have almost no contact with the surface, mostly there has been no exchange between water horizons [Gonzalez-Labajo et al,2001] and the radio-nuclides originate from the surrounding rocks and soils.

The correlations among radio-nuclides and major ions were not significant, suggesting independence among the variables. However the analyzed waters were from a bicarbonate type; the main influence on the radionuclide distribution is due to chloride ions. Some explanation is related to the surprising fact that most of the waters contain non negligible iron and manganese concentrations.

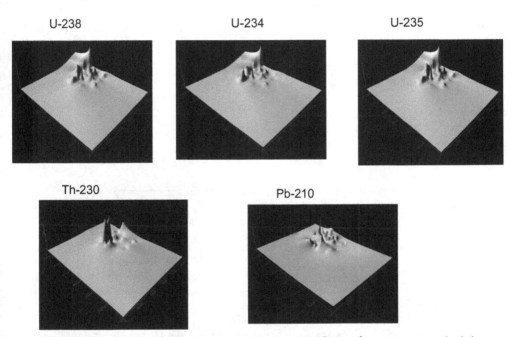

Fig. 4. Spatial distribution of uranium series isotopes over Luxembourg country. Activity concentration, represented on axe 'z' is in mBq/L

Correlation analysis show also no clear relation between calcium and magnesium content, however calcium is substituted well by strontium in almost all geological formations (figure 5). We obtain good correlation of strontium and radium radio-nuclides in Buntsandstein and Luxembourg sandstones formation where these form sulfates. In both formations reducing conditions exist in the deep wells.

In figure 6 are represented average values of main ions influencing radio-nuclide distribution over geological formations as well as the mean of different radioisotopes distributions over the same formations. The values are in 'ppm', except where for representation facility, they are converted in 'mg'. Radioactive elements are in 'mBq/L except radon that is in Bq/L. In Figure 6a are represented main cations and anions that may affect the solution of natural radio-nuclides, following the fact that generally Luxembourg waters are bicarbonate. In Figure 6b, after detailed analysis of the data are represented the main influencing radioactive distribution ions. Strontium and sulfates affect the radium distribution, as discussed previously; whereas iron and its supported copper and manganese are responsible for uranium and thorium distributions.

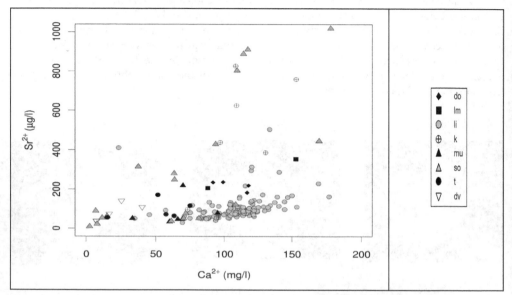

Fig. 5. Correlation between calcium and strontium content in analysed water vs. geological formation. The abbreviations are: do (dogger), km (keuper), li (liasic), mu (muschelcalc), so (buntsandstein 1), t (buntsandstein 2), dv (devonien)

In Figure 6c and 6d are represented the uranium and thorium series. The geological formations are also separated as for the Buntsandstein, where we observe a mixture of fissures and pores and permeability; besides an increasing mineralization from north to north-east. In the first ones we distinguish more sulfate and bicarbonate, while in the second more chlorides. This explains the low radium radio-isotopes and polonium contents in the bi-carbonate clay-sand and gypsum part, as well the presence of lead sulfate, and its dissolutions in mobile chlorides. Noticed is also the clear influence of iron and strontium in reducing conditions for the uranium and thorium presence. For similar reasons the sandstones of Luxembourg are presented into two different sub-formations. In the widely represented Luxembourg sandstones water circulation is favoured along vertically formed diaclases; the water table is charged continuously. Fractures are enlarged by water passages and by carbonate dissolutions; whereas the dissolved residue fraction in the sands constitutes a natural filter. As a consequence the reducing/oxidizing conditions change and induce the formation of sulfates where stable and radioactive thorium (80%) correlates with sodium, copper and manganese.

The $^{226}Ra/^{238}U$ ratios over the country range from the unique very low value of 0.3 to the values of 20, having a mean value not significantly different from 1. Similar is the range of ratios between ^{210}Po and ^{210}Pb. The highest observed values can be explained with the high solubility of radium and with polonium mobility in anoxic water.

Natural Radioactive Isotopes in Water in
Relation with Geology and Hydrological Investigations in the Territory of Luxembourg

11

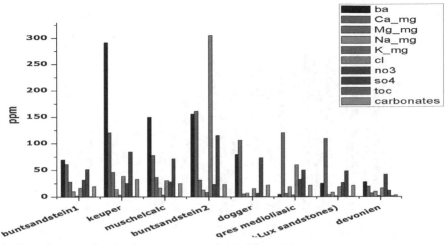

Fig. 6a. Distribution of major ions versus geological formation

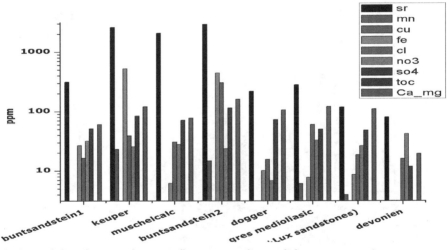

Fig. 6b. Spatial distribution of some influencing radio-nuclides occurrence ions versus geological formations

Analyzing the cross ratios $^{238}U/^{232}Th$ and $^{230}Th/^{232}Th$, it is observed that the first ones are higher than the second ones, but follow the same surface distribution. This fact again is explained with the specific chemistry of these radio-nuclides and should be attributed to the very low thorium solubility.

One can see that in many cases over the country, an opposite distribution between radon and radium is present. An analysis of the relation between radium and oxygen content shows higher radium associated with lower oxygen content. The initially anoxic water, when enriched in oxygen, deposits radium; this radium will produce elevated concentrations of radon in the source.

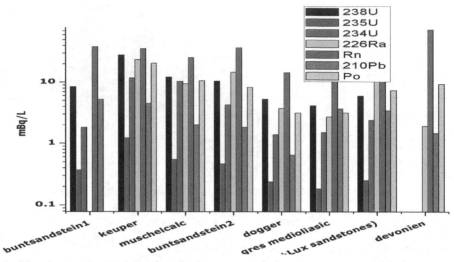

Fig. 6c. Uranium series isotopes distribution versus geology

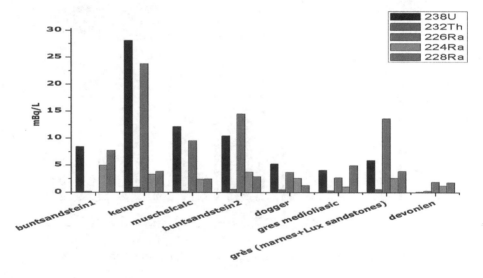

Fig. 6d. Distribution of main radio-nuclides versus geology

Another feature is the relatively high concentrations of [210]Po compared to the parent [226]Ra in the middle part of the country. One explanation is that, polonium product of ^{210}Pb is supplied to body waters from atmospheric input, run-off and the "in situ" decay of [226]Ra, its precursor in the water column. As the lead distribution follows the uranium one and it is 10 times lower than polonium the dependence of the [210]Po/[226]Ra ratio with the radon should be investigated in location with high values. Naturally produced 210Pb can exhale from the soils and can get transported through the lower atmosphere. It is a long-lived decay product of the noble gas radon from the uranium decay series, which reaches the atmosphere and is

Natural Radioactive Isotopes in Water in
Relation with Geology and Hydrological Investigations in the Territory of Luxembourg

13

transported with the atmospheric circulations in form of particles until deposition on the surface with the precipitations. Generally in keuper, muschelcalc and sandstone formations we observed some equilibrium between radon and radium. In these cases the 226Ra/210Po ratio is upper or near to 1. In the escape cases we had some spots of stable lead and it is related to Ba/Ca sulfates formations. Polonium values were high due to re-deposition of aerosols too.

In the devonien formation radon concentrations were higher than radium ones and 226Ra/210Po ratios lower than 1. In this case possible explanation of disequilibrium between polonium and lead were the exhalation to atmosphere processes.

For every analysed water sample the activity ratios $^{238}U/^{232}Th$ and $^{230}Th/^{232}Th$ are calculated. From the results is we obtain that these ratios may have significant values. In most cases, they can be attributed to the very low ^{232}Th concentrations and not to high ^{238}U. The mean ratio value of $^{238}U/^{232}Th$ for surface and soil infiltrated water is 2.4, whereas it exceeds 12 for underground water. It should be mentioned that in the investigated underground water with high $^{238}U/^{232}Th$ ratios, we remarked that the ^{210}Pb content is very low, and the ^{230}Th concentrations are twice as high as those of ^{232}Th [5]. This is associated with the higher solubility of uranium isotopes compared with thorium isotopes.

The obtained data also help to study the radioactive equilibrium disturbance within the uranium series. For each sample, the ratios $^{210}Pb/^{238}U$, $^{234}U/^{238}U$ are calculated. As expected for surface waters, the $^{238}U/^{234}U$ ratio is close to 1 and reveals a secular equilibrium (normal background). In the same samples, the lead content exceeds by more than 30 times the uranium content related to the normal hydrological cycle and atmospheric precipitations.

In all drilled wells the $^{234}U/^{238}U$ ratios range from the very low value of 1.3 to a value of 5, with a mean value of 1.8. A correlation with increasing depth or fault zones is presumed, which causes the irregularities.

As above -mentioned, the majority of the natural springs occurs in the outcropping region of the Luxembourg sandstone in the central part of the country. The normal background radon concentrations in water vary in the range of 2–60 Bq/L with an average value of 17 Bq/L. The distribution of the obtained data follows a positively skewed log- -normal dependence with the maximum at 15 Bq/L that is characteristic for deeper groundwater in this area. Only 3% of the obtained results are close to 100 Bq/L and are well explained with the high mineralization of these sources.

The comparison between indoor radon concentrations over the country and water radon concentrations show no direct relation. The reasons for this lack of relationship are not clear because soil air radon and, therefore, indoor radon concentrations are dependent on the geological substrates and their U-bearing minerals in general. It is supposed that the radon propagation in the water is more dependent of the local hydrology processes than from the geological structures.

The geographical distribution of radon shows two axes of elevated concentrations. One is located in the north-western region, followed by the second one in the central East of the country. They both are mainly situated on Lower Jurassic and Triassic formations. A radiation profile map of Luxembourg, prepared with aero-gamma measurements in 1995 [Geo-service of Luxembourg, 1997], also reveals that these parts of the country have increased radiation levels as a consequence of high geochemical distribution of 238U in the rocks of these formations. The test of radon dependency on main physical parameters, such as EC, temperature and pH, did not show any correlation for all sampled springs. The temperature of the waters varied in-between 8–16°C with average values in the range of 10–12°C. An anti-correlation of the radon and temperature in 65% of the cases is known from the literature, but this could not be observed in this case due to the small temperature fluctuations of the groundwater [Huyadi et al, 1999]. The same situation was found for the EC.

Radon concentrations in the north-central part show some differences in the 2003 and the 2007 campaigns. The latter campaign coincides with the recent construction of the northern highway in the time of this investigation. A hydro-geological and hydro-geochemical study of the Geological Survey of Luxembourg showed that the construction works influenced the flow regime and the hydro-geochemistry. A decreasing radon concentration could be observed in the same period. It is assumed that drilling and blasting works changed the flow path within the aquifer and, therefore, also changed the water chemistry. This was correlated with degreasing discharges in most of the monitored springs.

The spatial variations can be led back to the different geological materials and structural/tectonic features. Figure 6c shows that the radon activity changes with different geological units where the sources are located. Most aquifers in the region are double porosity aquifers, with proportion of matrix storage and fracture flow. Radon concentrations are influenced by the contact time between the water and the aquifer rock. Water from fast fracture flow is usually in contact with the rock matrix only for a short period of time and, therefore, cannot accumulate so much radon. On the other hand, radon with its short half-life reaches secular equilibrium after approximately 30 days. Hence, the groundwater flow should not influence its concentrations because groundwater flow is usually by magnitudes slower. Nevertheless, hydraulic conductivities of up to 100 m/d have been found along open fracture zones through tracer tests [Pistre et al, 2005], which support the results of this study.

The relation between radon and radium is of particular interest. Radon is a tracer, which can be used in a variety of groundwater studies, such as hydrograph separation, or any other type of groundwater/surface water interaction [Schmidt et al, 2009; Wu et al, 2004]. It is a daughter isotope of radium. Reasonable amounts of radium get into the hydrological cycle mainly through the dissolution of silicates and carbonates. It is not very mobile and can precipitate with other bivalent cation, such as Ba^{2+} or Ca^{2+}, as sulfates or carbonates. Nevertheless, considerable amounts of radium were measured in the sampled sources.

It is interesting to notice that the gamma airborne survey has shown a strong anti-correlation of radon and radium over Luxembourg, especially in the central-eastern part of the country. The radium concentrations in water are negatively correlated with radon. High levels of radium are associated with low levels of radon and vice versa. Radon activity in groundwater is controlled by the concentration of radium in the aquifer matrix [Vinson et al, 2009]. However, the amount of radon that can enter the groundwater depends on the emanating coefficient of the radium bearing material in the aquifer. This means that the radon emanation potential of the rocks is in some cases very low, especially for low permeable rocks. Then again, low radium content rock can nevertheless emit higher concentrations of radon, when their porosity and interconnected pore space is large. Both interpretations do not explain the negatively correlated behaviour in Luxembourg. The waters from the Devonian rocks, low porosity, show high concentrations in radon (Figure 6c), but low concentrations in radium, whereas the waters from the Lias and Buntsandstein, higher porosity, formations are low in radon, but high in radium. Hence, a possible interpretation is that in the first case the radon derives directly from emanating form the rock. The Devonian shits and sandstones are rich in uranium and radium, whereas the radon comes form dissolved radium in the water in the second case. The relation between radium and oxygen content shows higher radium concentrations associated with lower

Natural Radioactive Isotopes in Water in
Relation with Geology and Hydrological Investigations in the Territory of Luxembourg

15

oxygen content. The initially anoxic water, poor in oxygen, mobilizes only a little amount of radium. If this water is deposited on the surface layer, it will produce elevated concentrations of radon in the springs.

Besides, a dense spatial resolution we analyze temporal variations of radon as well. The 316 springs from 2007-2008 campaign were sampled twice, once during the winter month, and once during the summer month and therefore during base flow conditions. Furthermore, five of the selected springs have been equipped with continuous flow and electrical conductivity (EC) devices and have been sampled for radon on monthly basis over a whole year (figure 7). Data from a previous campaign in 2003 are also used for comparison.

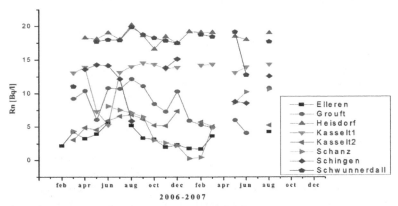

Fig. 7. Temporal variations of radon in sources of drinking water

Spatial heterogeneity was assumed in the rocks, with different correlation lengths. Long-scale effects were dominant for most of the sources.

The possible application of tritium is an essential tool for hydrological investigations. The most important applications are identifications of recharges in aquifers, estimation of hydraulic parameters related to pollutant transfer and determination of turnover time of groundwater. In the present work are analyzed for this isotope only 83 samples mainly for pollution investigation purposes this does not allow us to make significant conclusions. In all of them the concentrations are near to the capability of registration of used equipment.

Analyses of all obtained data over these two campaigns are in progress of elaboration.

5. Conclusions

The spatial distributions of the radio-nuclides as well the major ions are driven by two kinds of forces, anthropogenic (agricultural land use, iron mining etc.) and geogenic Uranium / thorium series isotopes are integrated in the crystal lattice of minerals, e.g. as it is the case with uranium in zircons, or substitute positively charged major ions, e.g. the substitution of calcium and magnesium by radium. Depending of local site geo-morphology this substitution is guided by carbonates (trias, buntsandstein), sulphates and chlorides (deep water in keuper, muschelcalc, lias) or nitrates and iron oxide / hydroxides (Luxembourg sandstones diaclases, land-used trias part).

Radon (in this case ^{222}Rn) constitutes a particular case, as it is a noble gas and therefore inert. It is the daughter nuclide of Radium (^{226}Ra decays to ^{222}Rn) and accumulates in open crevasses in the rock matrix or gets dissolved in the groundwater.

As the distribution of the analysed sources is much more concentrated in the part of Luxembourg sandstones and alluvial formation, more of the results are reported to trias and inferior lias formations.

In medio-liasic sandstones and supra-liasic formations (Minette), due to iron mining, showed a marked uranium/thorium disequilibrium.

In Muschelcalc formation radon concentrations are low due to low permeability, increased levels are measured in sources due to the existence of fractures. In Luxembourg sandstones due to changing oxidizing/reducing conditions variable radon concentrations are observed. Seasonal heterogeneities of radon are mainly relied to runoff production processes.

The response of radio-nuclides as tracers is more direct in sandstones compared to the schist in rain periods.

The investigation on spatial distribution dependency shows that the distribution in uranium series radio-nuclides, except of radium and polonium, are in correlation with local hydrogeology features.

Comparisons between two measurement campaigns are coherent and give us important information for the mechanism of distributions of the natural radio-nuclides in ground and surface water.

6. Acknowledgment

We want to thank the Water Service Department of Ministry of Interior and the Great Region of Luxembourg andespecially Dr. Denis Pittois form the Public Research Center Henri Tudor - CRTE for the chemical data and support. We would like to acknowledge the Service of Geology of Luxembourg for the guidelines and help during the analysis and whole project.

7. References

Colbach R (2005) Overview of the geology of the Luxem- bourg Sandstones. Scientific work of Museum of natural history of Luxembourg vol.44, pp. 155–160

EU Directive of drinking water (2000)

Tosheva Z., Taskaeva I., Kies A. (2006). International Journal of Environmental Analytical Chemistry, vol.86 No 9, pp.657-661

Tosheva Z., Hofmann H., Kies A. (2009). Journal of Radioanalytical Nuclear Chemistry, vol.282, pp.501-505

www.geology.lu

Berners (1983) Annales de la Société Geologique de Belgique 106:87-102

Porceli D., Swarzenski P (2003). Mineral Geochemistry, vol.52, pp.317-321

Oliveira J., Carvalho F (2006). Czechoslovak Journal of Physics, vol.56, No1, D545

Gonzales-Labajo J., Bolvar J.P., García-Tenorio R (2001). Radiation Physical Chemistry, vol.61, No3, pp.643-652

Geochemical map of G-D Luxembourg (1997). Geological Service Division of Luxembourg

Huyadi I., Csige I., Haki J., Baradacs E. (1999). Radiation and Measurements, vol.31, No116, pp.301-316

Pistre S., Marliak S., Joudre A., Bidoux P. (2005). Ground Watater, vol.40, pp.232-241

Schmidt A., Stringer Ch., Haferkorn U., Schubert M. (2009). Environmental Geology, vol. 56, No5, pp.855-863

Wu Y., Wen X., Zhang Y (2004). Environmental Geology, vol.45, No5, pp.647-653

Vinson D., Vengosh A., Hirschfeld D., Dwyer G. (2009). Chemical Geology, vol.260, No3/14, pp.159-171

Part 2

Minerology

2

Carbonate-Hosted Base Metal Deposits

Fred Kamona
University of Namibia
Namibia

1. Introduction

Carbonate-hosted base metal sulphide deposits are discussed with reference to ore deposit mineralogy and geology, isotope geochemistry, ore fluids and ore genesis of type examples. Three main carbonate-hosted deposit types are recognized: 1. Leadville-type mineralization (LTM); 2. Sedimentary-exhalative (SEDEX); and 3. Mississippi Valley-type (MVT) deposits. The three types constitute a distinct group of deposits characterized by orebodies within carbonate host-rocks containing Pb and Zn as the major metals with variable amounts of Cu and associated by-products, including Ag, Cd, Ge, V, Ga, As, Sb, Au and In.

LTM deposits form by magmatic-hydrothermal processes associated with igneous intrusions and correspond to the carbonate-hosted high-enthalpy type of Russell & Skauli (1991) and the chimney-manto type deposits of Hutchinson (1996). The SEDEX deposits are typically characterized by syngenetic to syndiagenetic processes of mineralization at the submarine surface within sedimentary basins prior to carbonate lithification and are spatially associated with faults located within, or at the margins of, the basins. They are also known as medium-enthalpy, Irish type or clastic dominated Pb-Zn deposits (Hitzman & Beaty, 1996; Russell & Skauli, 1991; R.D. Taylor et al., 2009). In contrast, typical MVT deposits (low-enthalpy type of Russell & Skauli, 1991) are epigenetic and form at the periphery of basins in lithified carbonate rocks spatially associated with faults, structural highs and facies changes. They are commonly referred to as MVT due to the occurrence of classic districts in the Mississippi River drainage basin of the USA (Leach et al., 2010). Both SEDEX and MVT deposits form from formation waters derived from sedimentary basins with high heat flows.

1.1 LTM deposits

LTM deposits are high temperature (>200°) carbonate-hosted Pb-Zn-Ag deposits distinctly different from MVT and SEDEX deposits by virtue of their temporal and spatial relationship to igneous intrusions. The fundamental diagnostic feature of this deposit type is its origin from magmatic-hydrothermal processes associated with felsic to intermediate magmatism and ore formation by replacement of carbonate rocks (Einaudi et al., 1981; Megaw et al., 1988; Thompson & Beaty, 1990). Although the type name originates in the Colorado mineral belt of the United States (Beaty et al., 1990), the type setting for these deposits is northern Mexico where the mineralization is hosted in carbonate-dominated portions of Jurassic-Cretaceous sedimentary sequences (Megaw et al., 1988).

The type of mineralization ranges from near-intrusion Cu and Zn-Pb skarns through distal Pb-Zn skarns to massive sulphide bodies hosted by limestone and dolostone cut by granite, quartz monzonite and other intermediate to felsic hypabyssal, porphyritic lithologies (Einaudi et al., 1981; Megaw et al., 1988). There is a continuous transition in most districts from skarn ores to massive orebodies beyond the skarn zones (Fig. 1).

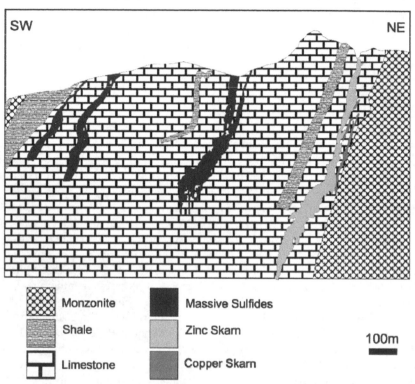

Fig. 1. Cross section of major orebodies of Providencia in Zacatecas, Mexico over a vertical interval of 1000m with concordant skarn and massive sulphide orebodies associated with a monzonite stock (modified from Megaw et al., 1988).

1.2 SEDEX deposits

Carbonate-hosted SEDEX Pb-Zn deposits are characterized by a stratiform morphology, an active tectonic setting during mineralization and low to moderate temperatures of formation (100° to 260° C). The Pb-Zn ores are associated with significant Ag and minor Cu (Hitzman & Large, 1986). The mode of occurrence of the mineralization includes cross-cutting and stratiform ores, such as in the classical Silvermines deposit in Ireland (Fig. 2), where epigenetic feeder zones occur below exhalative stratiform orebodies (Andrew, 1986a). The deposits are distinct from volcanogenic massive sulphide deposits because of a high Pb/Cu ratio and their lack of an intimate association with volcanic and volcaniclastic sequences (Turner & Einaudi, 1986). Some SEDEX deposits typically occur within marine shales or siltstones (Large, 1983) and are therefore distinct from carbonate-hosted Pb-Zn SEDEX deposits considered here.

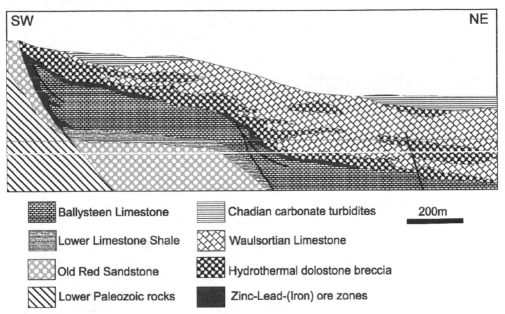

Fig. 2. Cross section of the Silvermines SEDEX deposit with an epigenetic feeder zone and stratiform ores (modified from Hitzman & Beaty, 1996).

The central Irish ore field may be considered as the type setting of carbonate-hosted SEDEX deposits. The ore field occurs in a first-order, intrashelf basin on the edge of the North Atlantic craton within Lower Carboniferous limestones (Andrew & Ashton, 1985; Hitzman & Large, 1986). The Alpine Triassic deposits of Central Europe are comparable to the Irish SEDEX deposits (Schneider, 1964; Sangster, 1976). They predominantly contain epigenetic replacement bodies and minor but extensive syngenetic stratiform orebodies in carbonate rocks (Schneider, 1964). Klau & Mostler (1986) regard the deposits as MVT based on the Pb-isotope data of Köppel (1983) which, according to them, shows no relation between magmatic rocks and mineralization. In addition, the ore lead is isotopically different from host-rock lead. However, according to Köppel (1983), basement rocks and upper Carboniferous to Permian magmatic rocks of the Southern Alps exhibit similar characteristics as the lead in the Triassic Pb-Zn deposits, whereas the lead of feldspars from Triassic volcanics shows significant differences. The syndiagenetic Pb-Zn deposits in Triassic carbonates could have obtained lead from lower crustal rocks, or material thereof, by leaching processes (Köppel, 1983).

1.3 MVT deposits
MVT deposits are stratabound, epigenetic orebodies that occur in clusters in carbonate formations of mineral districts distributed over large areas (Ohle, 1959, 1967; Heyl, 1967; Snyder, 1967). The orebodies are typically found at or near basin margins associated with domes and intrabasinal highs (Fig. 3). Like SEDEX deposits, MVT deposits are characterized by the absence of obviously associated igneous rocks, but the stratabound and often stratiform morphology of the former contrasts with the majority of MVT deposits whose morphology commonly crosscuts stratigraphy (Hitzman & Large, 1986; Sangster, 1990).

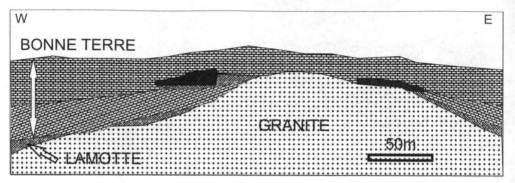

Fig. 3. Cross section through the Doe Run Mine, Southeast Missouri with ore zones (black) overlying a Precambrian granite "knob" (modified from Ohle, 1996).

Classical examples of MVT deposits are the districts of the mid-continent in the United States and the Pine Point district in Canada. However, they also occur in other districts, including the Appalachian in eastern United States, Silesia in Europe, southern and central Africa, and the Leonard shelf in Australia. They include districts varying from Zn- to Pb-dominant as well as Cu-rich deposits. With the exception of the Southeast Missouri district, which is dominantly Pb-rich, most districts are Zn-dominant and a few deposits (e.g. Kipushi and Tsumeb in Africa) contain significant amounts of Cu in addition to Zn and Pb.

2. Ore deposit geology

2.1 LTM deposits
This type of deposit is found on continental margins of orogenic belts and on the inner side of principal arcs where carbonate rocks representing miogeoclinal or stable platform depositional environments are widespread (Eunaudi et al., 1981; Sawkins, 1990). The deposits are associated with granodioritic to granitic magmatism which, together with the related metallogenesis in this environment, are believed to be a function of subduction (Einaudi et al., 1981; Bookstrom, 1990). In Mexico, the mineralized districts occur within or on the margins of a thrust belt with anticlinal ranges (the Sierra Madre Oriental). The thrust belt consists of Jurassic-Cretaceous sedimentary sequences of basal continental red beds with evaporites followed by a series of shale and carbonate facies which were deformed during the Laramide orogeny. The carbonate portions of these sequences are frequently more than 3 km thick and consist of limestone with minor dolomite (Megaw et al., 1988).
The mineralization-related intrusions range from batholiths and stocks to thin dikes and sills composed of diorite, granodiorite, quartz monzonite, monzogranite, quartz latite, and rhyolite (Einaudi et al., 1981; Megaw et al., 1988; Bookstrom, 1990). In Mexico the intrusions originated from the lower crust (Ruiz et al., 1988) and their age ranges from 47 to 26 Ma (Megaw et al., 1988). Mineralization occurred during a restricted interval of the Mid-Tertiary, well after evolution of the Laramide-aged Mexican thrust belt. Large deposits like Santa Eulalia and Naica grade outward from intrusion-associated mineralization to intrusion-free ores, suggesting that districts without intrusion relationships may not have been traced to their ends (Megaw et al., 1988).

The carbonate host rock may be predominantly limestone or dolomite and there are no consistent links between ore deposition and carbonate composition, facies, organic content, or insoluble components (Megaw et al., 1988). Many deposits contain mineralization in carbonate strata within or below relatively less permeable rocks such as cherts, shales, slates and volcanic rocks, and the formation of large orebodies appears to be related to structural enhancement of permeability in the carbonate rocks (Yun & Einaudi, 1982; Megaw, et al., 1988). Complex fold and fault structures are the dominant controls in strongly folded areas, whereas simple fault-related structures are more common on fold belt margins and radial and/or concentric structural patterns related to intrusions are important controls in domal areas (Megaw et al., 1988). The controlling structures apparently acted as conduits for channeling mineralizing fluids to sites of ore deposition.

The orebodies commonly display a combination of forms that include mantos, chimneys, pods and veins (Einaudi et al., 1981; Megaw et al., 1988; Beaty et al., 1990) and are temporally and spatially associated with igneous intrusions (Fig. 1). The change in orebody morphology is caused by local variations in stratigraphy, structural patterns, travel distance of hydrothermal fluids, and depth of intrusive emplacement (Megaw et al., 1988). Podiform skarn and sulphide bodies are the deepest modes of occurrence and are irregular or equant with no particular orientation or relationship to stratigraphy. They commonly occur along intrusive, fault, and lithological contacts (Megaw et al., 1988). Typical vein deposits include Bluebell in Canada and Uchucchacua in Peru (Einaudi et al., 1981). Vein mineralization is also dominant in some major skarn deposits such as San Martin (Rubin & Kyle, 1988).

Tonnage and grade data (Megaw et al., 1988) for skarn, chimney and manto ores in seventeen districts of Mexico show a wide range of values (Fig. 4). The ore districts have an average of 11.4 million tonnes (Mt) at 11.0 wt.% Pb, 8.5 wt.% Zn, 0.7 wt.% Cu, 243 ppm Ag and 1.8 ppm Au. Santa Eulalia is the largest district with 50 Mt of ore grading 5% wt. Pb, 6.7 wt.% Zn, 0.1 wt.% Cu and 242 ppm Ag. Other major ore districts include Providencia-Conception del Oro (25 Mt, 2 to 20 wt. % Pb+Zn, 30 to 500 ppm Ag) in Mexico (Megaw et al., 1988), and Leadville (24 Mt, 7 wt. % Pb+Zn, 320 ppm Ag) in Colorado (Beaty et al., 1990).

With the exception of the San Carlos and Los Lamentos manto districts which are Pb-rich and contain no Zn, the average Zn/(Zn+Pb) ratio of 0.5 in the other Mexican deposits indicates equal concentrations of Pb and Zn in most of the orebodies. However, Cu values are generally low (0.7 wt. %) and only range from 0.1 to 1.6 wt % Cu. The gold grades are also typically low and ranges from 0.2 to 6 ppm Au with an average of only 1.8 ppm Au for the 9 Mexican districts for which data is available (Megaw et al., 1988). In contrast, the Ag grades are high in the majority of deposits with an average of 243 ppm Ag and a range from 14 to 600 ppm Ag with most deposits (14 out of 17) having grades >100 ppm Ag.

2.2 SEDEX deposits

The central Irish ore field occurs in a first-order, northerly transgressive intrashelf basin on the edge of the North Atlantic craton (Andrew & Ashton, 1985). All the known major deposits lie along basin margins adjacent to historically active basement fault zones. The major stratiform deposits at Navan, Silvermines and Tynagh occur in early Carboniferous stratigraphic successions of shallow water carbonate sediments and argillites, and deep water Waulsortian limestones overlain by carbonates and argillites deposited in latest Courceyan to Chadian stages (Philips & Sevastopulo, 1986). The carbonate succession is underlain by basal Devonian sandstones (the Old Red Sandstones) which diachronously overlie a Precambrian-Lower Paleozoic basement.

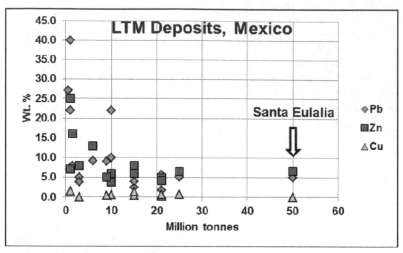

Fig. 4. Tonnage and grade data for 17 Mexican districts with skarn, chimney and manto ores based on data from Megaw et al. (1988).

There is no obvious spatial association between igneous rocks and ore deposits at Silvermines, Tynagh and Keel (Morrissey et al., 1971). A late Caledonian, post-tectonic syenite with a model lead isotopic age of 400 Ma intrudes Lower Paleozoic rocks beneath the Navan deposit, but it is not regarded as a mineralizing pluton as it is older than the mineralization age of 366 Ma (Andrew & Ashton, 1985; Ashton et al., 1986). Although thin pyroclastic beds occur at the same stratigraphic level as the most mineralized beds at Tynagh, the pyroclastic material is considered to have been derived from the Limerick volcanic centre 70 km away (Morrissey et al., 1971).

The mode of occurrence of SEDEX orebodies includes cross-cutting, epigenetic veins and breccia ores commonly regarded as feeders to overlying tabular stratiform lenses. Such a feeder-exhalative system is best developed at Silvermines (Fig. 2) where the upper stratiform ores occur at the base of and within Waulsortian carbonates (Andrew, 1986a). The mineralization is preferentially enriched immediately below siltstone and shale bands as well as in dolomitization voids and fracture systems. The thickness and distribution of the stratiform ore is related to the paleotopographic control of knolls of Waulsortian micrites (S. Taylor, 1984). Navan represents a different style of mineralization in that the orebody consists of five vertically superimposed stratiform lenses with cross-cutting veins and breccia zones within Lower Carboniferous limestones (Andrew & Ashton, 1985; Ashton et al., 1986). The mineralization is consistently restricted to non-argillaceous units, being generally best developed in micritic, oolitic, pelloidal or slightly arenaceous carbonate beds.

SEDEX deposits have a spatial relationship to fault structures active during mineralization, and, in the case of Silvermines and Tynagh, during host rock deposition (Hitzman & Large, 1986). All the ore zones at Silvermines are closely related to WNW or westerly faults (Andrew, 1986a). At Navan, the mineralization is grossly located adjacent to fault intersections on the flank of an anticline (Ashton, et al., 1986). The main orebody at Tynagh was a subhorizontal wedge that extended updip of the Tynagh fault in two elongated zones separated by a barren dolomitized reef (Boast et al., 1981a; Clifford et al., 1986). In addition, a residual deposit of oxidized and unoxidized mineralization occurred in the hanging wall of the fault in a deep post-karstification trench.

Tonnage and grade data for 28 carbonate-hosted SEDEX deposits (Fig. 5) from the data bank of Goodfellow & Lydon (2007) shows that the average deposit contains 8.9 Mt grading at 6.74 wt. % Zn, 2.51 wt. % Pb and 63 ppm Ag. The deposits are typically Zn-rich with a Zn/(Zn+Pb) ratio of 0.71.

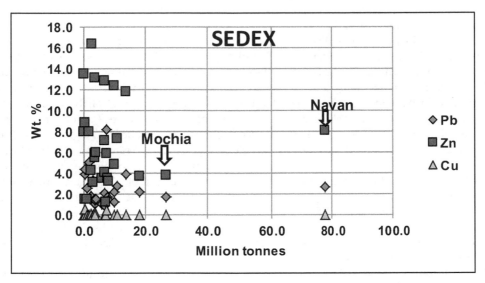

Fig. 5. Grade tonnage plot for carbonate-hosted SEDEX deposits based on data from Goodfellow & Lydon (2007).

With the exception of Navan (77.8 Mt at 8.04 wt. % Zn and 2.68 wt. % Pb) and the Palaeoproterozoic Mochia deposit in India (26.8 Mt at 3.79 wt. % Zn and 1.69 wt. % Pb) most carbonate-hosted SEDEX deposits contain less than 20 Mt of ore (Fig. 5). The Cu content of these deposits is also very low with an average of only 0.30 wt. % Cu in the four deposits for which Cu data is available. The Mehdiabad deposit in Cretaceous dolostone in Iran contains exceptionally large geological resources estimated at 218 Mt with 2.3 wt. % Pb, 7.2 wt. % Zn and 51 ppm Ag (Goodfellow & Lydon, 2007).

2.3 MVT deposits

MVT Pb-Zn deposits have low temperatures (100° to 150° C), uniform salinity, density and composition (Roedder, 1967, 1984) and typically occur in shallow-water platform carbonates peripheral to intracratonic basins (Sangster, 1990). The classical districts of the Upper Mississippi Valley, Tri-State and Southeast Missouri are found within the stable continental interior of the North American craton, whereas the Appalachian district occurs in an orogenic belt on the eastern margin of the continent (Hoagland, 1967, 1976; Thacker & K.H. Anderson, 1977). The mineral districts are distributed over thousands of square kilometres (Ohle, 1967) and the carbonate-hosted ore deposits occur over a stratigraphic interval of 200 million years ranging from Cambrian to Mississippian-Pennsylvanian (Ohle, 1980).

MVT orebodies are stratabound in essentially horizontal carbonates, chiefly dolomites and dolomitic limestones. They are epigenetic and typically occur within breccia bodies below unconformities (Callahan, 1967; Sangster, 1988). Zones of high permeability such as channels and troughs in algal reefs, collapse breccias, pinchouts, facies changes, bedding planes, faults and fractures are the focus of mineralization, particularly along flanks or on crests of basement highs (Fig. 3) or below unconformities. The ore-hosting breccias include discordant domes and columns that may be interconnected to concordant breccias. The breccia tops are vuggy and porous and contain mainly open-space filling mineralization while replacement mineralization is dominant near the base of the breccias (Rogers & Davis, 1977).

The regional distribution of the Southeast Missouri orebodies closely follow the pinchouts of the Lamotte Sandstone (Kisvarsanyi, 1977). Although algal reefs are also spatially associated with the mineralization, the reefs themselves are heavily mineralized only in some parts of the Viburnum Trend, and they are barren in many places. Faults seem to have been important in the Old Lead Belt where fault zone ores provided most of the lead (James, 1952). In the Tri-State district the ore occurs as elongate tabular bodies ("runs") of breccias up to 1 km or more in length which may form circular map patterns (Hagni, 1976). The dominant ore-bearing collapse breccias developed by solution enlargement of fractures along silicified dolomite-limestone contacts. In addition, flat "sheet" or blanket ore occurs in partly broken, stratified chert bodies. In the Upper Mississippi Valley the ore occurs dominantly as open-space filling in pitch-and-flat structures along bedding planes, joints and faults (McLimans et al., 1980; Heyl, 1983). The orebodies are linear, arcuate, or elliptical in plan and vary in length from 0.4 to 2 km. In the Appalachian district the ores are associated with karst breccias or reefs in dolomitic limestone at or adjacent to a dolomite-limestone interface (Hoagland, 1976).

The widespread regional distribution of MVT deposits is also observed in other districts, including the Otavi Mountainland in Namibia where regional scale ore fluid migration is indicated by the presence of Pb–Zn occurrences over 2500 km^2 within stratabound breccias of the Elandshoek Formation (Kamona & Günzel, 2007) as well as in central Africa (Fig. 6) where numerous prospects are associated with a few Neoproterozoic economic deposits such as Tsumeb, Kipushi, Kabwe and Berg Aukas (Kamona & Friedrich, 2007; Kampunzu et al., 2009).

Most MVT deposits are Zn-dominant as indicated by the Zn/(Zn+Pb) ratio of 0.7 in individual orebodies and districts. However, some deposits like the Viburnum No. 27 mine with 8 Mt grading 2.9 wt. % Pb and 0.2 wt. % Zn (Grundmann, 1977) and the Buick mine with >50 Mt, 8 wt. % Pb and 2 wt. % Zn (Rogers & Davis, 1977), indicate the Pb-rich nature of these ores. In contrast, sphalerite is the only ore mineral in Central Tennessee which contains 20 Mt with 5 wt. % Zn (Kyle, 1976).

Tonnage and grade data for 23 MVT deposits (Fig. 7) from Canada (Paradis et al., 2007) and Africa (Kamona & Günzel, 2007; Kamona & Friedrich, 2007; Kampunzu et al., 2009) indicate that the average MVT deposit contains 13 Mt with grades of 7.2 wt. % Zn and 3.4 wt. % Pb. The average Zn/(Zn+Pb) ratio of 0.69 is similar to that of carbonate-hosted SEDEX deposits (0.71). Copper averages 3.2 wt. % in nine of these deposits for which Cu grades are available with significant Cu grades in the Kipushi (8.0 wt. %), Tsumeb (4.42 wt. %) and Khusib Springs (10.06 wt. %) deposits. Silver contents can be considerable with a maximum of 584 ppm in the Khusib Springs deposit and an average of 96.8 ppm Ag in 11 of the deposits considered here.

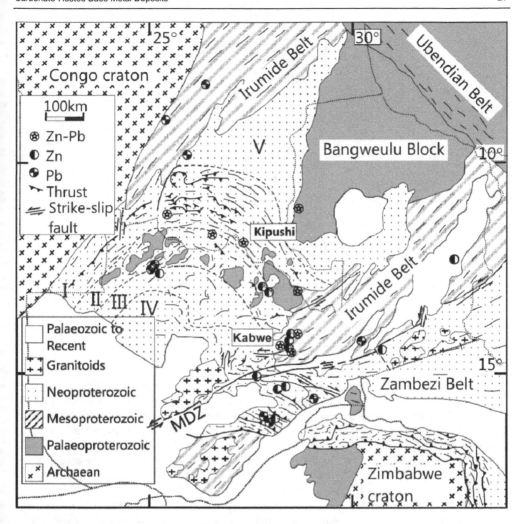

Fig. 6. Location of the Kipushi and Kabwe Pb-Zn deposits and Pb-Zn occurrences in the geotectonic framework of central Africa (modified from Kampunzu et al., 2009). I. External fold and thrust belt, II. Domes region, III. Synclinorial belt, IV. Katanga high, and V. Kundelungu aulacogen or palaeograben. MDZ (Mwembeshi Dislocation Zone).

The Kipushi (Democratic Republic of the Congo) and Gayna River (Canada) deposits are examples of large deposits of this type, each with 50 Mt of ore. The Zn-Cu dominant Kipushi deposit is characterised by zones of Cu, Zn+Cu and Zn+Pb along its length (Kampunzu et al. 2009). The 80 Mt Pine Point district, which contains 100 individual orebodies (Paradis et al., 2007) is included for comparison (Fig. 7). The data of Leach et al. (2005, 2010) and Paradis et al. (2007) indicate that MVT districts with carbonate-hosted orebodies in the USA and Canada contain an average of 233 Mt with grades of 3.3 wt. % Zn and 2.2 wt. % Pb with a Zn/(Zn+Pb) ratio of 0.72 (n=10).

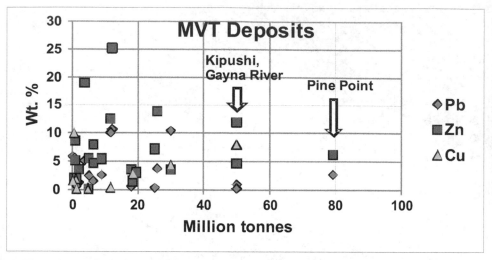

Fig. 7. Grade and tonnage plot for carbonate-hosted MVT deposits in Canada and Africa based on data from Kamona & Günzel (2007), Kamona & Friedrich (2007), Kampunzu et al. (2009) and Paradis et al. (2007).

3. Mineralogical and geochemical features

3.1 LTM deposits

The principal sulphides are galena and sphalerite with variable amounts of pyrite, pyrrhotite, marcasite, chalcopyrite and arsenopyrite (Einaudi et al., 1981; Megaw et al., 1988; Thompson & Arehart, 1990). The massive sulphide orebodies commonly exhibit banding due to textural, size or mineralogical differences between bands and they are composed of more than 65 wt. % sulphides. The sulphides usually exhibit a zonal distribution with galena being more abundant than sphalerite in mantos, but the latter is more abundant in chimneys. The chalcopyrite content may increase with depth, and districts that contain both pyrite and pyrrhotite show an increase of pyrrhotite with depth or an increase of pyrite with time.

Tetrahedrite-tennantite, chalcopyrite and arsenopyrite commonly occur near skarns at deep levels, and near mineralization-related intrusions. Acanthite, cinnabar, stibnite, realgar and silver sulfosalts are most common in the peripheral parts of districts. In addition, some skarns also contain bornite, covellite, molybdenite, scheelite, powellite, cassiterite, magnetite, and hematite. Silver occurs in solid solution in galena in most Mexican deposits (Megaw et al., 1988), whereas galena from the Leadville district is devoid of Ag, Sb and Bi, and the source of silver is tetrahedrite and electrum with the former occurring partly as disseminations within galena (Thompson & Arehart, 1990).

The gangue is dominated by carbonates, fluorite, or quartz with or without minor barite and anhydrite. Calc-silicates may be locally present and make up less than 5 % of the mineral content (Megaw et al., 1988). The main minerals of the skarns are pyroxene and garnet with associated wollastonite, bustamite, rhodonite, dannemorite, idocrase, olivine, ilvaite, chlorite, Mn-actinolite, epidote, fluorite and cummingtonite. Skarns are typically zoned from andraditic garnet to hedenbergitic-johannsenitic pyroxene to wollastonite-bustamite to marble (Einaudi et al., 1981; Shimizu & Iyama, 1982).

The contacts between mineralization and unaltered carbonate wall rocks are sharp, and where alteration exists it is variable in extent and may appear as zones of disseminated Mn-oxide mineralization, hydrothermal dolomitization and/or recrystallized carbonates (Megaw et al., 1988). Silicification or jasperoid formation varies from poorly to strongly developed peripheral to mineralization in some ore districts, but it may be totally lacking in other districts. Isotopic depletion halos of C and O surround orebodies at La Encantada and Santa Eulalia (Megaw et al., 1988). C and O isotopes in calcite veinlet stockworks above mineralization, and limestone wallrocks adjacent to mineralization show shifts to lighter values.

3.2 SEDEX deposits

Sphalerite is the major sulphide with lesser galena, pyrite and marcasite (S. Taylor & Andrew, 1978; Boast et al., 1981a; Andrew & Ashton, 1985; Andrew, 1986a; Ashton et al., 1986). In addition, tennantite is of economic importance at Tynagh (Boast et al., 1981a). Common accessories include chalcopyrite, bornite, pyrrhotite, arsenopyrite, freibergite, pyrargyrite, boulangerite, bournonite, semseyite, jordanite and cylindrite. Principal gangue minerals are barite (except at the Magcobar barite orebodies of Silvermines), calcite, dolomite, siderite, and quartz with fluorite as an accessory. In the upper stratiform ores at Silvermines Ag is directly proportional to Pb due to inclusions of boulangerite and jordanite in galena (S. Taylor & Andrew, 1978). In addition, an isolated lens of rich silver mineralization in barite contains prominent patches of discrete silver minerals such as proustite, xanthoconite, smithite, miargyrite, argentite-acanthite, and argyrodite as well as gersdorffite (S. Taylor, 1984).

The lateral zonation at Silvermines and Tynagh is Cu-Pb-Zn-Mn with Mn forming an extensive aureole of up to 7 km in the carbonates at Tynagh (Russell, 1978). The Magcobar barite body and associated pyrite-rich zones occur on the fringes of the stratiform sulphide ores, whereas the iron-oxide facies of the Iron Formation is lateral to the sulphide mineralization (Andrew, 1986a; Caulfield et al., 1986). Vertical zoning in feeder zones is characterized by increasing Pb/Zn ratios which decrease slightly in the upper parts of the ores as in the feeder zones at Silvermines (S. Taylor, 1984). Zoning patterns at Navan are complex due to variations in the metal content of deposited sulphides with time, resulting in superposition of differing zoning trends (Andrew & Ashton, 1985). Mn, As, Zn and Pb values are irregular, but they form an intensive halo over the mineralization at Navan.

Dolomitization and silicification of Waulsortian micrites is associated with the mineralization at Silvermines, Tynagh and Navan. At Silvermines, dolomitization occurred before silicification of the host rock and precipitation of laminated cherts which in turn preceded sulphide deposition (Andrew, 1986a). Dolomitization at Tynagh occurred before and during mineralization (Boast et al., 1981a; Clifford et al., 1986). In addition, alteration aureoles containing Na and K feldspars surround epigenetic Cu-Pb-Ba veins which represent the final stage of sulphide mineralization (Boast et al., 1981a). According to Clifford et al. (1986), silicification related to dolomitization post-dates the Cu-Pb-Ba veins. In rocks of the Navan Group, zones of complex dolomitization involving both ferroan and non-ferroan dolomite, as well as more coarsely crystalline dolomite, occur only in mineralized areas (Hitzman & Large, 1986).

Depletion in the heavier ^{13}C and ^{18}O isotopes was observed in mineralized limestones, ore-stage carbonates and post-ore carbonates with respect to unmineralized Waulsortian limestone at Tynagh (Boast et al., 1981a). The depletion in ^{13}C and ^{18}O is a result of isotopic exchange between host rock and mineralizing fluids. In general, the unmineralized micrites and diagenetic calcites define a heavy isotope end-member, whereas the post-ore carbonates tend to be most depleted in ^{13}C and ^{18}O. Carbonates coeval with sulphide mineralization have intermediate $\delta^{13}C$ and $\delta^{18}O$ values.

3.3 MVT deposits

Sphalerite is the main ore mineral of most MVT districts except Southeast Missouri where galena is more dominant. The gangue generally consists of dolomite, calcite, quartz, marcasite and pyrite (Hagni, 1976, 1983; Hoagland, 1976; Hagni & Trancynger, 1977; McLimans et al., 1980). Chalcopyrite and several Co-Ni sulphides such as siegenite, fletcherite, bravoite, vaesite and polydymite are abundant in Southeast Missouri (Jessey, 1983). Barite and/or flourite are associated with the mineralization in the Appalachian, Tri-State and Upper Mississippi Valley districts. Sphalerite in the upper Mississippi Valley is characterized by colour banding which can be correlated stratigraphically over several kilometers (Maclimans et al., 1980). Sphalerite in Appalachian deposits also shows fine laminar to single crystal banding with colour variations of light and dark zones (Craig et al., 1983). The dark zones contain numerous inclusions of fluid, vapour and solids, and the darkest bands are rich in hydrocarbons and Cd.

The paragenesis of the ores are complex and variable from deposit to deposit and usually consist of several generations due to repetitive sulphide formation over a long period of time (Hagni & Trancynger, 1977; McLimans et al., 1980; Sverjensky, 1981; Hagni, 1983). Pyrite and marcasite are generally early, followed by chalcopyrite, sphalerite and galena, in that order. According to Jessey (1983), the Ni-Co-Cu minerals in Southeast Missouri preceded Pb-Zn mineralization.

Dolomitization, silicification, recrystallization and associated isotopic changes in the host carbonates are the most significant alteration effects in MVT deposits. In the Southeast Missouri district, an early porosity network was enlarged through the corrosive action of the ore fluids with pre-ore dolomite (Type I) to produce Type II and Type III end-member dolomite compositions (Frank & Lohmann, 1986). The isotopic composition of the host dolomite exhibits trends from Type I dolomite to Type II (depletion of both $\delta^{13}C$ and $\delta^{18}O$ isotopic ratios) and Type III (depletion of $\delta^{13}C$ and enrichment of $\delta^{18}O$) dolomites. The isotopic trends suggest a fluid dominated system in which the host rock was pervasively altered in areas of fluid flow to compositions in equilibrium with mineralizing fluids.

The C and O isotopes of host rock limestones and dolomite as well as of gangue calcite in the Upper Mississippi Valley define a trend extending from the isotope ratios of unaltered dolomite, through those of limestone and dolomite with decreased $\delta^{13}C$ and $\delta^{18}O$ values, to the significantly lower values for calcite deposited late in the ore paragenesis (Sveryensky, 1981b). The significant decrease of $\delta^{13}C$ and $\delta^{18}O$ values toward the orebodies is attributed to recrystallization under conditions of progressively higher water to rock ratios near the orebodies.

In the Tri-State district the close association of ore with areas of coarsely crystalline dolomite constitutes the single most important guide to ore (Hagni, 1976). Most orebodies are located along one or both margins of dolomitic zones that are probably dolomitized bioherms (Hagni, 1982). Jasperoid with disseminated sulphides forms a zone around many dolomite zones and its lateral and vertical extent essentially coincides with sulphide mineralization. Recrystallization of limestone occurs for distances of 13 to 27 m beyond the margin of silicification.

Although minor local variations may exist, the following set of minerals is typical of massive Zn–Pb sulphide ores in the low grade metamorphic zone of the Copperbelt of Central Africa: sphalerite+pyrite+galena+chalcopyrite, with subsidiary ±arsenopyrite, ±gallite, ±tennantite, ±briartite, ±reniérite. Variations in the relative proportions of the three important sulphides (pyrite, galena and sphalerite) result in the identification of massive pyrite ore, galena-rich and sphalerite-rich ores, respectively (Kampunzu et al., 2009). In the Otavi Mountainland mineralogically simple Zn–Pb-dominated ores comprising sphalerite, galena, pyrite, minor chalcopyrite, bornite, colusite, reniérite (such as the Berg Aukas deposit) are distinguished from polymetallic ores of the Tsumeb-type which contain Cu, Pb, Zn and As, as well as trace elements such as Ag, Cd, Ga, Ge, Mo, and Sb. The sulphide assemblages in the latter ores are composed of variable amounts of galena, tennantite, chalcopyrite, sphalerite, chalcocite, enargite, bornite, pyrite, minor germanite, reniérite, briartite, Ge-bearing colusite, and Mo–W sulphides (e.g. Melcher et al., 2006; Kamona & Günzel, 2007).

4. Ore fluids

4.1 LTM deposits

These deposits are characterized by formation temperatures from 200° to 500° C (Megaw et al., 1988) as well as high chlorine concentrations of up to 21.2 wt. % Cl in fluid inclusions at Providencia (Rye & Haffty, 1969). In addition, Rye & O'Neil (1968) measured δD values in the range -68 and -83 ‰ and $\delta^{18}O$ from 5.8 to 6.2 ‰. At Leadville the ranges are from -45 to -70 ‰ δD and 5 to 8 ‰ $\delta^{18}O$, respectively (Thompson & Beaty, 1990). Similar values of δD= -45 to -60 ‰ and $\delta^{18}O$= 4 to 9 ‰ have been reported for the Gilman deposit (Beaty, ed. 1990). These values are comparable to those for waters which have equilibrated with magmas (Sheppard, 1986).

In general, the metasomatic fluid of skarn and massive sulphide ores has a low CO_2 content (X_{CO2} < 0.1) and moderate to high salinities (10 to 45 wt. % NaCl equivalent) (Einaudi et al., 1981). Fluid inclusion data from host minerals coeval with sulphide deposition or skarn formation (fluorite, sphalerite, quartz, pyroxene and garnet) indicate homogenization temperatures in the range 200° to 550° C and salinities with a total range from 1 to 60 wt. % NaCl equivalent (Einaudi et al., 1981; Megaw et al., 1988). The extreme variation in salinity may result from various mixtures of highly saline magmatic fluids and later circulating groundwaters (Roedder, 1984). Hotter (400° - 650° C), more saline (>26 wt. % NaCl equivalent) solutions are typical of the skarn zones, whereas the massive sulphide ores were deposited at temperatures between 200° and 450° C and variable salinities. Boiling appears to be characteristic of shallower environments (0.3 to 1 kb).

4.2 SEDEX deposits

Detailed fluid inclusion studies have been carried out at the Silvermines deposit (Samson & Russell, 1983, 1987). The mineralizing fluids have high Na concentrations with lesser, more variable K and Ca, and uniformly low Mg concentrations. Fluid inclusion waters in quartz, dolomite, sphalerite, galena and barite have δD values in the range of -23 to -58 ‰. The calculated $\delta^{18}O$ values of the mineralizing fluid, based on mineral values for quartz and dolomite, range from 1.1 to 7.7 ‰. The data represent aqueous fluids that equilibrated with the geosynclinal sequence and granites underlying the deposit (Samson & Russell, 1987).

The δD and $\delta^{18}O$ values of the ore fluid at Silvermines partly overlap that of both MVT and LTM ore fluids.

Homogenization temperatures and salinities in SEDEX deposits range from 100 to 260 °C and from 8 to 28 wt. % NaCl equivalent, respectively with deposition temperatures of Pb-Zn mineralization varying from 100° to 185° C at Navan, 150° to 195° C at Tynagh and 180° to 240° C at Silvermines (Andrew, 1986b). Ore stage fluids at Silvermines were more saline (12 to 22 wt. % NaCl equivalent) than those at Tynagh and Navan which had salinities ranging from 8 to 12 wt. % NaCl equivalent. According to Samson & Russell (1987), at Silvermines a negative correlation between the homogenization temperature and salinity of quartz-hosted fluid inclusions indicates mixing of high temperature, low salinity (8 to 12 wt. % NaCl equivalent) fluids with lower temperature, high salinity (18 to 22 wt. % NaCl equivalent) fluids.

4.3 MVT deposits

The ore fluids of MVT deposits have low temperatures (100° to 150° C), uniform salinity, density and composition (Roedder, 1967, 1984). The gross salinity of the fluid inclusions is usually >15 wt. % NaCl equivalent and frequently >20 % wt. % NaCl equivalent, but no daughter crystals are found, indicating the presence of cations other than Na (Roedder, 1984). Fluid inclusion investigations have shown that MVT deposits form from highly concentrated Na-Ca-Cl brines (Newhouse, 1932; Hall & Friedman, 1963; Roedder, 1967). Apart from Na and Ca chlorides, the solutions contain minor K and Mg with total sulphur (expressed as SO_4^{2-}) seldom exceeding a few thousand ppm. The density is always >1.0 g/cm^3 and frequently >1.1 g/cm^3 and the solute composition is uniform among various MVT districts. In addition, organic matter is frequently but not always observed as gases such as methane, as immiscible oil-like droplets, and as organic compounds in solution in the brines (Roedder, 1984).

Other regional fluid inclusion studies by Leach (1979), Leach & Rowan (1986) and Rowan & Leach (1989) have shown that sphalerite and associated sparry dolomite in the Ozark region have homogenization temperatures in the range 77° to 140° C with salinities from 14 to >23 wt. % NaCl equivalent. The narrow temperature range for sphalerite suggests that the ore fluid was near thermal equilibrium with rocks over a rather large geographic distribution. Even higher fluid inclusion temperatures have been found in sparry dolomite from the Reelfoot Rift Complex where values in the range of 150° to 280° C have been recorded (Leach et al., 1997). According to Leach et al. (1997), these high homogenization temperatures may be related to widespread igneous activity in the Reelfoot Rift Complex in Late Pennsylvanian to Permian times.

As discussed by Kampunzu et al. (2009), fluid inclusions studies in the Pb-Zn-Cu deposits of central Africa indicate at least two stages of mineralization: an earlier high temperature (220° to 390° C) and moderate to high salinity(15 to 43 wt. % NaCl equivalent) stage and a later stage associated with lower fluid temperature (≤80° to 180° C) and variable salinities (<12 to 31 wt. % NaCl equivalent).

The ore fluids have δD values ranging from 0 to -45 ‰ and $\delta^{18}O$ values from -3 to +4.7 ‰ (Hall & Friedman, 1963; Pinkney & Rye, 1972). Hall & Friedman (1963) established that the concentrations of deuterium in ore-stage fluids were relatively high in comparison to that in late gangue calcite which had low deuterium concentrations similar to the local groundwaters. In general MVT ore fluids are similar to SEDEX ore fluids in terms of salinity and homogenization temperatures as indicated above.

5. Sulphur and lead isotopes

5.1 LTM deposits

The isotope ratios of sulphides (including sphalerite, galena, pyrite, chalcopyrite and pyrrhotite) are homogeneous and cluster around 0 ±4 ‰ $\delta^{34}S$ in most LTM deposits (Sawkins, 1964; Rye, 1974; Gilmer et al., 1988; Thompson & Beaty, 1990; Beaty, 1990). Deposits with homogeneous isotope ratios clustering around 0 ±4 ‰ include Providencia, Velardeña, Leadville and Gilman. Other deposits have positive homogeneous values in the range +5.9 to +9.6 ‰ as at Temperino in Italy (Corsini et al., 1980) or from +18.9 to +24.1 ‰ as at Los Lamentos (Megaw et al., 1988). Isotopic equilibrium between sulphide species may or may not have been maintained during mineral deposition.

At Santa Eulalia, a wide spread is observed at the East Camp deposit with $\delta^{34}S$ ratios ranging from +4 to -11 ‰, whereas a much narrower spread occurs in the West Camp deposit where negative values from -8 to -15 ‰ are characteristic (Megaw et al., 1988).

The lead isotope ratios of galena from Mexico are relatively homogeneous within individual mineral districts, and deposits of similar type exhibit similar isotope ratios (Cumming et al., 1979). Lead isotope data for Cenozoic limestone replacement and vein deposits define a linear array which is regarded as a secondary isochron resulting from a two-stage process. The linear array partly coincides with the field of mature arc volcanics of Doe & Zartman (1979), but extends to more radiogenic compositions characterized by pelagic sediments and/or upper crustal material. Isotopic data for these carbonate replacement deposits are characterized by intermediate values between the less radiogenic SEDEX deposits and the more radiogenic MVT Appalachian subdistrict of Ivanhoe-Austinville.

5.2 SEDEX deposits

The $\delta^{34}S$ ratios of sulphide minerals from the epigenetic feeder zones at Silvermines and Keel cluster around 0 ± 10 ‰ (Coomer & Robinson, 1976; Caulfield et al., 1986). Other sulphides, particularly from the stratiform zones at Silvermines and most of the sulphides from Tynagh, are characterized by isotopically light sulphur generally ranging from -42.5 to -4 ‰. Sulphides with relatively narrower spreads occur within this wide range of $\delta^{34}S$ values, such as most of the stage one pyrite (-3.1 to -7.4 ‰) at Tynagh, stratiform ore sphalerite and galena (-20 and -18 ‰) at Silvermines, and sedimentary sphalerite (-15 and -22 ‰) from Navan (Coomer & Robinson, 1976; Boast et al., 1981a; I.K. Anderson et al., 1986). Pyrite from hydrothermal vents at Silvermines also exhibits relatively homogeneous $\delta^{34}S$ values in the range -42.5 to -31.3 ‰ (Type I pyrite vents) and -23.7 to -18.4 ‰ (Type II pyrite vents) (Boyce et al., 1983).

Sulphides with heavy $\delta^{34}S$ values of up to +14 ‰ include post-sedimentary galena at Navan (0 to +14 ‰), and sphalerite and galena from Tatestown (-3 to +14 ‰) (I.K. Anderson et al., 1986; Caulfield et al, 1986). Barite from the Irish deposits forms a fourth distinct isotopic group characterized by positive $\delta^{34}S$ values in the range 14.2 to 22.6 ‰ (Boyce et al., 1983; Andrew & Ashton, 1985; Caulfield et al., 1986).

Caulfield et al. (1986) observed that the lead isotope ratios of the Irish deposits depend on geographic location rather than on the mode of occurrence or age of host rocks, indicating direct involvement of local basement rocks. The [206/204]Pb isotope ratios are parallel to basement trends of Caledonide inheritance and closely follow geophysically identified basement structures. The isotope data defines a linear array of both Lower Paleozoic and Carboniferous-hosted mineralization on the lead-lead plot which reflects mixing between a less radiogenic source with a mantle type lead signature and a more radiogenic source with upper crustal type characteristics (O'Keefe, 1986).

With the exception of the Navan deposit, the lead isotope ratios of galena from Tynagh and Silvermines (Boast et al., 1981b; Boast, 1983), as well as minor deposits such as Keel, Tatestown, Ballinalack and Moyvoughly (Caulfield et al., 1986), are characterized by uniform values within a given deposit. All the lenses except the stratigraphically highest number 1 lens at Navan have uniform isotope ratios on a lens scale (Mills et al., 1987). In general, the average $^{206/204}$Pb ratio decreases from the stratigraphically lowest number 5 lens to the number 1 lens. The number 1 lens contains a less radiogenic component in addition to the more radiogenic lead characteristic of the other lenses.

5.3 MVT deposits

The isotope ratios of sulphur in MVT deposits are generally positive and uniform as in the Upper Mississippi Valley (Pinckney & Rafter, 1972; Maclimans, 1975), or positive with wide ranges as in Southeast Missouri (Sverjensky, 1981). Sulphur with negative isotope ratios has been reported from the Tri-State district (Ault & Kulp, 1960; Deloule et al., 1986).

The data of Pinckney & Rafter (1972) for the stratigraphically correlated bands in ores from four deposits in the Upper Mississippi Valley indicate that the total δ^{34}S range of the main stage sulphides is from 6.3 to 15.9 ‰, but individual sulphides have narrower spreads. The bands include early (bands 1 to 7), intermediate (bands 8 to 10), and late (bands 11 to 18) paragenetic stages. The latest stage is represented by post-sphalerite deposition of galena. The isotopic variations as deposition proceeded are small and largely due to equilibrium isotopic fractionation between sphalerite and galena with decreasing temperature from an ore fluid with essentially constant δ^{34}S$_{H2S}$ (Pinckney & Rafter, 1972; Maclimans, 1975).

An example of isotopic disequilibrium during sulphide precipitation is provided by sulphides from the Buick mine, Southeast Missouri (Sverjensky, 1981) where the successive stages include early (dark sphalerite), intermediate (cuboctahedral galena), late (cubic galena) and latest (pale sphalerite). Early sphalerite (δ^{34}S = 17.54 ‰) and intermediate galena (δ^{34}S = 16.53 ‰) have similar average values, but galena has a wider spread ranging from 13.0 to 21.1 ‰. Cubic galena of the late stage is isotopically distinct from the earlier phases as indicated by a spread from 0.2 to 9.9 ‰. The latest stage of sphalerite is a minor phase with isotope ratios overlapping those for cuboctahedral and cubic galena.

In the Tri-State district Deloule et al. (1986) found a total range of -6 to -12 ‰ δ^{34}S in a single galena crystal from the Pitcher orebody. The inner part of the crystal had a constant δ^{34}S ratio of -9 ‰ δ^{34}S, whereas the outer part had a value of -12 ‰ δ^{34}S with local variations. The main variation is spatially related to a major change in the lead isotope ratio of the crystal, indicating a common source for the lead and sulphur.

In southern and central Africa, the sulphur isotopes of MVT deposits fall into three categories: (1) heavy and homogeneous δ^{34}S values (e.g. Tsumeb δ^{34}S = 20.5 ‰, n= 35; Kamona & Günzel (2007)), (2) highly variable with a wide range of values (e.g. Kombat δ^{34}S = −11.3‰ to +26.2‰ with an average value of +5.2‰, n=19; Hughes (1987)) and (3) negative and homogeneous δ^{34}S values (e.g. Kabwe δ^{34}S = −15.8‰, n= 43; Kamona & Friedrich (2007)). The negative and homogeneous sulphur isotope ratios of ore sulphides (−18 to −12‰ δ^{34}S) from Kabwe are typical of sedimentary sulphides produced through bacterial reduction of seawater sulphate and suggest a sedimentary source for the sulphur, whereas the heavy δ^{34}S isotopes at Tsumeb and other deposits (Berg Aukas, Khusib Springs) indicate seawater as the ultimate source of the sulphur (Kamona & Günzel, 2007). A seawater sulphate source is also suggested for the Kipushi deposit (Kampunzu et al., 2009).

The seawater sulphate may have been initially concentrated in basinal brines and evaporite beds that later provided heavy sulphur to ore-forming hydrothermal solutions that eventually deposited the base metals in the carbonate host rocks. The large variation in the sulphur isotope ratios at Kombat is more enigmatic and could be due to thermochemical sulphate reduction (Innes & Chaplin, 1986), mixing of seawater with a fluid containing isotopically light sulphur derived by bacteriogenic reduction of seawater sulphate, variations in oxygen fugacity and/or pH with decreasing temperature (Hughes, 1987) or the involvement of a magmatic ore fluid (Innes & Chaplin, 1986).

Mississippi Valley type deposits of the United States are characterized by anomalous radiogenic linear trends that do not fit standard lead evolution models (Heyl et al., 1966). This anomalously radiogenic lead is traditionally termed "J-type" after the Joplin galena in the Tri-state district. Galena from the Appalachian district is the least radiogenic of all the MVT Pb-Zn districts (Heyl et al., 1966).

In contrast, the lead isotopes of galena from the Austinville-Ivanhoe subdistrict of the Appalachians (Foley et al., 1981) have a relatively narrow spread. According to Foley et al. (1981), galenas of different generations have distinctive lead isotope ratios, with galena becoming increasingly more radiogenic as ore deposition proceeded. In addition, smaller orebodies have a wider range in lead ratios and are more radiogenic than larger orebodies due to contamination by more radiogenic crustal lead.

Lead isotopic variations have also been observed in single galena crystals from MVT deposits (e.g. Sverjensky, 1981; Deloule et al., 1986; Crocetti et al., 1988). The data indicate frequent and rapid isotopic changes that show increasing radiogenic lead with time. In addition, the data shows several isotopic domains that suggest mixing between at least three end members with distinct isotope ratios such as different sedimentary layers and basement rocks (Deloule et al., 1986).

Sulphides from the Zn–Pb orebodies in southern and central Africa show small variations in their Pb isotopic compositions with values typical of the upper continental crust (Hughes et al., 1984; Kamona et al., 1999).

6. Discussion

6.1 LTM deposits

The δD and $\delta^{18}O$ ratios for waters in fluid inclusions at Providencia and Leadville (Rye & O'Neil, 1968; Thompson & Beaty, 1990), as well as the close spatial and temporal relationship between ores and intrusives, suggest a magmatic origin for the ore fluids and metals (Shimazaki, 1980; Einaudi et al., 1981). The deposits may be associated with subduction-related, magnetite-series granitoids (Einaudi et al., 1981) or with magnetite- and ilmenite-series granitoids as in Japan (Shimazaki, 1980).

The lead isotope data of Cumming et al. (1979) indicates that in the case of the Mexican deposits, the lead was derived from the Cenozoic mantle as well as from the Precambrian crust by magma contamination. The sulphur isotope values do not indicate a unique magmatic source for all the deposits since the $\delta^{34}S$ values of granitoid source-magmas are controlled by geographic location and may therefore reflect sulphur isotope ratios of local country rocks (Ohmoto, 1986). Nevertheless, a predominantly magmatic sulphur source is indicated in deposits such as Providencia (Rye, 1974; Sawkins, 1964), Velardeña (Gilmer et al., 1988), and Leadville (Thompson & Beaty, 1990).

The high concentrations of chlorine in fluid inclusions at Providencia (Rye & Haffty, 1969) indicate that the major metals are transported as chloride complexes to the deposition site. The general lack of sulphates suggests that H_2S is the dominant sulphur species (Ohmoto, 1986). The solutions were moving up fast enough to maintain uniformly heated wall rocks adjacent to the fluid conduits as indicated by the low vertical temperature gradient of less than 50° C/km at Providencia (Sawkins, 1964). Initial skarn formation by infiltrational metasomatism and diffusion in limestones occurs between 400° and 650° C and precedes sulphide deposition (Einaudi et al., 1981; Einaudi & Burt, 1982). Sulphide deposition takes place as a consequence of declining temperature, local oxidation-reduction reactions, or neutralization of the ore fluid at the marble contact. A generalized reaction between metal chlorides and hydrogen sulphide leading to sulphide deposition is (Barnes, 1979):

$$MeCl_{2(aq)} + H_2S_{(aq)} = MeS + 2H^+ + 2Cl^- \qquad (1)$$

where Me is metal (e.g. Zn or Pb). From reaction (1) deposition of sulphides could be caused by an increase in reduced sulphur (due to sulphate reduction or leaching of previously formed sulphide minerals), an increase in pH (e.g. acid neutralization by carbonate dissolution), and a decrease in chloride concentration (by removal of Cl^- ions through pairing by ions such as Ca^{2+} and Mg^{2+}). A decrease in temperature of the chloride-sulphide bearing fluid stabilizes H_2S relative to SO_2 and thus promotes deposition of metal sulphides (Rubin & Kyle, 1988).

6.2 SEDEX deposits

In the case of SEDEX deposits there is no general consensus about fluid and metal sources (White, 1986). The problem is caused by the fact that although the lead isotope data indicates the involvement of basement rocks, it does not distinguish between lead derived directly from the basement and lead from disaggregated basement rocks. O'Keefe (1986) established that epigenetic lead veins hosted in Lower Paleozoic rocks in Ireland are of the same age and have been derived from the same lead sources as the surrounding Carboniferous-hosted deposits, suggesting a pre-Carboniferous source for the mineralization. The relative abundance of Pb over Cu indicates silicic continental-derived rocks and sediments which are relatively rich in Pb rather than deep ocean rifts and basaltic rocks with high Cu and Zn but low Pb contents (White, 1986).

A basement source for the metals in the Irish SEDEX deposits is postulated by the deep con-vection cell model of Russell (1978) and Russell et al. (1981) in which fluid convection within deep fault structures in upper crustal rocks leads to ore fluid generation. In the model, the ore solution evolves from seawater into a brine capable of leaching metals from underlying rocks during fluid-rock interaction with increasing depth down to the base of the brittle zone in the upper crust where temperatures approach 250° C.

Objections raised against the deep fluid convection cell model concern the lack of an intimate water-rock contact (Cathles, 1986) and the absence of suitable rock types such as basalts susceptible to mineral:solution disequilibrium (Lydon, 1986). Lydon (1986) proposed a stratal aquifer model in which the ore fluid is generated from connate waters within an arenaceous aquifer. The metals are derived mostly from clays and iron-oxide pigments in the aquifer itself with minor contributions from faulted basement rocks during reflux convection to depths of 1-3 km and not up to 15 km as in the Russell model. Dense chloride rich brines were produced by evaporitic conditions during the early part of a marine transgression and collected as connate pore waters in Old Red Sandstone lithologies within a basinal structure. The brines became geopressured and heated to over 200° C due to hydraulic sealing and thermal insulation by about 500 m of semi-compacted Lower Carboniferous shale and argillaceous limestone. Heat to the stratal aquifer is supplied directly by conduction from basement rocks and by reflux convection of brines along dilatant fault zones to depths of 3 km.

The sources of sulphur are less problematic. The sulphur isotope ratios of sulphides from the epigenetic feeder zones of the Irish SEDEX deposits are similar to those of diagenetic and vein sulphides in underlying Lower Paleozoic strata from which the sulphur was presumably leached (I.K. Anderson et al., 1989). An ultimate source for the sulphur may be Caledonian granites in the age range 390 to 435 Ma which have a mean $\delta^{34}S$ value of +0.7 ± 2.6 ‰ (Laouar, et al., 1990). Isotopically light sulphur with ratios from -42.5 to -4 ‰ is considered to have originated through bacteriogenic reduction of Lower Carboniferous seawater sulphate and was the dominant sulphur source in the Pb-Zn deposits (Hitzman & Large, 1986; I.K. Anderson et al., 1989). A third group of sulphides in the Irish SEDEX ores is characterized by $\delta^{34}S$ values in the range -3 to +14 ‰. The heavy sulphur component in these sulphides is believed to be of evaporitic origin (Caulfield et al., 1986), but it may represent deep-seated sulphur or low temperature chemical reduction of seawater at the deposition site (I.K. Anderson et al., 1986). Barite formed by reaction of exhaled Ba^{2+} with Lower Carboniferous seawater sulphate (Boyce et al., 1983).

It is generally agreed that stratiform sediment-hosted ores are deposited with or soon after the enclosing sediments and are therefore syngenetic or syndiagenetic (White, 1986). In the deep fluid convection model, circulation of seawater is initiated during rifting and is driven by a high geothermal gradient as indicated by local volcanic flows and/or shallow intrusives in the rock successions at Navan, Tynagh and Silvermines (Russell et al., 1981). The modified seawater with metals and H_2S in solution returns to surface along convective updrafts such as intersecting major fractures. Ore deposition takes place as a result of fluid mixing with alkaline groundwaters and seawater brines. In contrast, in the stratal aquifer model, geopressured formation waters are released intermittently to the surface along fault zones during periods of tectonic activity (Lydon, 1983) and ore precipitation takes place from a brine pool in a third-order basin due to cooling and/or bacteriogenic reduction of sulphate to sulphide within the water column.

6.3 MVT deposits

The origin of MVT deposits is regarded as a normal part of sedimentary basin evolution (Jackson & Beales, 1967). According to the basinal brine model of Beales & Jackson (1966, 1967), ore-forming brines are produced during burial of sediments. Heat, metals and other solutes are acquired during brine migration along aquifers. The similarity of the fluid inclusion data in minerals from MVT deposits to typical oil-field brines with respect to element compositions, high salinities, and hydrogen and oxygen isotopes has been noted (Hall & Friedman, 1963; Roedder, 1967). The main difference is the much lower Na/K ratio of MVT fluids (Sawkins, 1968; White, 1968). This difference may be due to water-rock interactions in aquifers during fluid migration (Sverjensky, 1984; Roedder, 1984), or to contributions of interstitial fluids from evaporite beds enriched in residual potassium (G.M. Anderson & Macqueen, 1982). According to Crocetti & Holland (1989), the lower Na/K ratio of MVT fluids results from equilibration with K-feldspar-albite assemblages in basement and/or sedimentary rocks, whereas higher ratios are due to equilibrium reactions with clay minerals.

The much higher salinities of the brines compared to seawater result from dissolution of salt beds by surface-derived groundwaters or by expulsion of interstitial fluids from evaporite beds (G.M. Anderson & Macqueen, 1982). The chloride-rich brines are able to leach metals from rocks during fluid migration by desorption of loosely bound metals, or release of metals from metal-organic complexes through thermal alteration or destruction of such complexes.

The isotope ratios of sulphur are generally positive and may be uniform or have wide ranges, indicating a crustal, ultimately seawater origin involving sulphate reduction (G.M. Anderson & Macqueen, 1982; Sverjensky, 1986). Multiple sources for the lead have been suggested, including a radiogenic lead component from the basement rocks, lead from carbonate cements and feldspars in sandstones, and normal lead from oil-field brines (Brown, 1967; Doe & Delevaux, 1972). The isotopic studies of Deloule et al. (1986) indicate that numerous and discontinuous inputs of Pb and S from isotopically distinct sources such as different sedimentary layers and basement rocks in the same basin are involved.

In the basinal-brine model, fluids are expelled from strata by sediment compaction during early stages of diagenesis. However, Bethke (1986) observes that fluids displaced from deep basins by compaction-driven flow move too slowly to avoid conductive cooling and cannot account for the deposition temperatures of 100° to 150° C. Episodic dewatering from overpressured aquifers (Cathles & Smith, 1983) also appears to be incapable of adequate heat transfer because of the small volumes required and a narrowly defined aquifer (Leach & Rowan, 1986). In addition, significant overpressures did not develop in basins such as the Illinois basin in the Upper Mississippi Valley (Bethke, 1986). A gravity-driven groundwater flow system due to topographic differences across basins with high heat flows and adequate aquifer permeabilities is considered more likely for the mid-continent districts (Bethke, 1986; Leach & Rowan, 1986).

The fluids are expelled from basins that have undergone deformation or uplift as indicated by the proximity of MVT deposits to tectonically deformed or uplifted basin margins, a coincidence which may be an important unifying factor for MVT deposits in diverse geologic provinces (Leach & Rowan, 1986). Other mechanisms include topography-driven, sediment compaction and overloading, overpressured gas reservoirs, and thermal and density reflux drives (e.g. Leach et al., 2010; Vearncombe et al., 1996)

Due to the low solubilities of sphalerite and galena in the presence of significant H_2S in solutions at 100° to 150° C, the metals and H_2S are not transported together (G.M. Anderson, 1983). The sulphide is instead supplied at the deposition site by reduction of sulphate already in the brine or by adding H_2S to the brine, resulting in sulphide deposition. The source of reduced sulphur is sulphate, either from locally available sulphate minerals or dissolved brine-transported sulphate. This is the so-called "mixing-model" which has been applied with success at the Pine Point deposit where H_2S was derived locally through thermochemical reactions involving sulphate, bitumen and H_2S (Macqueen & Powell, 1986). Alternatively, transport of metals and H_2S in the order of 1 ppm is possible under conditions of acidic pH and high total dissolved CO_2 concentrations (Sverjensky, 1986), as indicated by the correlation of Pb and S isotopes in galena in the Southeast Missouri district and the district wide sphalerite color banding in the Upper Mississippi Valley. This is the "non-mixing" or "reduced-sulphur" model in which deposition of metal sulphides may be caused by pH change, cooling or dilution with groundwater.

The Zn–Pb–(Cu) deposits of southern and central Africa formed from basinal brines during two main mineralizing events that characterize syntectonic deposits (e.g., Kabwe, Tsumeb) and post-tectonic deposits (e.g., Kipushi). These deposits exhibit many of the typical characteristics of Mississipi Valley-type ore-forming systems, including their classical stratabound, epigenetic nature and occurrence in clusters within platform marine carbonates overlying continental crust (e.g. Frimmel et al., 1996; Kampunzu et al., 2009). A sialic crustal origin of the metals from basinal sediments and basement rocks is supported by lead and strontium isotope data, whereas the sulphur isotope compositions are compatible with a sedimentary origin of sulphur from seawater via evaporites and/or diagenetic sulphides. The main ore-forming fluids were saline with moderate to high temperatures and could have been produced by normal geothermal gradients during basin evolution. Ore deposition occurred in carbonate rocks with favourable permeable structures including faults, veins, breccias and hydrothermal karsts as a result of cooling, fluid mixing, pH change or addition of H_2S.

Although the one universal and fundamental characteristic of MVT deposits is the absence of igneous rocks that are potential sources of the ore solutions (Ohle, 1959), intermittent igneous activity occurred in the Southeast Missouri district from Cambrian to Cretaceous time (Thacker & K.H. Anderson, 1977). In particular, Paleozoic explosive igneous activity and intrusion of dikes occurred during Upper Cambrian and Devonian times in the district (Gerdemann & Myers, 1972). In addition, a volcanic tuff bed of up to 1 m or more in thickness occurs over a limited area near the top of a sandy transition zone at the Lamotte-Bonneterre contact. Although the igneous activity was localized and short-lived to have had much direct bearing on the mineralization, it provides evidence that the region had a high geothermal gradient (Gerdemann & Myers, 1972).

7. Conclusions

Carbonate-hosted Pb-Zn deposits range from high temperature, intrusion –related LTM ores through moderate temperature, syngenetic to syndiagenetic SEDEX ores to low and moderate temperature MVT ores characterized by a lack of associated igneous rocks. Metal ratios expressed as Zn/(Zn+Pb) vary from 0.5 in LTM districts to 0.7 in both SEDEX and MVT deposits. Pyrite, sphalerite and galena are the major sulphides, but the mineral paragenesis and relative proportions of these sulphides vary from deposit to deposit, even within the same district. On average Ag grades are relatively high in LTM (>200 ppm) compared to SEDEX (63 ppm) and MVT (96.8 ppm) ore deposits. Copper may be significant in some MVT deposits, especially those in which sulphosalts of tennantite and tetrahedrite form part of the mineral paragenesis as at the Tsumeb and Kipushi deposits.

There are numerous mechanisms that may be responsible for MVT ore fluid migration in sedimentary basins, including topography-driven, sediment compaction, orogenic squeezing, overpressured gas reservoirs, as well as thermal and density reflux mechanisms. However, no single mechanism can be applied to all districts. In the case of the LTM deposits metal and sulphur sources include mantle and crustal rocks, whereas for SEDEX and MVT deposits the metals may have been derived from basement rocks as well as basin sediments during fluid-rock interactions over a prolonged period of time in sedimentary basins with a high heat flow. In all three cases the formation of large deposits requires focusing of ore fluids along conduits that include faults and folds, stratigraphic aquifers, discontinuities and contacts, breccias, pinch-outs and reefs in carbonate host rocks.

8. References

Anderson, G.M. (1983). Some geochemical aspects of sulphide precipitation in carbonate rocks. In: *International Conference on Mississippi Valley-type lead-zinc deposits.* Kisvarsanyi, G., Grant, S.K., Pratt, W.P. & Koenig, J.W. (Eds.), Proceedings Vol., pp. 61-76, University of Missouri, Rolla.

Anderson, G.M. & Macqueen, R.W. (1982). Mississippi Valley-type lead zinc deposits. *Geoscience Canada,* Vol. 9, No. 2, pp. 108-117, ISSN 0315-0941.

Anderson, I.K., Andrew, C.J., Ashton, J.H., Boyce, A.J., Caulfield, J.B.D., Fallick, A.E. & Russell, M.J. (1989). Preliminary sulphur isotope data of diagenetic and vein sulphides in the Lower Paleozoic strata of Ireland and northern Scotland: implications for Zn + Pb + Ba mineralization. *Journal of the Geological Society of London,* Vol. 146, No. 4, pp. 715-720.

Anderson, I.K., Boyce, A.J., Russell, M.J., Fallick, A.E., Hall, A.J. & Ashton, J. (1986). Textural and sulphur isotopic support for Lower Carboniferous sedimentary exhalative base metal deposition at Navan, Ireland. *Terra Cognita,* Vol. 6, p.133.

Andrew, C.J. (1986a). The tectono-stratigraphic controls to mineralization in the Silvermines area, County Tipperary, Ireland. In: *Geology and genesis of mineral deposits in Ireland,* Andrew, C.J., Crowe, R.W.A., Finlay, S., Pennell, W.M. & Pyne, J.F. (Eds.), pp. 377-417, Irish Association for Economic Geology, Dublin.

Andrew, C.J. (1986b). Sedimentation, tectonism, and mineralization in the Irish orefield. In: In: *The genesis of stratiform sediment-hosted lead and zinc deposits: conference proceedings,* Turner, R.J.W. & Einaudi, M.T. (eds.), Vol. XX, pp. 44-56, Stanford University Publications.

Andrew, C.J. & Ashton, J.H. (1985). Regional setting, geology and metal distribution patterns of the Navan orebody, Ireland. *Transactions of the Institute of Mining and Metallurgy,* Vol. 94, B66-B93, ISSN 0371-7453.

Ashton, J.H., Downing, D.T. & Finlay, S. (1986). The geology of the Navan Zn-Pb orebody. In: *Geology and genesis of mineral deposits in Ireland,* Andrew, C.J., Crowe, R.W.A., Finlay, S., Pennell, W.M. & Pyne, J.F. (Eds.), pp. 243-280, Irish Association for Economic Geology.

Barnes, H.L. (1979). Solubilities of ore minerals. In: *Geochemistry of hydrothermal ore deposits,* Barnes, H.L. (Ed.), pp. 404-460, Wiley-Interscience, 2nd. ed., New York.

Beales, F.W. & Jackson, S.A. (1966). Precipitation of lead-zinc ores in carbonate rocks as illustrated by Pine Point ore field. *Transactions of the Canadian Institution of Mining and Metallurgy,* Vol. 75, pp. B278-B285, ISSN 0371-5701.

Beaty, D.W., Landis, G.P. & Thompson, T.B. (1990). Carbonate-hosted sulfide deposits of the Central Colorado Mineral Belt: Introduction, general discussion and summary. *Economic Geology Monograph 7,* pp. 1-18.

Beaty, D.W. (Ed.) (1990). Origin of the ore deposits at Gilman, Colorado. *Economic Geology Monograph 7,* pp. 193-265.

Bethke, C.M. (1986). Hydrologic constraints on the genesis of the Upper Mississippi Valley mineral district from Illinois basin brines. *Economic Geology,* Vol. 81, No. 2, pp. 233-249, ISSN 0361-0128.

Boast, A.M. (1983). Discussions and contributions. *Transactions of the Institute of Mining and Metallurgy,* Vol. 92, p. B101, ISSN 0371-7453.

Boast, A.M., Coleman, M.L. & Halls, C. (1981a). Textural and stable isotopic evidence for the genesis of the Tynagh base metal deposit, Ireland. *Economic Geology*, Vol. 76, No. 1, pp. 27-55, ISSN 0361-0128.

Boast, A.M., Swainbank, I.G., Coleman, M.L. & Halls, C. (1981b). Lead isotope variation in the Tynagh, Silvermines and Navan base-metal deposits, Ireland. *Transactions of the Institute of Mining and Metallurgy,*Vol. 90, pp. B115-B119, ISSN 0371-7453.

Bookstrom, A.A. (1990). Igneous rocks and carbonate-hosted ore deposits of the Central Colorado Mineral Belt. *Economic Geology Monograph 7*, pp. 45-65.

Boyce, A.J., Coleman, M.L. & RusselL, M.J. (1983). Formation of fossil hydrothermal chimneys and mounds from Silvermines, Ireland. *Nature*, Vol. 306, pp. 545-550.

Brown, J.S. (1967). Isotopic zoning of lead and sulfur in Southeast Missouri. In: *Genesis of stratiform lead-zinc-barite-fluorite deposits in carbonate rocks*, Brown, J.S. (Ed.),. Economic Geology Monograph 3, pp. 410-426.

Callahan, W.H. (1967). Some spatial and temporal aspects of the localization of Mississippi Valley-Appalachian type ore deposits. In: *Genesis of stratiform lead-zinc-barite-fluorite deposits in carbonate rocks*, Brown, J.S. (Ed.), Economic Geology Monograph 3, pp. 14-19.

Cathles, L.M. (1986). A tectonic/hydrodynamic view of basin-related mineral deposits. In: *The genesis of stratiform sediment-hosted lead and zinc deposits: conference proceedings*, Turner, R.J.W. & Einaudi, M.T. (Eds.), Vol. XX, pp. 171-176, Stanford University Publications.

Cathles, L.M. & Smith, A.T. (1983). Thermal constraints on the formation of Mississippi Valley-type lead-zinc deposits and their implications for episodic basin dewatering and deposit genesis. *Economic Geology*, Vol. 78, No. 5, pp. 933-1002, ISSN 0361-0128.

Caulfield, J.B.D., LeHuray, A.P. & Rye, D.M. (1986). A review of lead and sulphur isotope investigations of Irish sediment-hosted base metal deposits with new data from the Keel, Ballinalack, Moyvoughly and Tatestown deposits. In: *Geology and genesis of mineral deposits in Ireland*. Andrew, C.J., Crowe, R.W.A., Finlay, S., Pennell, W.M. & Pyne, J.F. (Eds.), pp. 591-615, Irish Association for Economic Geology, Dublin.

Clifford, J.A., Ryan, P. & Kucha, H. (1986). A review of the geological setting of the Tynagh orebody, County Galway. In: *Geology and genesis of mineral deposits in Ireland. Irish Association forEconomic Geology*, Andrew, C.J., Crowe, R.W.A., Finlay, S., Pennell, W.M. & Pyne, J.F. (Eds.), pp. 419-439, Irish Association for Economic Geology, Dublin.

Coomer, P.G. & Robinson, B.W. (1976). Sulphur and sulphate-oxygen isotopes and the origin of the Silvermines deposits, Ireland. *Mineralium Deposita*, Vol. 11, No. 2, pp. 155-169, ISSN 0026-4598.

Corsini, F., Cortecci, G., Leone, G. & Tanelli G. (1980). Sulfur isotope study of the skarn-(Cu-Pb-Zn) sulfide deposit of Valle del Temperino, Campiglia Marittima, Tuscany, Italy. *Economic Geology*, Vol. 75, No. 1, pp. 83-96, ISSN 0361-0128.

Craig, J.R., Solberg, T.N. & Vaughan, D.J. (1983). Growth characteristics of sphalerites in Appalachian zinc deposits. In: *International Conference on Mississippi Valley type lead-zinc deposits*, Proceedings Vol., Kisvarsanyi, G., Grant, S.K., Pratt, W.P. & Koenig, J.W. (Eds.), pp. 317-327, University of Missouri, Rolla.

Crocetti, C.A. & Holland, H.D. (1989). Sulfur-lead isotope systematics and the composition of fluid inclusions in galena from the Viburnum Trend, Missouri. *Economic Geology*, Vol. 84, No. 8, pp. 2196-2216, ISSN 0361-0128.

Crocetti, C.A., Holland, H.D. & Mckenna, L.W. (1988). Isotopic composition of lead in galenas from the Viburnum Trend, Missouri. *Economic Geology*, Vol. 83, No. 2, pp. 355-376, ISSN 0361-0128.

Cumming, G.L., Kesler S.E. & Krstic D. (1979). Isotopic composition of lead in Mexican mineral deposits. *Economic Geology*, Vol. 74, No. 6, 1395-1407, ISSN 0361-0128.

Deloule, E., Allegre, C. & Doe, B. (1986). Lead and sulfur isotope microstratigraphy in galena crystals from Mississippi Valley-type deposits. *Economic Geology*, Vol. 81, No. 6, pp. 1307-1321, ISSN 0361-0128.

Doe, B.R. & Delevaux, M.H. (1972). Source of lead in southeast Missouri galena ores. *Economic Geology*, Vol. 67, No. 4, pp. 409-425, ISSN 0361-0128.

Doe, B.R. & Zartman, R.E. (1979). Plumbotectonics, the Phanerozoic. In: *Geochemistry of hydrothermal ore deposits*, Barnes, H.L. (Ed.), 2nd. ed., pp. 22-70, Wiley-Interscience, New York.

Einaudi, M.T., Meinert, L.D. & Newberry, R.J. (1981). Skarn deposits. In: *Economic Geology 75th Anniversary Volume 1905-1980*, Skinner, B.J. (Ed.), pp. 317-391, Economic Geology Publishing Company, Catalog Card No. 81-84611, New Haven.

Einaudi, M.T. & Burt, D.M. (1982). Introduction - Terminology, classification, and composition of skarn deposits. *Economic Geology*, Vol. 77, No. 4, pp. 745-754, ISSN 0361-0128.

Foley, N.K., Sinha, A.K. & Craig, J.R. (1981). Isotopic composition of lead in the Austinville-Ivanhoe Pb-Zn district, Virginia. *Economic Geology*, Vol. 76, pp. 2012-2017, ISSN 0361-0128.

Frank, M.H. & Lohmann, K.C. (1986). Textural and chemical alteration of dolomite: interaction of mineralizing fluids and host rock in a Mississippi Valley-type deposit, Bonneterre Formation, Viburnum Trend. In: *Process Mineralogy VI: conference proceedings, Metallurgical Society Annual Meeting*, pp. 103-116, Louisiana, March 2-6, 1986, 103-116.

Frimmel, H.E., Deane, J.G. & Chadwick, P.J. (1996). Pan-African tectonism and the genesis of base metal sulphide deposits in the northern foreland of the Damara Orogen, Namibia. In: *Carbonate-hosted lead-zinc deposits*, Sangster, D.F. (Ed.), pp. 204-217, Society of Economic Geologists, 1-887483-95-0, Littleton.

Gerdemann, P.E. & Myers, H.E. (1972). Relationships of carbonate facies patterns to ore distribution and to ore genesis in the southeast Missouri lead district. *Economic Geology*, Vol. 67, No. 4, 426-433, ISSN 0361-0128.

Gilmer, A.L., Clark, K.F., Conde, J.C., Hernandez, I.C., Figueroa, J.I.S. & Porter, E.W. (1988). Sierra de Santa Maria, Velardeña Mining District, Durango, Mexico. *Economic Geology*, Vol. 83, No. 8, pp. 1802-1829, ISSN 0361-0128.

Goodfellow, W.D. & Lydon, J.W. (2007). Sedimentary exhalative (SEDEX) deposits. In: *Mineral Deposits of Canada: A Synthesis of Major Deposit Types, District Metallogeny, the Evolution of Geological Provinces, and Exploration Methods*. Goodfellow, W.D., (Ed.), Geological Association of Canada, Mineral Deposits Division, Special Publication No. 5, pp. 163-183.

Grundmann, W.H. Jr. (1977). Geology of the Viburnum No. 27 mine, Viburnum Trend, Southeast Missouri. *Economic Geology*, Vol. 72, No. 3, pp. 349-364, ISSN 0.61-0128.

Hagni, R.D. (1976). Tri-State ore deposits: the character of their host rocks and their genesis. In: *Handbook of stratabound and stratiform ore deposits*, Wolf, K.H. (Ed.), Vol. 6, pp. 457-494. Elsevier, New York.

Hagni, R.D. (1982). The influence of original host rock character upon alteration and mineralization in the Tri-State district of Missouri, Kansas, and Oklahoma, U.S.A. In:

Ore genesis: the state of the art, Amstutz, G.C. et al., (Eds.), pp. 97-107, Springer-Verlag, Heidelberg.

Hagni, R.D. (1983). Ore microscopy, paragenetic sequence, trace element content, and fluid inclusion studies of the copper-lead-zinc- deposits of the Southeast Missouri lead district. In: *International Conference on Mississippi Valley type lead-zinc deposits.* Kisvarsanyi, G., Grant, S.K., Pratt, W.P. & Koenig, J.W. (Eds.), Proceedings Vol., pp. 243-256, University of Missouri, Rolla.

Hagni, R.D. & Trancynger T.C. (1977). Sequence of deposition of the ore minerals at the Magmont mine, Viburnum Trend, Southeast Missouri. *Economic Geology*, Vol. 72, No. 3, pp. 451-464, ISSN 0361-0128.

Hall, W.E. & Friedman, I. (1963). Composition of fluid inclusions, Cave-in-Rock fluorite district, Illinois, and Upper Mississippi Valley zinc-lead district. *Economic Geology*, Vol. 58, No. 6, pp. 886-911, ISSN 0361-0128.

Heyl, A.V. (1967). Some aspects of genesis of stratiform zinc-lead-barite-fluorite deposits in the United States. In: *Genesis of stratiform lead-zinc-barite-fluorite deposits in carbonate rocks*. Brown, J.S. (Ed.), pp. 20-32, Economic Geology Monograph 3.

Heyl, A.V. (1983). Geologic characteristics of three major Mississippi Valley districts. In: *International Conference on Mississippi Valley type lead-zinc deposits*, Kisvarsanyi, G., Grant, S.K., Pratt, W.P. & Koenig, J.W. (Eds.), Proceedings Vol., pp. 27-60, University of Missouri, Rolla.

Heyl, A.V., Delevaux, M.H., Zartman, R.E. & Brock, M.R. (1966). Isotopic study of galenas from the Upper Mississipppi Valley, the Illinois-Kentucky, and some Appalachian Valley mineral districts. *Economic Geology*, Vol. 61, No. 5, 933-961, ISSN 0361-0128.

Hitzman, M. W. & Beaty, D.W. (1996). The Irish Zn-Pb-(Ba) orefield. In: *Carbonate-hosted lead-zinc deposits*, Sangster, D.F. (Ed.), pp. 112-143, Society of Economic Geologists, 1-887483-95-0, Littleton.

Hitzman, M. W. & Large, D. (1986). A review and classification of the Irish carbonate-hosted base metal deposits. In: *Geology and genesis of mineral deposits in Ireland*, Andrew, C.J., Crowe, R.W.A., Finlay, S., Pennell, W.M. & Pyne, J.F. (Eds.), pp. 217-238, Irish Association for Economic Geology, Dublin.

Hoagland, A.D. (1967). Interpretations relating to the genesis of East Tennessee zinc deposits. In: *Genesis of stratiform lead-zinc-barite-fluorite deposits in carbonate rocks*, Brown, J.S. (Ed.), pp. 52-58, Economic Geology Monograph 3.

Hoagland, A.D. (1976). Appalachian zinc-lead deposits. In: *Handbook of stratabound and stratiform ore deposits*, Wolf, K.H. (Ed.), Vol. 6, pp. 495-534, Elsevier, New York.

Hutchinson, R.W. (1996). Regional metallogeny of carbonate-hosted ores by comparison of field relationships. In: *Carbonate-hosted lead-zinc deposits*, Sangster, D.F. (Ed.), pp. 08-17, Society of Economic Geologists, 1-887483-95-0, Littleton.

Jackson, S.A. & Beales, F.W. (1967). An aspect of sedimentary basin evolution: the concentration of Mississippi Valley-type ores during the late stages of diagenesis. Bulletin of the Canadian Society of Petroleum Geologists, Vol. 15, No. 4, pp. 393-433.

James, J.A. (1952). Structural environments of the lead deposits in the southeastern Missouri mining district. *Economic Geology*, Vol. 47, No. 6, pp. 650-660, ISSN 0361-0128.

Jessey, D.R. (1983). The occurrence of nickel and cobalt in the southeast Missouri mining district. In: *International Conference on Mississippi Valley type lead-zinc deposits*,

Kisvarsanyi, G., Grant, S.K., Pratt, W.P. & Koenig, J.W. (Eds.), Proceedings Vol., pp. 145-154, University of Missouri, Rolla.

Kisvarsanyi, G. (1977). The role of the Precambrian igneous basement in the formation of the stratabound lead-zinc-copper deposits in Southeast Missouri. *Economic Geology*, Vol. 72, No. 3, pp. 435-442, ISSN 0361-0128.

Kamona, F. & Friedrich, G.H. (2007). Geology, mineralogy and stable isotopes geochemistry of the Kabwe carbonate-hosted Pb–Zn deposit, Central Zambia. *Ore Geology Reviews*, Vol. 30, Nos. 3-4, pp. 217–243, ISSN 0169-1368.

Kamona, A.F. & Günzel, A. (2007). Stratigraphy and base metal mineralization in the Otavi Mountain Land, Northern Namibia. *Gondwana Research*, Vol. 11, No. 3, pp. 396–413, ISSN 1342-937X.

Kamona, A.F., Lévêque, J., Friedrich, G., Haack, U. (1999). Lead isotopes of the carbonate hosted Kabwe, Tsumeb, and Kipushi Pb–Zn–Cu sulphide deposits in relation to Pan African orogenesis in the Damaran–Lufilian fold belt of Central Africa. *Mineralium Deposita*, Vol. 34, No. 3, pp. 273–283, ISSN 0026-4598.

Kampunzu, A.B., Cailteux, J.L.H., Kamona, A.F., Intiomale, M.M. & Melcher, F. (2009). Sediment-hosted Zn–Pb–Cu deposits in the Central African Copperbelt. *Ore Geology Reviews*, Vol. 35, Nos. 3-4, pp. 263-297, ISSN 0169-1368.

Klau, W. & Mostler, H. (1986). On the formation of Alpine Middle and Upper Triassic Pb-Zn deposits, with some remarks on Irish carbonate-hosted base metal deposits. In: *Geology and genesis of mineral deposits in Ireland*, Andrew, C.J., Crowe, R.W.A., Finlay, S., Pennell, W.M. & Pyne, J.F. (Eds), pp. 663-675, Irish Association for Economic Geology, Dublin.

Köppel, V. (1983). Summary of lead isotope data from ore deposits of the Eastern and Southern Alps: some metallogenic and geotectonic implications. In: *Mineral deposits of the Alps and of the Alpine Epoch in Europe*, Schneider, H.-J. (Ed.), pp. 162-168, Springer-Verlag, 0387122311, Heidelberg.

Kyle, J.R. (1976). Brecciation, alteration, and mineralization in the Central Tennessee zinc district. - *Economic Geology*, Vol. 71, No. 5, pp. 892-903, ISSN 0361-0128..

Laouar, R., Boyce, A.J., Fallick, A.E. & Leake, B.E. (1990). A sulphur isotope study on selected Caledonian granites of Britain and Ireland. *Geology Journal*, Vol. 25, pp. 359-369.

Large, D. E. (1983). Sediment-hosted massive sulphide lead-zinc deposits: an empirical model. In: *Sediment-hosted stratiform lead-zinc deposits*, Sangster, D.F. (Ed.), Short Course Handbook, Vol. 8, SC9, pp. 1-29, Mineralogical Association of Canada.

Leach, D.L. (1979). Temperature and salinity of the fluids responsible for minor occurrences of sphalerite in the Ozark region of Missouri. *Economic Geology*, Vol. 74, No. 4, pp. 931-937, ISSN 0361-0128.

Leach, D.L. & Rowan, E.L. (1986). Genetic link between Ouachita foldbelt tectonism and the Mississippi Valley-type lead-zinc deposits of the Ozarks. *Geology*, Vol. 14, No.11, 931-935, ISSN 0091-7613.

Leach, D.L., Apodaca, L.E., Repetski, J.E., Powell, J.W. & Rowan, E.L. (1997). Evidence for hot Mississippi Valley-type brines in the Reelfoot rift complex, south-central United States, in late Pennsylvanian-Early Permian. U.S. Geological Survey professional paper 1577, Accessed 09/04/2003, Available from: http://www.pubs.usgs.gov/puprod.

Leach, D.L., Sangster, D.F., Kelley, K.D., Large, R.R., Garven, G., Allen, C.R., Gutzmer, J., & Walters, S. (2005). Sediment-hosted lead-zinc deposits: a global perspective. Economic Geology One Hundredth Anniversary Volume 1905-2005, pp. 561-607, *Society of Economic Geologists*, ISSN 0361-0128.

Leach, D.L., Taylor, R.D., Fey, D.L., Diehl, S.F., & Saltus, R.W. (2010). A deposit model for Mississippi Valley-Type lead-zinc ores. In: *Mineral deposit models for resource assessment*. Accessed 15 January 2011, Available from; http://www.usgs.gov/pubprod.

Lydon, J.W. (1983). Chemical parameters controlling the origin and deposition of sediment-hosted stratiform lead-zinc deposits. In: *Sediment-hosted stratiform lead-zinc deposits*, Sangster, D.F. (Ed.), Short Course Handbook, Vol. 8, SC9, pp. 175-250, Mineralogical Association of Canada.

Lydon, J.W. (1986). Models for the generation of metalliferous hydrothermal systems within sedimentary rocks and their applicability to the Irish Carboniferous Zn-Pb deposits. In: *Geology and genesis of mineral deposits in Ireland*, Andrew, C.J., Crowe, R.W.A., Finlay, S., Pennell, W.M. & Pyne, J.F. (Eds.), p. 555-577, Irish Association for Economic Geology, Dublin.

Macqueen, R.W. & Powell, T.G. (1986). Origin of the Pine Point lead-zinc deposits, N.W.T., Canada: organic geochemical and isotopic evidence. *Terra Cognita*, Vol. 6, p. 134.

McLimans, R.K. (1975). Systematic fluid inclusion and sulfide isotope studies of the Upper Mississippi valley Pb-Zn deposits. Economic Geology, Vol. 70, No. 7, p. 1324, ISSN 0361-0128.

McLimans, R.K., Barnes, H.L. & Ohmoto, H. (1980). Sphalerite stratigraphy of the Upper Mississippi Valley zinc-lead district, Southwest Wisconsin. *Economic Geology*, Vol. 75, No. 3, pp. 351-361, ISSN 0361-0128.

Megaw, P.K.M., Ruiz, J. & Titley, S.R. (1988). High-temperature, carbonate-hosted Ag-Pb-Zn(Cu) deposits of Northern Mexico. *Economic Geology*, Vol. 83, No. 8, pp. 1856-1885, ISSN 0361-0128.

Melcher, F., Oberthur, T & Rammlmair, D. (2006). Geochemical and mineralogical distribution of germanium in the Khusib Springs Cu-Zn-Pb-Ag sulphide deposit, Otavi Mountain Land, Namibia. *Ore Geology Reviews*, Vol. 28, No. 1, pp. 32-56, ISSN 0169-1368.

Mills, H., Halliday, A.N. Ashton, J.H. Anderson, I.K. Russell, M.J. (1987). Origin of a giant orebody at Navan, Ireland. *Nature*, Vol. 327, No. 6119, pp. 223-226, ISSN 0028-0836.

Morrissey, C.J., Davis, G.R., & Steed, G.M. (1971). Mineralization in the Lower Carboniferous of Central Ireland. *Transactions of the Institute of Mining and Metallurgy*, Vol. 80, pp. B174-B184, ISSN 0371-7453.

Mouat, M.M. & Clendenin, C.W. (1977). Geology of the Ozark Lead Company mine Viburnum Trend, Southeast Missouri. *Economic Geology*, Vol. 72, No 3, pp. 398-407, ISSN 0361-0128.

Newhouse, W.H. (1932). The composition of vein solutions as shown by liquid inclusions in minerals. *Economic Geology*, Vol. 27, No. 5, pp. 419-436, ISSN 0361-0128.

Ohle, E.L. (1959). Some considerations in determining the origin of ores of the Mississippi Valley-type. *Economic Geology*, Vol. 54, No. 5, pp. 769-789, ISSN 0361-0128.

Ohle, E.L. (1967). The origin of ore deposits of the Mississippi Valley type. In: *Genesis of stratiform lead-zinc-barite-fluorite deposits in carbonate rocks*, Brown, J.S. (Ed.), pp. 33-39, Economic Geology Monograph 3.

Ohle, E.L. (1980). Some considerations in determining the origin of ores of the Mississippi Valley-type, Part II. *Economic Geology*, Vol. 75, No. 2, pp. 161-172, ISSN 0361-0128.

Ohle, E.L. (1996). Significant events in the geological understanding of the southeast Missouri district. In: *Carbonate-hosted lead-zinc deposits,* Sangster, D.F., pp. 1-7, Society of Economic Geologists, 1-887483-95-0, Littleton.

Ohmoto, H. (1986). Stable isotope geochemistry of ore deposits. In: *Stable isotopes in high temperature geological processes,* Valley, J.W., Taylor, H.P. & O'Neil, J.R. (Eds.), Reviews in Mineralalogy, Vol. 16, pp. 491-559 Mineralogical Society of America.

O'Keefe, W.G. (1986). Age and postulated source rocks for mineralization in Central Ireland, as indicated by lead isotopes. In: *Geology and genesis of mineral deposits in Ireland,* Andrew, C.J., Crowe, R.W.A., Finlay, S., Pennell, W.M. & Pyne, J.F. (Eds.), pp. 617-624, Irish Association for Economic Geology, Dublin.

Paradis, S., Hannigan, P. & Dewing, K. (2007). Mississippi Valley-type lead-zinc deposits. In: *Mineral Deposits of Canada: A Synthesis of Major Deposit-Types, District Metallogeny, the Evolution of Geological Provinces, and Exploration Methods,* Goodfellow, W.D., (Ed.), pp. 185-203, Geological Association of Canada, Mineral Deposits Division, Special Publication No. 5.

Phillips, W.E.A. & Sevastopulo, G.D. (1986). The stratigraphic and structural setting of Irish mineral deposits. In: *Geology and genesis of mineral deposits in Ireland,* Andrew, C.J., Crowe, R.W.A., Finlay, S., Pennell, W.M. & Pyne, J.F. (Eds.), pp. 1-30, Irish Association for Economic Geology, Dublin.

Pinckney, D.M. & Rafter, T.A. (1972). Fractionation of sulfur isotopes during ore deposition in the Upper Mississippi Valley zinc-lead district. *Economic Geology*, Vol. 67, No. 3, 315-328, ISSN 0361-0128.

Roedder, E. (1967). Environment of deposition of stratiform (Mississippi Valley-type) ore deposits, from studies of ore inclusions. In: *Genesis of stratiform lead-zinc-barite-fluorite deposits in carbonate rocks,* Brown, J.S. (Ed.), pp. 349-362, Economic Geology Monograph 3.

Roedder, E. (1984). Fluid inclusions. Mineralogical Society of America, Reviews in Mineralogy, Vol. 12, 644 p.

Rogers, R.K. & Davis, J.H. (1977). Geology of the Buick mine Viburnum Trend, Southeast Missouri. *Economic Geology*, Vol. 72, No.3, 372-380, ISSN 0361-0128.

Rowan, E.L. & Leach, D.L. (1989). Constraints from fluid inclusions on sulphide precipitation mechanisms and ore fluid migration in the Virburnum Trend Lead District, Missouri. *Economic Geology*, Vol. 84, No. 7, pp. 1948-1965, ISSN 0361-0128.

Rubin, J.N. & Kyle, J.R. (1988). Mineralogy and geochemistry of the San Martin skarn deposit, Zacatecas, Mexico. *Economic Geology*, Vol. 83, No. 8, pp. 1782-1801, ISSN 0361-0128.

Ruiz, J., Patchett, P.J. & Ortega-Gutierrez, F. (1988). Proterozoic and phanerozoic basement terranes of Mexico from Nd isotopic studies. Geological Society of America Bulletin, Vol. 100, pp. 274-281.

Russell, M.J. (1978). Downward-excavating hydrothermal cells and Irish-type ore deposits: importance of an underlying thick Caledonian prism. *Transactions of the Institute for Mining and Metallurgy*, Vol. 87, pp. B168-B171, ISSN 0371-7453.

Russell, M.J. & Skauli, H. (1991). A history of theoretical development in carbonate-hosted base metal deposits and a new tri-level enthalpy classification. In: *Historical perspectives of genetic concepts and case histories of famous discoveries, Economic Geology*

Monograph 8, Hutchinson, R.W. & Grauch, R.I., (eds.), pp. 96-116, Economic Geology Publishing Company, No. 91-073091, New Haven.

Russell, M.J., Solomon, M. & Walshe, J.L: (1981). The genesis of sediment-hosted, exhalative zinc + lead deposits. *Mineralium Deposita*, Vol. 16, No. 1, 113-127, ISSN 0026-4598.

Rye, R.O. (1974). A comparison of sphalerite-galena sulfur isotope temperatures with filling temperatures of fluid inclusions. *Economic Geology*, Vol. 69, No. 1, pp. 26-32, ISSN 0361-0128.

Rye, R.O. & Haffty, J. (1969). Chemical composition of the hydrothermal fluids responsible for the lead-zinc deposits at Providencia, Zacatecas, Mexico. *Economic Geology*, Vol.64, No. 6, pp. 629-643, ISSN 0361-0128.

Rye, R.O. & O'Neil, J.R. (1968). The O^{18} content of water in primary fluid inclusions from Providencia, North-Central Mexico. *Economic Geology*, Vol. 68, No. 3, pp. 232-238, ISSN 0361-0128.

Samson, I.M. & Russell, M.J. (1983). Fluid inclusion data from Silvermines base metal-baryte deposits, Ireland. Transactions of the Institute for Mining and Metallurgy, Vol. 92, pp. B67-B71, ISSN 0371-7453.

Samson, I.M. & Russell, M.J. (1987). Genesis of the Silvermines zinc-lead-barite deposit, Ireland: Fluid inclusion and stable isotope evidence. *Economic Geology*, Vol. 82, No. 2, pp. 371-394, ISSN 0361-0128.

Sangster, D.F. (1976). Carbonate-hosted lead-zinc deposits. In: *Handbook of stratabound and stratiform ore deposits*, Wolf, K.H. (Ed.), Vol. 6, pp. 447-456, Elsevier, New York.

Sangster, D.F. (1988). Breccia-hosted lead-zinc deposits in carbonate rocks. In: *Paleokarst*, James, N.P. & Choquette (Eds.), pp. 102-116, Springer-Verlag, New York.

Sangster, D.F. (1990). Mississippi Valley-type and sedex lead-zinc deposits: a comparative examination. *Transactions of the Institution of Mining and Metallurgy*, Vol. 99, pp. B21-B42, ISSN 0371-7453.

Sawkins, F.J. (1964). Lead-zinc ore deposition in the light of fluid inclusion studies, Providencia Mine, Zacatecas, Mexico. Economic Geology, Vol. 59, 883-919.

Sawkins, F.J. (1968). The significance of Na/K and Cl/SO_4 ratios in fluid inclusions and subsurface waters, with respect to the genesis of Mississippi Valley-type ore deposits. - Economic Geology, Vol. 63, 935-942.

Sawkins, F.J. (1990). *Metal deposits in relation to plate tectonics* (2nd edition), Springer-Verlag, 0-387-50920-8, Heidelberg.

Schneider, H.-J. (1964). Facies differentiation and controlling factors for the depositional lead-zinc concentration in the Ladinian geosyncline of the eastern Alps. In *Sedimentology and ore genesis*, Amstutz, G.C. (Ed.), pp. 29-45, Elsevier, Amsterdam.

Sheppard, S.M.F. (1986). Characterization and isotopic variations in natural waters. In: *Stable isotopes in high temperature geological processes*, Valley, J.W., Taylor, H.P. Jr. & O'Neil, J.R. (Eds.), Reviews in Mineralogy, Vol. 16, pp. 165-183, Mineralogical Society of America.

Shimazaki, H. (1980). Characteristics of skarn deposits and related acid magmatism in Japan. *Economic Geology*, Vol. 75, No. 2, pp. 173-183, ISSN 0361-0128.

Shimizu, M. & Iiyama, J.T. (1982). Zinc-lead skarn deposits of the Nakatsu Mine, Central Japan. *Economic Geology*, Vol. 77, No. 4, pp. 1000-1012, ISSN 0361-0128.

Snyder, F.G. (1967). Criteria for origin of stratiform ore bodies with application to Southeast Missouri. In: *Genesis of stratiform lead-zinc-barite-fluorite deposits in carbonate rocks*, Brown, J.S. (Ed.), pp. 1-12, Economic Geology Monograph.

Sverjensky, D.A. (1981). The origin of a Mississippi Valley-type deposit in the Viburnum Trend, Southeast Missouri. *Economic Geology*, Vol. 76, No. 7, pp. 1848-1872, ISSN 0361-0128.

Sverjensky, D.A. (1984). Oil field brines as ore-forming solutions. *Economic Geology*, Vol. 79, No. 1, pp. 23-37, ISSN 0361-0128.

Sverjensky, D.A. (1986). Genesis of Mississippi Valley-type lead-zinc deposits. *Annual Reviews of Earth and Planetary Science*, Vol. 14, pp. 177-199.

Sweeney, P.H., Harrison, E.D. & Bradley, M. (1977). Geology of the Magmont mine, Viburnum Trend, Southeast Missouri. *Economic Geology*, Vol. 72, No. 3, pp. 365-371, ISSN 0361-0128.

Taylor, R.D., Leach, D.L., Bradley, D.C., & Pisarevsky, S.A. (2009). Compilation of mineral resource data for Mississippi Valley-type and clastic-dominated sediment-hosted lead-zinc deposits: U.S. Geological Survey Open-File Report 2009–1297, pp. 42, U.S. Geological Survey, Reston, Virginia.

Taylor, S. (1984). Structural and paleotopographical controls of lead-zinc mineralization in the Silvermines ore bodies, Republic of Ireland. - Economic Geology, Vol. 79, 529-548.

Taylor, S. & Andrew, C.J. (1978). Silvermines orebodies, County Tipperary, Ireland. - Trans. Inst. Min. Met., Vol. 87, B111-B124.

Thacker, J.L. & Anderson, K.H. (1977). The geologic setting of the Southeast Missouri lead district - regional geologic history, structure and stratigraphy. *Economic Geology*, Vol. 72, No. 3, 339-348, ISSN 0361-0128.

Thompson, T.M. & Arehart, G.B. (1990). Geology and the origin of ore deposits in the Leadville District, Colorado: Part I. Geologic studies of orebodies and wall rocks. *Economic Geology Monograph 7*, pp. 130-155.

Thompson, T.M. & Beaty, D.W. (1990). Geology and the origin of ore deposits in the Leadville District, Colorado: Part II. Oxygen, hydrogen, carbon, sulfur, and lead isotope data and development of a genetic model. Economic Geology Monograph 7, pp. 156-179.

Turner, R.J.W. & Einaudi, M.T. (1986). Introduction. In: *The genesis of stratiform sediment-hosted lead and zinc deposits: conference proceedings*, Turner, R.J.W. & Einaudi, M.T. (Eds.), Vol. XX, pp. 1-2, Stanford University Publications.

Vearncombe, J.R., Chisnall, A.W., Dentith, M.C., Dorling, S.L., Rayner, M.J., & Holyland, P.W. (1996). Structural controls on Mississippi Valley-type mineralization, the southeast Lennard Shelf, Western Australia. In: *Carbonate-hosted lead-zinc deposits*, Sangster, D.F. (Ed.), pp. 74-95, Society of Economic Geologists, 1-887483-95-0, Littleton.

White, D.E. (1968). Environments of some base-metal ore deposits. *Economic Geology*, Vol. 63, No. 4, pp. 301-335, ISSN 0361-0128.

White, D.E. (1986). Environments favorable for generating sediment-hosted Pb-Zn deposits. In: *The genesis of stratiform sediment-hosted lead and zinc deposits: conference proceedings*, Turner, R.J.W. & Einaudi, M.T. (Eds.), Vol. XX, pp. 177-180, Stanford University Publications.

Yun, S. & Einaudi, M.T. (1982). Zinc-lead skarns of the Yeonhwa-Ulchin district, South Korea. *Economic Geology*, Vol. 77, No 4, pp. 1013-1032, ISSN 0361-0128.

3

Simulation of Tunnel Surrounding Rock Mass in Porous Medium with Hydraulic Conductivity Tensor

Lin-Chong Huang and Cui-Ying Zhou
Sun Yat-sen University
China

1. Introduction

The deformation of surrounding rock in tunnel is a comparatively complex process, because of the heterogeneous and discontinuous characters in deformation, which belongs to a highly nonlinear problem.

In recent years, many researchers have paid a lot of attention to the study of deformation in the soft surrounding rock. Sulem(1987) and Stille(1989) got the analytic solution of displacement under the hydrostatic pressure state. In fact, this result is based the linear yield criterion in the condition of small deformation. According the experiments, Lade(1977), Agar(1985), and Santarelli(1987) pointed out that, in soft surrounding rock, especially in the soft soil, the relationship between the maximum principle and the minimum stresses is nonlinear, and the linear relationship is only just the special case. In 1966, Hobbs proposed the Power law nonlinear criterion for the fist time, and then Ladanyi suggested a new nonlinear criterion from the crack theory of Griffith in 1974. Kennedy and Linderg studied it using the segment linear theory in 1978, and Brown got the Hoek-Brown nonlinear failure criterion based on predecessors.

With the rapid development of computing power, extensive research has been done on the 3D modeling of tunnel construction. In addition to the special issue on tunneling mentioned earlier, Shahrour and Mroueh(1997) performed a full 3D FEM simulation to study the interaction between tunneling in soft soils and adjacent surface buildings. Their analysis indicated that the tunneling-induced forces largely depended on the presence of the adjacent building and neglecting of the building stiffness in the tunneling-structure analysis yielded significant over-estimation of internal forces in the building members. Tsuchiyama et al. (1988) analysed the deformation behaviour of the rock mass around an unsupported tunnel intersection in the construction of a new access tunnel to the existing main tunnel using 3D linear elastic FEM and found that the influence area along the main tunnel was on the order of one tunnel diameter on the obtuse angle side and about three times the tunnel diameter on the acute angle side from the point of intersection. Rowe and Kack(1983) carried out a numerical analysis based on the finite element method and compared the results with case histories for predicting and designing the settlement above tunnels constructed in a soft ground. Kasper and Meschke(2004) developed a 3D finite element model for a shield-driven tunnel excavation in a soft ground and reproduced settlement, pore pressure distribution, stress levels, and deformations in the lining and in the soil.

Constitutive modeling is an important aspect for the analysis of the displacement and stress fields. A review of existing literature suggests that elastoplastic theory seems to be the most popular framework for constitutive modeling. Borja et al. (1997) formulated the problem of elastoplastic consolidation at finite strain. They showed that for saturated soil media with incompressible solid grains and fluids, balance of energy suggests that Terzaghi's effective stress was the appropriate measure of stress for describing the constitutive response of the soil skeleton. Consequently, the formulation had the advantage of being able to accommodate a majority of the effective stress-based models developed in geotechnical engineering for describing the deformation behaviour of compressible clays. Elastoplastic analyses of circular tunnels excavated in Mohr-Coulomb media have been performed by numerous investigators. For elastic brittle plastic case, Brown et al. (1983) presented the closed-form solution for stress and radial displacement in the plastic zone. However, they did not consider the variation of elastic strain, resulting in the neglect of the influence of the unloading in the plastic zone. Recently, several improved solutions were provided. An analytical solution for Hoek – Brown rock mass given by Sharan(2005) was not exact in calculating displacements in plastic zone, as it was assumed that the elastic strain field in the plastic zone was the same as that of thickwall cylinder problem. Solutions by Park and Kim (2006) offered an exact expression for displacement in the plastic zone.

To sum up, most of the research efforts on tunneling to date have focused on the assessment of the ground surface settlement although some have begun to pay attention to the interaction between tunneling and existing surface structures such as adjacent buildings. Relatively, little research work can be found in the literature on the particular constitutive model for soft soil, especially for the tunnel in porous medium. Also, in the numerical modeling, the Finite Difference Method (FDM) is seldom used in the simulation. The purpose of this paper is to present an elastoplastic model with hydraulic conductivity tensor perform utilizing FLAC3D, and monitor the mechanical behaviour of the tunnel deformation response in the porous medium tunnel during construction.

2. Constitutive model framework

There is the need to establish a link between the state of stresses and the deformations, also a link between the flow vector and the fluid pressure in soft argillaceous shale tunnel. In these links, the stresses are assumed to be a nonlinear function of the deformation via an elastoplastic constitutive response.

The deformation of the surrounding rock includes elastic and plastic deformation in elastoplastic model theory, thus we should set up the constitutive model by combining elastic and plastic theory. In the constitutive relation, there are Mohr-Coulomb and Drucker-Prager yield criterions for the surrounding rock material.

As showed in Fig. 1, when friction angle $\varphi > 0$, the yield surface of the Drucker-Prager criterion is a conical surface in the principal stress space, which is inscribed at the Mohr-Coulomb yield surface; while if $\varphi = 0$, the Drucker-Prager criterion is exactly the Mises criterion.

3. Deformation numerical simulation in soft argillaceous shale tunnel

The Guan Kouya tunnel is taken as the research objection here, which is a typical soft argillaceous tunnel located at Hunan province in China. It is a four-lane bidirectional separated tunnel with 880m in length. The geologic investigation data show that the rock is mainly soft argillaceous shale with distinct stratification structure, and the surrounding rock is classified as IV and V. The structure is supervised by NATM, and the support obtains the composite lining, with anchor and sprayed concrete to be the primary support together with the reinforced concrete as the secondary lining.

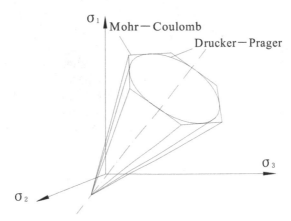

Fig. 1. Mohr−Coulomb & Drucker−Prager yield surface

3.1 Computational model
In order to obtain the deformation characters and compare them with the measurement results in site, using the software FLAC3D, the typical cross section (YK73+839.5, classified as V) is simulated and computated, which is exactly the section measured in site.

3.1.1 Computational bound
The boundary of the computational model is more than three times of the cavity width in each direction, so that the adverse influnce caused by the boundary constraint condition can be reduced sharply in the process of computation. To be specific, the computational zone includes $100\,m$ in the horizontal direction, $30\,m$ from the arch head to the ground in the vertical direction, and $30\,m$ along with the route direction, which can be clearly observed in Fig. 2. The water height is 5.6m over the arch crown.

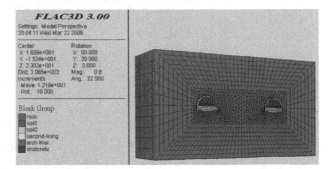

Fig. 2. Numerical computation model

3.1.2 Anchor simulation

It is unrealistic to simulate each anchor in computation, because there are so many anchors in the three-dimensional model. There is an effective method to solve this problem shown in [Huang, 2009] by enhancing the cohesive force instead of the effection of anchoring. The equation of the cohesive force in the anchoring rock is shown as:

$$C = C_0 \left[1 + \frac{\eta}{9.8} \cdot \frac{\tau S_m}{ab} \times 10^4 \right] \tag{1}$$

Where C_0 and C are the cohesive forces before and after adding the anchor(MPa), respectively; τ is the maximum shear stress of the anchor(MPa); S_m is the acreage of the anchor (m^2); a and s are the distances between each other in the longitudinal and lateral directions, respectively (m); η is a empirical coefficient, and it equals 4.0 in this project. The material parameters utilized in the simulation is summarized in Table 1.

Items	R (kN/m^3)	E (GPa)	μ	C (kPa)	ϕ (°)
soft argillaceous shale	2.0	1	0.45	10	31
C25 concrete	23	29.5	0.2	—	—
C30 concrete	23	31	0.2	—	—

Table 1. Summary of simulation papameters

3.1.3 Construction procedure

This structure is simulated as two saparate sigle cavities. We obtain the "bench method" in computational simulation, just as the method used in the pratical construction phases, and the procedures are followed as:

step 1→excavating 5 m in the upper bench and setting primary support; Step 2→excavating 5 m in the upper bench additionally and setting primary support, at the same time, excavating 5 m in the lower bench and setting primary support; Step 3→setting the secondary support; Step 4→excavating 5 m in the upper bench and setting primary support, at the same time, excavating 5 m in the lower bench and setting primary support. The construction simulation is done according to this flow operation.

In order to compare the computational results with the measurement data in site conveniently, some typical locations are chosen to be computated and analyzed, and these locations are the same as those measured in site. The construction procedure and these typical locations are shown in Fig. 3.

3.2 Deformation analysis

The maximum computational displacement results of the typical locations around the cavity are summarized in Table 2, and the displacement convergence of the arch crown at cross section YK73+839.5 is shown in Fig. 4.

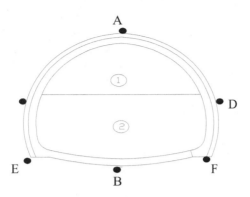

Fig. 3. Typical location & construction procedure

Construction phase	Vertical displacement of the typical locations（mm）					
	A	C	E	B	F	D
Step 1	-37.853	-22.046	-13.091	11.276	-13.581	-24.482
Step 2	-38.518	-12.091	-13.528	11.143	-13.170	-27.905
Step 3	-39.175	-22.975	-14.039	10.791	-14.786	-31.044
Note: minus means the displacement direction downwards.						

Table 2. Displacement computation results of the typical locations around the cavity

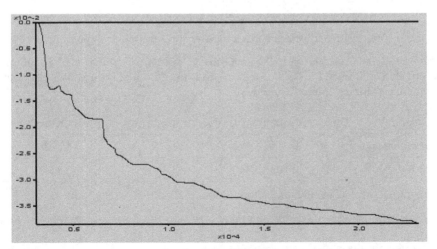

Fig. 4. Displacement convergence of arch crown (Unit:m)

The relative displacements at the arch crown (point A) in section YK73+839.5 are plotted in Figure 5.

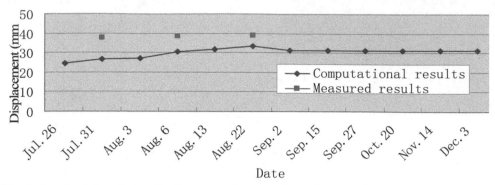

Fig. 5. Graph of the relative displacement change at the arch crown (point A)

While the relative displacements at the right of arch springing (point F) in section YK73+839.5 are plotted in Figure 6.

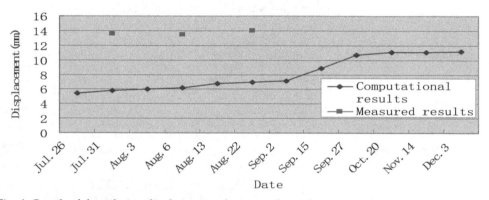

Fig. 6. Graph of the relative displacement change at the arch spinging (point F)

The computational results and the measured results of some typical points are comparatively summarized in Table 3.(Noting that the Graph of the relative displacements at the left arch springing (point E) is elided)

Result	Left arch springing（E）	Arch crown (A)	Right arch springing (F)
Computational value	14.039	39.175	14.786
Measured value	10.02	31.16	11.19

Table 3. Internal displacement value of the section YK73+839.5 (Unit : mm)

The comparative analysis about the computational and the measured results are followed below.

Computational results analysis:

1. The deformaion velocity is much high in the beginning, compared with the late phase, while the time that the deformation get steady is much long. These laws are coincident with the data measured in site.

2. The displacement convergence at the arch crown is 0.51576% after the construction of the secondary lining, while it is 0.6504% at the arch. Spingring. Both of the convergence values are much large, but are not beyond the allowable values, which are less than 0.4%~1.2%. At this point, evaluating with the displacement convergence, this tunnel is steady under constructrue howbeit the convergence values are large. These laws are also coincident with the data measured in site.

Measured results analysis:

1. The deformation velocity rises rapidly in the first 6 days, and goes to steady state after 40 days.
2. The displacement at the arch crown get steady after 30 days, which shows that the deformaion velocity is much high, and the time that the deformation get steady is also much long
3. The displacement at the arch spinging goes to steady state after 40 days, at which point the time is longer than the arch crown.
4. The vertical displacement of the arch spinging increases at all time, and the displacement at the arch crown are always larger than the arch spinging. This rule is also coincident with the computational results.

3.3 Plastic zone analysis

Figure 7 and 8 show the plastic zone of the computational section after excavating the upper bench and the lower bench, respectively, where "shear-n" means failure, while "shear-p" means yielding but no failure.

The plastic zone is small after step 2, in which tension and shearing yield appears in few zone. The whole initial support is almost in a state of "yield" before the construction of the secondary lining, which takes the form of the so-called "pulling yield" from the middle to the bottom of the arch, whereas it appears as the so-called "shear yield" in the 3~4 m near the excavating surface and the arch spinging. These suggest that the secondary lining and the Invert should be set in time during the construction of soft surrounding rock tunnel, so that the closed support can be formed in time, which is especially important to insure the tunnel safety, which are shown in Figure 7 and 8.

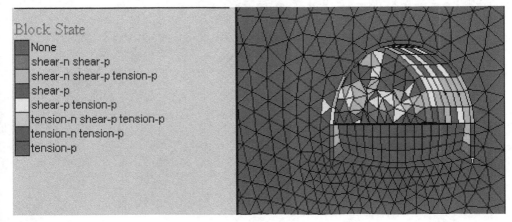

Fig. 7. Distributing of the plastic region

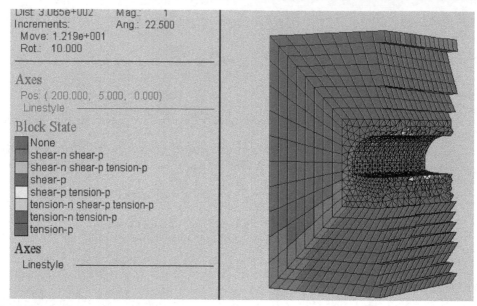

Fig. 8. Portrait distributing of the plastic region

3.4 Secondary lining stress analysis

Figure 9 and 10 show the maximum and minimum principle stresses of the computational values, respectively, while the the maximum and minimum principle stresses of the typical points can be clearly observed in Table 4.

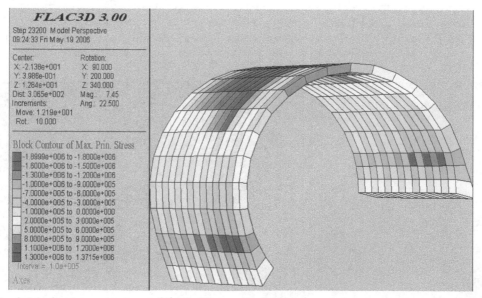

Fig. 9. Maximum main stress of the 2ed lining (Unit:Pa)

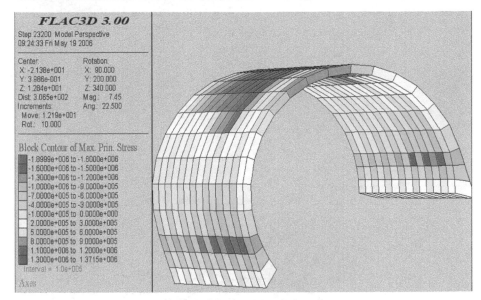

Fig. 10. Maximum main stress of the 2ed lining (Unit:Pa)

Item	A	C	E	B	F	D
Maximum principle stress（MPa）	6.50	4.16	9.91	None	10.76	10.06
Minimum principle stress (MPa)	-0.11	-0.19	-0.39	None	-1.34	-1.76
Note: The minus means tension stress.						

Table 4. The max. & min. principle stress of the typical locations at the secondary lining

Some results can be obtained from the computation such that: There are mainly pression stress in secondary lining, and the maximum pression stress is 10.76 $_{MPa}$, which is located at the middle of the arch, at the same time, which is 44.46% of the designed concrete compressive strength. At this point, the secondary lining is steady. However, there are few tension stresses near the arch spinging, and the maximum is 1.76 $_{MPa}$, 88% of the designed concrete tensile strength, which suggests some reinforcing steel bars should be configured.

Table 5 shows the internal force values of the locations at secondary lining according to the measrued data, from which, we get the shematic diagram of the axial force and moment, shown as Figure 11 and 12, respectively.

Some results can also be suggested from the measured data, i.e.:

1. In the secondary lining, the initial stresses are mostly the tension, because of the contractive stress of the concrete in the prophase. However, they will trun to be the compressive stresses about one month later, in particular, these compressive stresses increace slowly with the time, which will be steady about two months later.

2. The secondary lining moment is much high at the arch crown and the arch spinging, which is basiclly coincident with the mechanical characteristic of the mold casting lining.

location	axial force (KN)	moment (KN·m.)
A	542	11
C	262	4.6
E	678	-9.5
D	433	1.3
F	981	-5.3

Table 5. Internal force results of the measure locations at the secondary lining

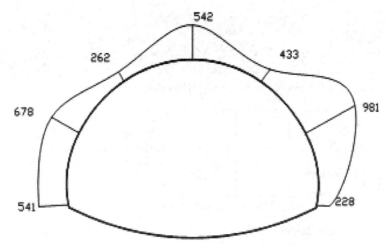

Fig. 11. Sketch map of axial force in 2ed lining (unit: KN)

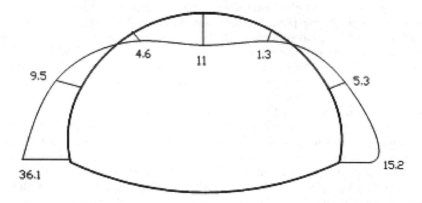

Fig. 12. Sketch map of moment in 2ed lining (unit: KN· m)

4. Conclusion

We have presented a nonlinear finite element model for the simulation of the tunnel in porous medium with hydraulic conductivity tensor. Using the FLAC3D code, the numerical simulation analysis is done to the Guan Kouya soft argillaceous shale tunnel .Some results are obtained, compared with the measurement data in site, such that:

1. The deformation velocity is fast during the prophase of the tunnel excavating, compared with the later phase, but the steady time remains long;
2. Regarding the vertical displacement, the displacement of the arch crown appears bigger than that of the middle, while the displacement of the arch springing is the least obvious, and the horizontal displacement always remains small;
3. The whole initial support is almost in a state of "yield" before the construction of the secondary lining, which takes the form of the so-called "pulling yield" from the middle to the bottom of the arch, whereas it appears as the so-called "shear yield" in the 3~4m near the excavating surface and the arch. These suggest that the secondary lining and the Invert should be set in time during the construction of soft surrounding rock tunnel, so that the closed support can be formed in time, which is especially important to insure the tunnel safety;
4. The surrounding rock stress releases rapidly after the excavating of the cavity, at this point, some assistant methods, especially the lock-foot anchor should be applied to enhance the steady stability due to the big compressive stress located at the arch of the two sides in the tunnel.

5. Acknowledgment

This work has been supported by China Postdoctoral Science Foundation (No. 2009046082724 & 201003386), the National Natural Science Foundation of China (No.51108472),the Natural Science Foundation of Guangdong Province, China (No.S2011040005172),and the State Key Program of National Natural Science Foundation of China(No.41030747), these supports are gratefully acknowledged.

6. References

Sulem, J. ; Panet, M. & Guenot, A (1987). An analytical solution for time-dependent displacement in a circular tunnel. *Int J. Rock Mech Mi Sci & Geomech Abstr*, Vol.24, No.3,pp. 155-164.

Stille, H. ; Holmoery, M. & Mord, G(1989). Support of week rock with grouted bolts and shoterete. *Int J Rock Mech Mi Sci & Geomech Abstr*, Vol.26, No.1,pp. 99-103.

Lade, P. V. (1977). Elasto plastic stress strain theory for cohesionless soil with curved yield surface. *Int J Skilds Struct*, Vol.13, pp. 1019-1035.

Agar, J. G. ; Morgensteren, N. R. & Scott, J. Shear strength and stress strain behaviour of Athabasca oil sand at elevated temperatures and pressure (1985). *Can. Geotech. J.*, Vol.24, No.1, pp. 1-10.

Santarelli, F. Theoretical and experimental investigation of the stability of the axisy metric borehole(1987). University of London.

Shahrour, I. & Mroueh, H.(1997). Three-dimensional non linear analysis of a closely twin tunnels. *Sixth International Symposium on Numerical Models in Geomechanics*, Montreal, pp. 481-487.

Tsuchiyama, S. ; Hayakawa, M. ; Shinokawa, T. & Konno, H.(1988). Deformation behaviour of the tunnel under the excavation of crossing tunnel. *Proceedings of the 6th International Conference on Numerical Methods in Geomechanics*, Innsbruck, pp. 1591-1596.

Rowe, R.K. & Kack, G.J(1983). A theoretical examination of the settlements induced by tunneling four case histories. *Can. Geotech. J.*, Vol.20, pp, 299-314.

Kasper, T. & Meschke, G.(2004). A 3D finite element simulation model for TBM tunneling in soft ground. *Int. J. Numer. Anal. Meth. Geomech.*, Vol.28, pp, 1441-1460.

Borja, R.I. & Andrade, J.E.(2006). Critical state plasticity, Part VI: Meso-scale finite element simulation of strain localization in discrete granular materials. *Computer Methods in Applied Mechanics and Engineering*, Vol.195, pp, 5115-5140.

Borja, R.I.; Tamagnini, C. & Amorosi, A.(1997). Coupling plasticity and energy conserving elasticity models for clays, *J. Geotech. Geoenviron. Engrg.*, Vol.123, pp, 948-957.

Brown, E.T. ; Bray, J.W ; Ladanyi, B. & Hoek, E.(1983). Ground response curves for rock tunnels. *J. Geotech. Eng., ASCE*, Vol.109, pp, 15 – 39.

Huang, L.C, Xu, Z.S, Wang, L.C. Constitutive equations and finite element implementation of strain localization in sand deformation(2009). Journal of Central South University of Technology. Vol.16,No.3, pp, 482-487

Sharan, S.K.(2005). Exact and approximate solutions for displacements around circular openings in elastic – brittle – plastic Hoek – Brown rock. *Int. J. Rock Mech. Min. Sci.* Vol.42, pp, 542-549.

Park, K.H. & Kim, Y.J.(2006). Analytical solution for a circular opening in an elasto-brittle-plastic rock. *Int. J. Rock Mech. Min. Sci.* Vol.43, pp, 616-622.

Kumar, P. (2000). Infinite Elements for Numerical Analysis of Underground Excavations[J]. Tunneling and Underground Space Technology, Vol.15, No.1, pp, 117-124.

Part 3

Soil

4

Evaluation of Soil Hydraulic Parameters in Soils and Land Use Change

Fereshte Haghighi[1], Mirmasoud Kheirkhah[2] and Bahram Saghafian[2]

[1]Soil Conservation and Watershed Management Institute, Tehran,
[2]Soil Conservation and Watershed Management Research Institute, Tehran,
Iran

1. Introduction

The knowledge of soil water properties and land-use effects on these properties are important for efficient soil and water management. Furthermore, the use of the pedotransfer functions (PTFs) to estimate soil water content (θ_h) is important to assess. The loosening effect of dryland farming on soil water retention is known. In this chapter we review soil water content, pedotransfer functions and some infiltration models applicability for two land-use types. The land-use effect on soil water retention may be significant at water potentials of -33 kPa and 0 kPa in the soil. At the -1500 kPa pressure head, water content may not be affected by cultivation of rangeland at different soil depths. In addition, pedotransfer functions can be used as a physically based model for soil water retention characterization in the various areas. Moreover, it is essential to evaluate the infiltration models applicability for different soils and various land-uses.

1.1 Definition
1.1.1 Soil hydraulic properties
Soil hydraulic properties govern transport processes and water balance in soils. Water retention capacity, infiltration rate, and saturated hydraulic conductivity are important soil hydraulic properties. Soil water retention and saturated hydraulic conductivity (K_s) are necessary input data for the simulations of water flow in soil and water engineering. Characterizing hydrological behavior of catchments requires knowledge of hydraulic parameters.

1.1.2 Soil water retention
Soil water retention at field capacity (FC) and permanent wilting point (PWP) are used to estimate the water depth applied by irrigation (Hansen et al., 1980), and to calculate water availability, as a crucial factor to assess the land area suitability for crop producing (Sys et al., 1991).

2. Soil water retention capacity and land use

One important soil hydraulic property is water retention capacity, which affects soil productivity and management. Soil water content (θ_h) governs the transport characteristics of water and solutes in soils. The knowledge of water retention capacity and land use effects on this property is important for efficient soil and water management. Upon conversion of natural lands to cultivated fields, water retention capacity is strongly influenced (Schwartz et al., 2000; Bormann and Klaassen, 2008; Zhou et al., 2008). Soil water retention at field capacity (FC) and permanent wilting point (PWP) are important to estimate the irrigation water depth which may be affected by land use change. Soil water retention characteristic, is affected by soil organic matter (SOM) content and porosity, which are significantly influenced by land use type (Zhou et al., 2008).

We conducted a study to evaluate, document, and quantify the effect of cultivation of rangeland on soil water retention in field capacity (FC), permanent wilting point (PWP), and to test the use of the van Genuchten equation to estimate θ_h in cultivated and natural lands in the same soils of the Taleghan watershed in Iran.

Significant differences in the OM and bulk density (BD) were observed between dryland farming and rangeland at both depths of 0 cm - 15 cm and 15 cm - 30 cm. Soil sample water contents at different pressure heads under both land use types are presented in Figures 1. The overall measured and fitted soil water retention curves did not show significant difference within the selected water potentials for both land use types in this study. However, measured θ_s (0 kPa) values were found to be significantly lower for dryland farming when compared with rangeland at depths of 0 cm - 15 cm and 15 cm - 30 cm, respectively. Moreover, the land use effect on soil water retention was significant at a water potential of -33 kPa (FC) based on laboratory measurements only at the top (15 cm depth). The results indicated that the conversion of rangeland to dryland farming led to a significant decrease (16.56% on average) in the FC at a depth of 0 cm - 15 cm. The mean -1500 kPa (PWP) water content was not affected by the land-use type. Figure 1 indicates that the mean total field capacity (FC) was significantly greater in rangeland when compared with dryland farming at a depth of 0 cm - 15 cm. In this study, there were not statistically significant differences in water content at other potentials (-50 kPa, -100 kPa, -500 kPa, and -1000 kPa pressures) between the two types of land use presented in Figure 1. At those pressure heads and at a -1500 kPa water content, the amount of micropores were not affected by cultivation of rangelands (Fig. 1). Overall, the results showed that the soil pore system and reduced total porosity under dryland farming can decrease water storage capacity at water potentials of -33 kPa and 0 kPa. Ndiaye et al., (2007) has shown that improper soil management decreases the soil macroporosity in the long-term affecting the θ_s. The data obtained in our study demonstrated the loosening effect of dryland farming on soil water retention. Previous studies on the effect of land use have demonstrated clear changes in soil physical properties, such as soil porosity, SOM, and BD, in relation to hydraulic properties (Bormann and Klassen 2008; Haghighi et al., 2010b).

3. Pedotransfer functions (PTFs)

Determination of soil water properties required as input data for simulation models is time consuming and relatively costly (Wösten et al., 1995). Thus, indirect estimation of these characteristics has been proposed as one alternative to direct estimation of the soil hydraulic parameters based on the measured water retention data. Pedotransfer functions (PTFs) are emerged as the relationship between soil hydraulic and other more available measured properties (Bouma, 1989) which can be used to estimate hydraulic parameters. PTFs are useful tools for modeling applications.

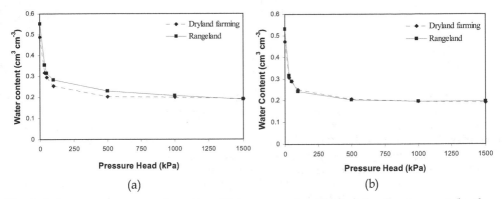

Fig. 1. Soil water content as a function of the pressure head for two landuse types at depths of (a) 0 cm -15 cm and (b) 15 cm - 30 cm.

4. Pedotransfer functions (PTFs) and different land uses

To estimate the land use effects on soil water retention, van Genuchten model (Van Genuchten, 1980) may be applied. Some researches have correlated van Genuchten parameters with soil organic matter, bulk density (BD), and soil particle size distribution and many researchers have estimated the water retention curve using soil texture, bulk density, and porosity.

Many statistical equations (pedotransfer functions) characterizing the water retention curve have been presented (Kutilek and Nielsen., 1994). PTFs are useful tools for modeling applications. Such analytical functions are derived involving various soil data. Such data are measured in the field and laboratory analysis. Soil hydraulic parameters derived through PTFs can be used to express soil hydraulic properties and water retention (Brooks and Corey, 1964). Consequently, physically based models such as van Genuchten representing a pedotransfer function may be considered as a valuable tool to simulate the soil water properties in different land uses.

The $\theta_{(h)}$ data may be fitted to van Genuchten equation to derive retention curves and parameters (a, n, and θ_r), using the RETC (RETention curve) optimization computer code (Van Genuchten et al., 1991). The van Genuchten model (van Genuchten, 1980) is defined as:

$$\theta(h) = \theta_r + \frac{\theta_s - \theta_r}{(1 + |\alpha h|^n)^m} \qquad \theta_{(h)} = \theta_s \qquad h \geq 0 \tag{1}$$

Where $\theta_{(h)}$ (cm³ cm⁻³) is the volumetric water content (for h<0), θ_r (cm³ cm⁻³) is the residual water content, and θ_s (cm³ cm⁻³) is the saturated water content. Here, m is $1-(1/n)$ with $n>1$. a (cm⁻¹) and n are empirical parameters determining the shape of the curve which were obtained for each core. Parameter n is related to steepness of the water retention curve.

$$K(S_e) = K_s S_e [1 - (1 - S_e^{(1/m)})^m] \tag{2}$$

Where K_s (mm/h) is saturated hydraulic conductivity and S_e is the effective saturation expressed as:

$$S_e = \frac{\theta - \theta_r}{\theta_s - \theta_r} \tag{3}$$

The effect of landuse type on soil water retention and PTF applications have not been documented for different land-uses to the best of our knowledge. In developing countries, there is a lack of large databases that are needed to develop PTFs. Thus, in many developing countries, the use of available PTFs can cause errors for estimating soil hydraulic properties. This encourages further investigations of the model applications and development of suitable point and parametric PTFs for estimating soil hydraulic properties in the studied area. The selection of more suitable PTFs for application where there are not developed PTFs caused by a lack of large databases is difficult. Consequently, it is essential to evaluate the model applicability and to develop point and parametric PTFs for estimating soil hydraulic properties for the soils in various sites. Thus, the estimates may be improved by comprehensive local studies.

Location (site)	Depth: 10–20 cm					Depth: 25–50 cm				
	K_{sat} (cm/day)	θ_{sat} (−)	α (1/cm)	l (−)	n (−)	K_{sat} (cm/day)	θ_{sat} (−)	α (1/cm)	l (−)	n (−)
A	16.0	0.43	0.007	0.944	1.37	23.1	0.43	0.014	− 0.350	1.41
B	16.0	0.42	0.007	0.331	1.36	15.8	0.40	0.009	− 0.448	1.33
C	17.4	0.40	0.008	0.093	1.35	14.7	0.38	0.010	− 0.482	1.33

Table 1. Mualem-van Genuchten parameters calculated for old grassland (site A), recently reseeded grassland (site B) and previous maize cultivated land (site C) (Sonneveld et al, 2003)

5. Evaluation of common infiltration models for different land-uses

The evaluation of infiltration characteristics as a hydrologic process in soils is necessary in agricultural studies. The knowledge of final steady infiltration rate is important for irrigation water efficiency, designing desirable irrigation systems, and loss of water. Thus, infiltration rate is important factor in sustainable agriculture, effective watershed management, surface runoff, and retaining water and soil resources. Since measuring the final infiltration rate is time consuming, several physical and empirical models have proposed to determine it. The empirical models such as Kostiakov (1932) and Horton (1940), and physical model such as Philip (1957) are the most common models to estimate infiltration rate of the soils.

5.1 Kostiakov-Lewis model
The model of Kostiakov modified for long times as follows:

$$f = at^{-b} + f_c \tag{4}$$

Where a and b are the equation's parameters ($a > 0$ and $0 < b < 1$). i_c is the steady infiltration rate (LT^{-1}).

5.2 Horton's model

The Horton's infiltration model (Horton, 1940) is expressed as follows:

$$f = (f_0 - f_c)e^{-kt} + f_c \tag{5}$$

Where i_c is the presumed final infiltration rate (LT⁻¹), i_0 is the initial infiltration rate (LT⁻¹) and t is time (T). k is the infiltration decay factor.

5.3 Philip two-parameter model

The Philip two-term model is expressed as (Philip, 1957):

$$f = \frac{1}{2}St^{-0.5} + A \tag{6}$$

Where f is the infiltration rate (LT⁻¹) as a function of time.
A= Transmissivity factor (LT⁻¹) as a function of soil properties and water contents, S = Sorptivity that is function of soil matric suction (LT⁻⁰·⁵).
t= time (T)
Singh (1992) expressed that the various models can estimate different values of the final infiltration rate in a soil which seems to be uncorrect, because of the final infiltration rate is a soil-dependent factor. Compared to the previous investigations on soil infiltration properties and models, studies on soil infiltration modelling depending on land use are scarce. Nevertheless, it can be assumed that landuse type have a significant impact on soil infiltration and infiltration models performance. Machiwal et al (2006) observed the infiltration process was well described by the Philip's model in a wasteland of Kharagpur, India. However, different soil management that influences the final infiltration rate is a major reason for different applicability of these models. Long-term effects of land use changes on soil infiltration and infiltration models (e.g., Horton, Kostiakov, and Philip models) can be observed (Navar and Synnott, 2000; Shukla et al, 2003).
Thus, the variability of soil infiltration characteristics and goodness of fit of the infiltration models for different land-uses should be considered during infiltration modelling studies helping on correct predictions of final infiltration for different land uses. Ability of these models for estimating the infiltration rate in different land-uses and soil management has been examined by some researchers. Gifford (1976) observed among the Horton, Kostiakov and Philip's models, the Horton's model was the best model to fit the infiltration data in mostly semi-arid rangelands from the Australia, but only under specific conditions. Shukla et al (2003) evaluated some of the infiltration models at different soil management and land-use systems in Ohio and observed among infiltration models, the Swartzendruber model was the best ones and fitted the observed infiltration data with lower sum of squares and higher model efficiency. Davidoff and Selim (1986) examined the goodness of fit for eight infiltration models on a Norwood soil with four winter cover crop treatments and results of their study showed that the Philip, Kostiakov and Horton's models had best predictions than the other models. Haghighi et al (2010b) evaluated the effects of rangeland and dryland farming land uses on performance of some infiltration models to estimate the final infiltration rate of soils. The study was conducted on some soils of Taleghan watershed, Iran. According to reports (Taleghan watershed study report, 1993), investigated soils are calcareous and classified as Typic Xerorthents. Mean annual rainfall alters from 464 to 796 mm and lands slope is by 15 %. The soil texture varied from clay-loam to silty clay loam.

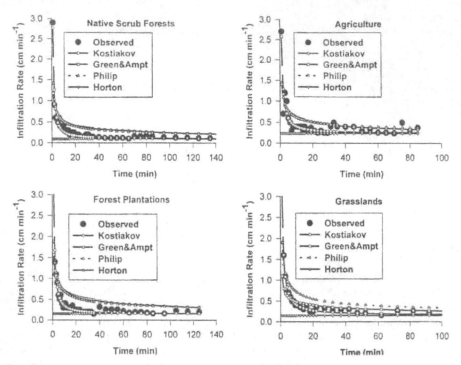

Fig. 2. Effects of land use changes on soil infiltration and infiltration models (Navar and Synnott, 2000)

In our study, the goodness of the fit of selected models and ability of them for estimating the final infiltration rate of rangeland and dryland farming soils was evaluated using the root mean squared errors (RMSE). The values of R^2 were determined high (0.99) and equal for all sites and land-uses, but the values of RMSE and the final steady infiltration showed that the estimated infiltration rates by the infiltration model of Horton, approached more closely to the measured ones at the selected area [Table 1]. The Horton's model was the best model selected for both of land-uses. It can be expressed that various models can suppose different final infiltration (f_c) values for a soil, which seems to be not practical, because f_c is a soil-dependent parameter, in general. Common changes in land-use negatively affect soil physical properties and decrease soil infiltration rate and could change modelling performance. Effect of land-use should be well documented aiming on good predictions in the studied areas and elsewhere.

The infiltration models can be used for estimating the infiltration rate in soils, well. But only, one or some of these models are better and appropriate for a specific site. Thus, the infiltration models should be analyzed for their ability to estimate the infiltration rate of each location. The investigation of Haghighi et al (2010a) showed that the Horton's model is the best ones selected for rangeland and dryland farming and land-use type is not an important factor to affect infiltration models efficiency. Due to a few number of investigation in this field of research, there is a need for further investigation on land-use effect on infiltration modelling and for the impact of land-use on soil infiltration characteristics, as well.

Land use	Kostiakov-Lewis model			Philip two-term model		Horton's model			Observed final infiltration rate (cm min^{-1})
	c	b	a	S	A	β	f_0	f_c	
Rangeland 1	0.2052	0.543	1.136	1.217	0.223	0.130	0.894	0.299	0.2803
Rangeland 2	0	0.770	1.058	1.322	0.235	0.021	0.596	0.190	0.2441
Rangeland 3	0.0870	0.606	0.839	0.970	0.127	0.090	0.556	0.183	0.1613
Rangeland 4	5.13×10^{-14}	0.855	0.5076	0.543	0.208	0.045	0.387	0.225	0.2285
Dryland farming 1	0.0585	0.636	1.475	1.741	0.162	0.085	0.909	0.258	0.2347
Dryland farming 2	0	0.522	2.863	2.984	0.018	0.120	1.601	0.200	0.1863
Dryland farming 3	1.37×10^{-12}	0.781	0.4857	0.592	0.119	0.029	0.290	0.120	0.1098
Dryland farming 4	0.0452	0.681	0.2555	0.315	0.073	0.071	0.198	0.089	0.0869

Table 2. Parameters of the selected infiltration models in both of land use types (Haghighi et al, 2010a)

6. Conclusions

Soil management and land use change may affect soil water retention at a -33 kPa (FC) potential in the soil based on laboratory measurements and model simulations. Lower water content at the -33 kPa potential would be expected upon conversion of natural lands to cultivated lands. In addition, the saturated soil water content (θ_s) may be affected by cultivation of rangeland. Moreover, because cultivation of natural lands affects soil macroporosity, we suggest measuring soil water retention at higher suction heads to document the land use effect on soil water retention properties in relation to soil macropores. Appropriate technology for dryland farming and suitable measures are necessary to improve soil water retention where cropping is required.

The findings show that the van Genuchten model is useful in describing soil water retention. Thus, use of this model may be considered as a valuable tool to gain more knowledge of hydraulic properties for various soil types. The effect of land use type on soil water retention and PTF applications have not been documented for dryland farming to the best of our knowledge. In many developing countries, such as Iran, the use of available PTFs can cause errors for estimating soil hydraulic properties. This review encourages further investigations of the model applications and development of suitable point and parametric PTFs for estimating soil hydraulic properties. The selection of more suitable PTFs for application where there are not developed PTFs caused by a lack of large databases is difficult. Consequently, it is essential to evaluate the model applicability and to develop point and parametric PTFs for estimating soil hydraulic properties for different land uses.

7. References

[1] Bormann, H., Klaassen, K., 2008. "Seasonal and land use dependent variability of soil hydraulic and soil hydrological properties of two Northern German soils." Geoderma. 145, 295-302.

[2] Brooks, R.H., Corey, A.T., 1964. "Hydraulic properties of porous media." Civil Engineering Dept., Colorado State University, Fort Collins, CO.

[3] Davidoff, B. and Selim, H.M., 1986. Goodness of fit for eight water infiltration models, Soil Sci. Soc. Am. J. 50, 759-764

[4] Gifford, GF. 1976. Applicability of some infiltration formulae to rangeland infiltrometer data. Journal of Hydrology. 28, 1-11.

[5] Haghighi F, Gorji M, Shorafa M, Sarmadian F, Mohammadi MH. 2010. "Evaluation of some infiltration models and hydraulic parameters." *Spanish Journal of Agricultural Research (INIA)*, 8 (1), 210-217.

[6] Haghighi F, Gorji M, Shorafa M. 2010. "A study of the effects of land use change on soil physical properties and organic matter." Land Degradation & Development Journal. *In press* 10.1002/ldr.999.

[7] Horton, R.E. 1940. An approach toward a physical interpretation of infiltration-capacity. Soil Sci. Soc. Am. Proc.5, 339-417.

[8] Kostiakov, A. N. 1932. On the dynamics of the coefficient of water percolation in soils and on the necessity of studying it from a dynamic point of view for the purposes of amelioration.

[9] Kutílek M. 2004. Soil hydraulic properties as related to soil structure, *Soil Till. Re* 79, 175–184.

[10] Machiwal, D., K.J.., and B.C. Mal. 2006. Modelling infiltration and quantifying spatial soil variability in a watershed of Kharagpur, India. Biosystems Engineering, 95(4), 569-582.

[11] Mishra, s.k, J.V.Tyagi, and V.P.Singh. 2003. Comparison of infiltration models. Hydrological processes. 17, 2629-2652 .

[12] Navar.J, and T.J. Synnote, 2000. Soil infiltration and land use in Linares, N.L., Mexico, Terra Latinoamericana. 18 (3), 255-262.

[13] Ndiaye, B., Molenat, J., Hallaire, V., Hamon, C.G.Y., 2007. "Effects of agricultural practices on hydraulic properties and water movement in soils in Brittany (France)". J. Soil & Tillage Res. 93, 251-263.

[14] Philip, J.R. 1957. The theory of infiltration: 4. Sorptivity and algebric infiltration equations. Soil Sci. 84, 257-264.

[15] Schwartz, R.C., Unger, P.W., Evett S.R., 2000. "Land use effects on soil hydraulic properties." ISTRO.

[16] Shukla, M.K., R.Lal., and P.Unkefer. 2003. Experimental evaluation of infiltration models for different land use and soil management systems. Soil Sci. 168 (3), 178-191.

[17] Singh V.P. 1992. Elementary Hydrology. Prentice Hall: Englewood Cliffs, NJ.

[18] Sonneveld MPW., Bachx MAHM., Bouma.J. 2003. Simulation of soil water regimes including pedotransfer functions and land use related preferential flow. *Geoderma*, 112, p: 97-110.

[19] University of Tehran. "Taleghan Watershed study report"., 1993. Irrigation Engineering Department, Iran.

[20] Van Genuchten, M.Th., 1980. "A closed-form equation for predicting the hydraulic conductivity of unsaturated soils." Soil Sci. Soc. Am. J. 44, 892–898.

[21] Van Genuchten, M.Th., Leij, F. J., Yates, S.R., 1991. "The RETC code for quantifying the hydraulic functions of unsaturated soils." EPA/600/2-91/065. R. S. Kerr, Environmental Research Laboratory. U. S. Environmental Protection Agency, Ada, OK. p: 83.

[22] Wösten, J.H.M., Finke, P.A., Jansen, M.J.W., 1995. "Comparison of class and continuous pedotransfer functions to generate soil hydraulic characteristics." Geoderma. 66, 227–237.

[23] Zhou, X., Lin, H.S., White, E.A., 2008. "Surface soil hydraulic properties in four soil series under different land uses and their temporal changes." Catena. 73, 180-188.

5

Soil Contamination by Trace Metals: Geochemical Behaviour as an Element of Risk Assessment

Monika Zovko and Marija Romić
University of Zagreb, Faculty of Agriculture
Croatia

1. Introduction

Trace metals occur naturally in rocks and soils, but increasingly higher quantities of metals are being released into the environment by anthropogenic activities. Metals are chemically very reactive in the environment, which results in their mobility and bioavailability to living organisms. People can be exposed to high levels of toxic metals by breathing air, drinking water, or eating food that contains them. As a consequence, metals get into the human body by different routes - by inhaling, through skin, and via ingestion of contaminated food.

Every decision on the application of any measures in the environment related to soil quality and management, whether statutory regulations or practical actions, must be based on reliable and comparative data on the status of this part of environment in the given area. Various aspects have to be considered by the society to provide a sustainable environment, including a soil clean of heavy metal pollution. The first step is to identify environments (or areas) in which anthropogenic loading of heavy metals puts ecosystems and their inhabitants at a health risk.

Long-term and extensive use of land for agriculture with frequent application of agri-chemicals is one of the major causes of trace metal, such as copper, nickel, zinc and cadmium, accumulation in soil. Accumulation of Cu in agricultural soil is a consequence of the century – old practice of using copper-sulphate (Bordeaux mixture) and other copper containing fungicides to control vine downy mildew. It is estimated that every time vines are sprayed with copper-containing solutions, some 2 to 5 kg ha-1 of copper enter the soil (Romić & Romić, 2003).

Widespread distribution of Cd and its high mobility makes it a potential contaminant in a wide range of natural environments. Generally, soil Cd concentrations exceeding 0.5 mg kg-1 are considered evidence of soil pollution (McBride, 1994). Phosphatic fertilizers are one of the most ubiquitous sources of Cd contamination in agricultural soils throughout the world. Total Cd inputs to soils through fertilizers in the countries of the European Union have been estimated to be around 334 t yr-1 (Alloway & Steinnes, 1999). Long-term investigations worldwide have shown that application of phosphate fertilizers has resulted in soil enrichment with Cd (0.3-4.4 g ha-1 year-1), depending on the rates and kinds of fertilizers applied (Singh, 1994). Andrews et al. (1996) and Gray et al. (1999) also determined a highly significant correlation between total concentrations of Cd and phosphorus in agricultural soils of New Zealand, which they attributed to long-term application of phosphate fertilizers. Chen et al. (2008) observed significant correlations between Cd and Pb and soil phosphorus in California vegetable croplands, indicating the application of P-fertilizers contributes significantly to the accumulation of Cd and Pb in soils. Zinc belongs to a group of trace metals that are potentially most dangerous for the biosphere. The main sources of the pollution are industry and use of liquid manure, composted materials and agrochemicals such as fertilizers and pesticides in agriculture.

Beside anthropogenic sources, trace metals can be also found in the parent material from which the soils developed. Whether the said inputs will become toxic and to what degree mobile depends on a number of factors: specific chemical and physical trace metal characteristics, soil type, land use, geomorphological characteristics within the soil type and exposure to emission sources. Processes that control the mobility, transformation and toxicity of metals in soil are of special importance in the soil root developing zone – the rhizosphere. For this reason, there is a considerable interest in understanding trace metals behaviour in soil, with special emphasis on the way they build-up in soil and on processes of by which plants take up metals.

2. Factors controlling trace metals behaviour in soil

2.1 Trace metal characteristics

Two thirds of all elements found in nature are metals. According to their chemical definition, metals are elements and as such cannot be synthesized or degraded by biological or chemical processes, though these processes can change chemical forms of metals. Metals are contained in the Earth's crust and in parent rocks, by whose weathering soils are formed, so their presence differs in different geographic regions. Terms like heavy metals, metalloids and microelements are the most commonly encountered in ecological studies. Among the 96 known metals, 17 are semimetals or metalloids (e.g., B, Si, Ge, As, Sn, Te, Po ...). The term heavy metal refers to a group of 53 metals with density higher than 5 g/cm^3. From the geochemical point of view, trace elements are metals whose percentage in rock composition does not exceed 0.1% (e.g., Cu, Cr, F, Fe, Mo, Ni, Se, Zn, As, Cd, Hg, Pb). In very small amounts, some of these elements are essential for normal growth and development of living organisms and they are, from the physiological point of view, called micronutrients or microelements (e.g., Fe, Mn, Zn, Cu, Mo, Ni, Se), while others are toxic even in small concentrations. The issue of toxicity is usually merely a matter of quantity, with the range varying for each element.

Regardless of whether soil metals originate from a natural source or are a consequence of anthropogenic activity, metals appear in groups; one element by itself is rarely the source of contamination. Hence, synergistic and antagonistic interactions of metals should not be disregarded in assessing bioavailability. For example, Zn and Cd are usually present together in ores and have similar physical and chemical properties.

Distribution, mobility, bioavailability and toxicity of metals depend not only on metal concentration but also on the form in which metal exist. Full understanding and prediction of chemical behaviour of an element in the environment is possible only by identification of all forms in which that element can be found under different environmental conditions. Metal speciation is one of the most important properties that determine the behaviour and toxicity of metals in the environment. Chemical speciation of an element refers to its specific form characterized by a different isotopic composition, molecular structure, and electronic or oxidation state (Manouchehri et al., 2006). Speciation is the process of identification and determination of different chemical and physical forms of elements present in a sample (Wang et al., 2006). Metals that occur in cationic forms have a higher ability of binding to negatively charged soil colloids, and are thus less bioavailable, but more easily accumulate in soil, unlike the anionic forms that are mainly present in soil solution and are more bioavailable, but are more readily leached from the soil.

2.2 Soil properties
Because of their very low abundance, trace elements are particularly sensitive to surrounding environmental conditions, which influence their physico-chemical speciation and their behaviour in the ecosystems (N'guessan et al., 2009). In agricultural environment soil is the main sink and source of trace metals. Not all soil properties have equal influence on the mobility and availability of a particular metal. For each metal it is therefore important to know the dominant soil property that will control the behaviour of that metal in that particular soil. With regard to bioavailability, the following metal fractions have an important role in soil: metals in soil solution; precipitated metals; metals bound to clay minerals, oxides and hydroxides, organic matter, and metals in the soil mineral matrix. All fractions of metals are in dynamic equilibrium, and only metals in aqueous soil solution are directly available to plants. Soil solution is in direct contact with the soil solid phase and transformations going on in it are a consequence of mineral equilibrium, exchange processes and sorption processes in the soil mineral phase and organic matter, as well as complexation with organic matter in the solid phase and in solution (figure 1). Major soil properties that affect changes in metal speciation, and thereby also their fractionation, are soil reaction, redox-potential, and existence of different organic and inorganic reactants – ligands.

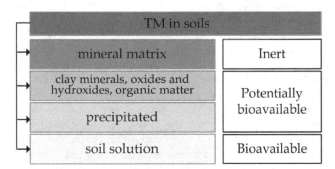

Fig. 1. Simplified schematic presentation of different trace metal (TM) fractions in soil

Whether soil metals will form complexes with organic matter, mineral colloids, inorganic complexes or exist as free ions depends on the soil solution reaction. Constant changes in soil pH are a consequence of cation/anion balance, leaching of organic acids, root respiration and oxidation-reduction reactions. Each change in pH alters the adsorption capacity of minerals and organic colloids. Positive charge prevails at low pH conditions under which anion exchange capacity is dominant, while at high pH negative charge prevails and cation exchange capacity predominates. Solubility of Mn and Zn is strongly dependent on soil reaction. The lower the pH, the stronger the Mn and Zn competition with H^+ and Al^{3+} ions for places in the soil cation exchange complex, and higher the Mn and Zn solubility and proportion of readily available free ions in soil solution. Soil reaction will affect Cu speciation, its solubility and adsorption; however, a weak correlation was determined between soil reaction and Cu concentration in soil solution (Wang et al., 2002) since dissolved Cu has high affinity for organic matter, that is, for phenolic groups and groups of weak acids of humic compounds (Cornu et al., 2007). Gummuluru et al. (2002) showed that Cu complexes with dissolved organic matter are the most dominant species, averaging 97.1, in the neutral and mildly acidic soils.

The presence of carbonate, high contents of organic matter and percentages of clay my suggest that trace metals could be retained in these soils, as these properties increase the adsorption capacity of metals by soil (Peris et al., 2008). Organic substances and oxyhydroxides have the highest surface charge density and are thus the most important and strongest sorbents of trace metals. Carboxylic groups (-COO-) of organic compounds can form stable complexes with metal ions, which leads to changes in the metal fraction available to biota. Surface charge of organic matter and oxyhydroxides is strongly dependent on soil pH, contrary to surface charge of clay (except kaolinite), which does not depend on pH (Reichman, 2002). Organic matter affects the soil redox potential. Fe and Mn hydroxides are especially susceptible to changes in soil redox potential. Electrons are released through organic matter decomposition, which results in a decrease in redox potential and reduction of Fe(III) into the easily soluble Fe(II) form. Manganese ions display similar behaviour. Organic matter can also influence temporary immobilization of Zn. Available Fe and Mn ions inhibit Zn uptake, which can affect precipitation of dissolved Zn into the mineral franklinite ($ZnFe_2O_4$).

Selective sequential extraction was applied to investigate the potential of mobilizing trace elements in agricultural soils of northwestern Croatia (Romić & Romić, 2003). In alluvial soils developed on Quaternary (Upper Pliocene to Holocene) deposits, extraction with 1M $Mg(NO_3)_2$ (pH 7) (Shuman, 1985) indicated possible remobilization of elements from the solid phase into soil solution, particularly in the case of copper. There are several possible mechanisms that increase solubility of metals in the surface layer: 1) the soil mineral component is more susceptible to weathering in shallower than in deeper soil layers owing to faster infiltration of precipitation, higher biological activity and greater changes in temperature; 2) shallower soil horizons are richer in organic matter, which can stimulate metal desorption by formation of soluble organic complexes; and 3) exchangeable complex of shallower soil layers contains more basic cations (Na^+, K^+, Ca^{2+}, Mg^{2+}), which can also reduce sorption of metals by increasing the competition for exchange sites. All these processes are even more pronounced in the anthropogenic horizon of arable soils.

For trace metals to be translocated by water through the soil profile, they have either to be in the soluble phase or bound to mobile particles. Metals can form complexes with particles of organic matter in topsoil and as such can be translocated vertically along the profile depth. It is generally taken that the water-soluble and exchangeable fractions, and sometimes also the organic fraction, are bioavailble forms. Investigations dealing with the leaching of metals from soil revealed a marked vertical distribution of organometallic compounds that can be leached up to 3 times deeper compared to metals not complexed with organic matter. This phenomenon is particularly pronounced in application of organic soil amendments. The observed enhanced mobility of metals in soils improved by different organic conditioners (Al-Wabel et al., 2002) is a consequence of metal binding to dissolved organic carbon (DOC) formed through decomposition of such conditioners. An increase in DOC concentration increases metal complexation with organic matter, that is, the proportion of metals in the soil liquid phase, and thereby also their mobility. In their investigations, Al-Wabel et al.

(2002) assessed that more than 99% of Cu and Zn in soil solution and about 90% of Pb were complexed with dissolved organic carbon or mineral colloids without being in the form of free ions or inorganic complexes. In contrast to Cu and Zn, speciation of Cd in soil solution was not strongly influenced by organic matter, but Cd existed in solution as a free divalent cation (Cd^{2+}) or with inorganic complexes Cl^-, SO_4^{2-} or HCO_3. It has been well documented in the literature that complex formation between metals and inorganic ligands Cl^- and SO_4^{2-} inhibits the adsorption of Cd on soil and soil constituents due to the formation of cadmium complexes that were not strongly adsorbed by soil (McLean & Bledsoe, 1992). In many arid and semiarid regions, including European Mediterranean region, saline irrigation water containing high level of chloride might aggravate trace metals pollution problem, as a result of metals mobilisation due to the formation of metals-chloride complexes (Ghallab & Usman, 2007). Soil salinity strongly increased the concentration of dissolved chloride (Cl^-) ligands, and significantly influenced Cd solubility and thereby its bioavailability and phytoaccumulation. Exposure to increasing NaCl salinity in the rhizosphere environment increased accumulation of Cd in muskmelon and radish leaves (Ondrašek et al., 2009).

The mobilization potential of trace metals, like Cd, Cu, Pb and Zn, may get increased in dredged (oxidised) sediments compared to submerged (reduced) sediments (Tack et al. 1998; Vandecasteele et al., 2007). Indeed, an important factor in the metal mobilization is changing redox conditions. In that sense, the Mn behavior is particularly varying. In oxic environment Mn precipitates as oxides with large mineral surfaces entering into the reactions with both trace metals and soil organic matter (Kaiser & Guggenberger, 2003).

3. Assessment of soil contamination by trace metals – geochemical methods

Various chemical methods, geochemical models and biotests are used for assessment of the bioavailable metal fraction in soil. However, these methods are not universally applicable for all elements and different soil characteristics. Chemical methods for assessment of metal bioavailability are commonly grouped within methods for assessment of total metal content in soil, methods for assessment of currently available and potentially available fractions, methods for assessment of metal speciation in soil solution.

Determination of the total content of metals in soils is an important step in estimating the hazards to the vital roles of soil in the ecosystem, and also in comparison with the quality standards in terms of the effects of pollution and sustainability of the system. Methods for assessment of total metal content are based on soil digestion with strong acids such as HNO_3, HF, $HClO_4$ and aqua regia. Although the total metal content in soil does not show a good correlation with the bioavailable fraction, it is still used in most countries as a statutory regulation for assessment of soil contamination.

From the ecotoxicological aspect, however, it is equally important to determine the bioavailabilty of trace metals accumulated in agricultural soils. Single and sequential extraction methods are applied for assessment of currently available and potentially available metal fractions in soil. There are several kinds of extraction solutions, the most commonly used being 0.001 to 1 M salt solutions ($CaCl_2$, $Ca(NO_3)_2$, $NaNO_3$), weak acids (acetic acid, citric acid) and strong complexes (EDTA, DTPA). Extraction methods are based on complexometric reactions between extractants and metals. In complexometric titrations use is made of suitable indicators, commonly compounds that can produce a less stable coloured product with a free (hydratized) cation. Extraction methods are extensively applied in bioavailability investigations, but are not acceptable for all kinds of metals; for example, EDTA is not suitable for metal bioavailability assessment in contaminated soils, particularly for Cu assessment (Brun et al., 2008). Available Cu in topsoils of the wine-growing regions in north-western Croatia was evaluated with DTPA extraction and calcium chloride extraction methods. Highly significant positive correlation was determined between total copper content and DTPA–extractable copper contents. Such strong correlation indicates that the DTPA extraction method is not suitable for assessing copper availability to plants (Romić et al., 2003).

There are two approaches to determining metal speciation in solution: analytical determination and chemical balance models. Direct measurements of metal ion speciation in soil solution are rare. Measurements are mostly conducted in solutions extracted with dilute salts (0.005 M HNO_3) of soil metals. Precise measurement of different forms of metals present in aqueous solution is the most demanding procedure, since it requires analytical methods of high selectivity and sensitivity; hence, different computer models (GEOCHEM, SOILCHEM) are mostly used for metal speciation determinations. These models are based on geochemical thermodynamic principles (Peijnenburg et al., 2003). Despite their high efficiency, computer models still have shortcomings that must not be disregarded in interpretation of the results obtained. The problem in the application of geochemical models for calculation of element speciation is the modelling of organic matter-metal complexation, since there are no reliable values for stability constants of dissolved organic matter, the properties of which can vary considerably in dependence on environmental conditions.

The main flaw of all chemical methods for bioavailability assessment is that they invalidate all complexity of the mechanism of metal uptake by plants and neglect the very important role of root metabolism and of the microorganisms that surround it. For this reason, these tests are not adequate for assessment of plant available metals in soil (Chaignon et al., 2003). An ideal method for metal phytoavailability assessment should simulate soil-plant interactions as closely as possible (Fang et al., 2007). Further, such a method should extract the amount of metal that corresponds to metal concentration taken up by the plant. Along these lines, the RHIZO - method (rhizosphere based method) was developed. It is based on the application of extraction of 0.01 M solution of organic acids of low molecular mass in wet rhizosphere soil. In laboratory investigations, Feng (2005) demonstrated the efficiency of the RHIZO method application in acid, neutral and alkaline soils, which makes this method more suitable than other bioavailability assessment methods. However, the results were satisfactory only for Cr, Cu, Zn, and Cd, while the method was not a good indicator of Pb and Ni phytoavailability.

Application of bioassays with plants offers a different approach to assessment of metal bioavailability in contaminated soils. Biotests can be done in nutrient solutions or in soil, depending on the research goal. Nutrient solutions can be considered as models of the soil system, more precisely, soil solution. Whether a nutrient solution will be a good soil solution model depends on the experimental design, but it should still be pointed out that this is a simplified soil system and that metal-plant interactions are different in solution and in soil. Thus, the interaction between Zn and Cu is of synergistic character in soil, and of antagonistic character in solution. It is also important to know metal rhizotoxicity because some metals, e.g., Cu, accumulate much more in roots than in stalks (Chaingon et al., 2003). Therefore it is essential for bioavailability assessment to develop biotests that will enable an unobstructed approach to root and rhizosphere studies.

4. Assessment of soil contamination by trace metals: A case of NW Croatia

4.1 Environmental soil functions

Soil plays many important roles in the environment. As being situated at the interface between the atmosphere and the lithosphere it acts as a filter and a buffer: it may weaken and degrade environmentally harmful compounds protecting the air quality. It also has an interface with hydrosphere and therewith it affects surface and groundwater quality. Furthermore, soil, as a part of biosphere, provides nutrient-bearing environment that sustains the growth of plant and animals. As a habitat and protecting media of flora and fauna it contributes to the maintaining of the global nutrient cycling as well as biomass production, whether by natural vegetation growing or plant cultivation. Beside these ecological functions, soil is ground to build and live on, raw material and reserve of cultural heritage.

Because soil quality and its utilization are directly linked, each of above mentioned functions or use mode requires a certain soil quality level. Otherwise, any change of soil quality may affect its utilization potential. Soil is a natural resource essential for the food production and global economy. The way and rate of soil degradation on the global scale point out the importance of the sustainable land use. Harris et al. (1996) define the soil quality as a capacity of the certain soil volume in given conditions (land use, relief, and climate) to protect water and air quality, to sustain plant and animal growth, promoting thus the human health. Out of total degraded land on the global scale that are estimated to 1,965 mha, about 55% was water eroded, about 28% wind eroded, and about 12% is polluted by chemicals (Adriano et al., 1995). Land degradation caused by physical, biological and chemical processes runs up the changes of the key soil properties that have a pivotal role in geochemical cycling.

From the standpoint of soil degradation, the presence of some trace elements in a toxic concentration may be due to both natural and anthropogenic factors. Therefore, it may become quite difficult to discriminate among the different causes. The parent material largely influences trace metals content in many soil types, with concentration sometimes exceeding the critical values (Palumbo et al., 2000; Romić & Romić, 2003; Salonen & Korkka-Niemi, 2007). Some metals, such as Ni, Cr and Mn, are contained as trace elements in some rock types of volcanic and metamorphic origin (Alloway, 1995). During weathering processes the primary crystalline structures of some rock minerals are completely broken, relevant chemical elements may be thus either adsorbed in the topsoil or transported towards surface water or ground water targets. The transformation of metals and metalloids is influenced significantly by adsorption-desorption reaction in soil environments; these reactions are affected by physicochemical and biological interfacial interactions, which should be especially important in rhizosphere (Huang, 2008). Soil buffer capacity may be defined as its ability to postpone the negative effects of more or less continuous input of toxic substances by inactivation of contaminants (Moolenar & Lexmond, 1999). This inactivation can generally be reached by effective binding of contaminant and soil particles, or by forming of insoluble complexes. When the contaminant input exceeds the level of so-called «critical content», their buffer capacity is getting overcome as well and then the soil is characterized as polluted (de Haan, 1996). So that, buffer capacity of the diverse soil types defers considerably regarding the soil characteristics reflecting thus its vulnerability or resistance.

Generally, two main types of pollution may be distinguished: diffuse pollution or non-point source, and point source (O'Shea, 2002). The example of the non-point source is atmospheric deposition as a result of urban, transport and construction activities, as well as mineral fertilizer or sewage sludge application in agriculture. Diffuse sources of pollution are not easy to control, and the best methods for soil pollution control often depend on the legal regulations and management strategies. It becomes easier to control point sources of pollution, because it usually refers to the single source that is easy to identify (local pollution caused by chance, accidentally or undertaking prohibited activities). Sources of agricultural land contamination, especially if places near urban or industrial area, certainly have a diffuse nature.

Geochemical maps are good visual demonstration of contaminant changes in the space, and enable the identification of the areas that are likely to contain harmful substances (Goodchild et al., 1993). The knowledge on spatial variability becomes equally important both for the assessment of the study site and for the prediction of the possible risks. The procedure of the geochemical surveys, including exhausting field work, long-lasting and expensive chemical analysis, require the sampling scheme optimization for the efficient interpolation and mapping.

4.2 Soil pollution assessment

A soil pollution assessment becomes very difficult to carry out when different sources of contamination are present and their products are variably distributed. In these cases the spatial variability of the trace metal concentrations in soils is basic information for identifying the possible sources of contamination and to delineate the strategies of site remediation. An approach was described that interpolate sampled trace metal concentrations using numerous environmental predictors and then represent the overall pollution by using the continuous limitation scores, as proposed by Romić et al. (2007). Such visualizations can supplement maps of separate trace metals so that the areas of high overall pollution can be more easily delineate presenting the basis for further studies on risk assessment or decision making.

In Europe, decision makers and spatial planners more and more require information on soil quality for different purposes: to locate areas suitable for organic (ecologically clean) farming and agro-tourism; to select sites suitable for conversion of agricultural to non-agricultural land, particularly for urbanization; setting up protection zones for groundwater pumped for drinking water; to estimate costs of remediation of contaminated areas and similar. Every decision on the application of any measures in the environment relating to soil quality and management, whether statutory regulations or practical actions, must be based on reliable and comparable data on the status of this part of environment in the given area. Various aspects must be considered by the society to provide a sustainable environment, including a soil clean of heavy metal pollution. The first among them is to identify environments (or areas) in which anthropogenic loading of heavy metals puts ecosystems and their inhabitants at health risk. Maps indicating areas with pollution risks can provide decision-makers or local authorities with critical information for delineating areas suitable for the planned land use or soil clean up (Van der Gaast et al., 1998; Broos et al., 1999). Maximum permissible concentrations of trace metals in soil are now regulated by law in many countries.

In urban and industrial environments there are numerous potential sources of contamination with harmful substances, including trace metals, mainly combustion processes in industry and transportation. Lead and cadmium are the main trace elements arising from combustion and are often associated with zinc owing to tyre wear on the roads. Waste water from industrial processes may contain an important load of zinc, copper, chromium and nickel. Moreover, mining activities for extraction and manufacturing of metal products may result in a large amount of pollutants to be released into the atmosphere and, secondly, in the adjoining soils and waters.

Long-term and extensive use of agricultural land with frequent application of growing practices and use of pesticides (Nicholson et al., 2003) may cause heavy metals such as copper, nickel, zinc and cadmium to be strongly accumulated in the topsoil.

The estimation of the total trace metals content and spatial variability of these elements in soil is the main indicator of the degree of contamination, but is not sufficient for establishing the relevant guidelines or decision making. GIS-based mapping techniques in conjunction with statistical and geostatistical analysis of the data are widely used to highlight the influence of human activities on the trace metals content of topsoils in urban and sub-urban areas (Kelly et al., 1996) and to assess the transport pathways, sinks, and impact of particulate associated trace metals in the various spheres making up the urban environment (Charlesworth et al., 2011). Intensive urbanisation of the Croatian capital of Zagreb has led to a situation where very good agricultural soils, particularly for vegetable production, are entrapped within urban and suburban areas. On the example of the Zagreb region (Northwest Croatia), different approaches to the assessment of the soil trace metals spatial variability and level of contamination will be demonstrated.

4.2.1 Data acquisition

The research on pollution in agricultural soils of the Zagreb region and establishment of the monitoring on a regional scale has started in 1997. At the beginning, the survey was carried out on approximately 860 km² of urban and suburban areas of the city of Zagreb. Later on, the research was extended for about 3000 km², covering agricultural soils of the City of Zagreb and the surrounding Zagreb County.

A total of 916 topsoil (0-20 cm) samples were collected using a systematic sampling on a 2-km grid, with sampling density increasing to 1 km within the industrial and residential area and wine-growing areas (Fig. 2). The observation sites were spatially referenced using GPS and data were stored in different GIS layers. Site survey, carried out during sampling process, provided site-specific information related to land use and other human activities near the sampling points.

For the determination of soil properties, the surface soil samples (average weight of 2 kg) were air-dried and mixed well. A subsample of about 1 kg soil was sieved through a 0.5-mm mesh. Digestion in aqua regia (HRN ISO11466, 2004) was done by the microwave technique on a PerkinElmer Multiwave 6MF 100 (1000W) apparatus in closed TFM vessels and with automatic pressure and temperature regulation. Heavy metal concentrations in soil digests were determined by inductively coupled plasma optical emission spectroscopy (ICP-OES) on a VistaMPX AX (Varian).

Before any solution for the problem of soil heavy metal pollution can be suggested, a distinction needs to be made between natural anomalies and those resulting from human activities. Namely, it often happens that also natural concentrations and distribution of potentially toxic metals could present health problems, like in the case of chromium, cobalt, and particularly nickel in ultramafic soils (Proctor & Baker, 1994). Rock type and geological-geochemical processes can change markedly in a relatively small area, resulting in great spatial variability in the soil content of elements. The region exhibits a variety of soils developed on diverse lithologies described in detail by Sollito et al. (2010). The oldest stratigraphic units are represented by the Paleozoic magmatic and metamorphic complex of the deep earth crust (Fig.

3a), mostly comprising diabases, gabbros, greenshist and blueshist facies rocks outcropping at Mt Medvednica (Belak & Tibljaš, 1998). These types of rocks are known to have a chemical composition characterized by high content of heavy metals, such as Ni and Cr, which are accumulated during the weathering processes in the soil (Alloway, 1995). Mesozoic calcareous rocks (mostly dolomite and limestone) outcrop at NW. Paleozoic–Mesozoic massif is rimmed by Miocene sediments due to the presence of tectonic structures. These deposits are composed by a transgressive sequence of calcareous breccias and conglomerates, marls, clays, sands and silts (Vrsaljko et al., 2005). The main geo-lithological features are the Pliocene and Quaternary alluvial sediments of the Sava River basin in the central and southern parts of the studied area. Terraced sediments outcrop mainly in the western and southwestern sectors of the region and consist of gravel and sands, and secondarily of sandy and silty clays. The sediments in the floodplain area and in the recent stream beds consist of coarse grained sandy-clayey silts and silty clays, with thin layers of charcoal that were flooded from the Slovenian coal mines. Moreover, Pleistocene deposits are made of pond sediments and non-carbonate loess mixed with sand and gravel. Quartz is the main component of the light mineral fraction of these sediments.

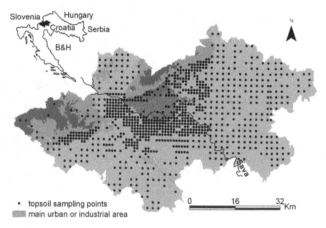

Fig. 2. Study area and sampling locations

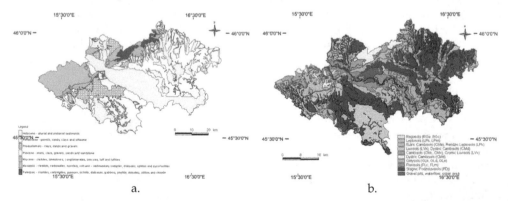

a.   b.

Fig. 3. a) Simplified geologic map of the studied area; b) Simplified soil map of the study area

The large heterogeneity of the parent material, combined with the climate and geomorphology of the Zagreb region, has caused the development of a wide variety of soil types (Fig. 3b). Since the floodplain soils are mixture of the material eroded from the geological units present in the catchment basin, the mineralogical and chemical composition as well as the textural properties of the sedimentary deposits in the alluvial region are largely influenced by the dynamic of the Sava River flow and its solid transportation. Hydromorphic soils prevail in the Sava River valley, where the recent deposits form fluvial terraces. Molic Fluvisols, Calcaric Fluvisols, Eutric Cambisols, Eutric and Calcic Greysols are developed on the Holocene deposits. In the area of Pleistocene terraces, Stagnic Podzoluvisols prevail on the plateau, whereas Stagnic Podzoluvisols and Glayic Podzoluvisols are prevalent on the slopes (FAO, 1998). More than 50% of the land is used for agriculture: soils developed on loamy aeolian materials are mostly used for intensive cultivation of field crops (like cereals), but some of them are under permanent grassland. Anthropogenic vineyard soils, classified as Aric Anthrosols (FAO, 1998), of the Mt Medvednica foothills were formed on Tertiary carbonate deposits of marl and limestone. Moreover, as the Žumberačka Gora piedmont spreads perpendicularly to the mountains, some slopes are firmly interlinked by ridges, forming well protected, amphitheatre-shaped vineyard areas. The presence of intense agricultural activities imposes the risk of soil contamination due to the use of pesticide, which sometimes may constitute a diffuse source over large regions. In addition to agricultural land use other main usage classes are forest, covering the mountain areas and some parts of the terraces, pasture and orchard. Local source of pollution may be related to industrial areas and urban networks, which are scattered distributed over the floodplain, as well as to mining activities which are mainly located in the upstream regions.

Soils in the vicinity of urban areas and industry are exposed to input of potentially toxic elements, and the situation of agricultural soils gets additionally complicated due to continuous application of agrochemicals.

In practice, soil pollution by heavy metals is commonly assessed by interpolating concentrations of heavy metals sampled at point locations, so that each heavy metal is represented in a separate map (Fig. 4).

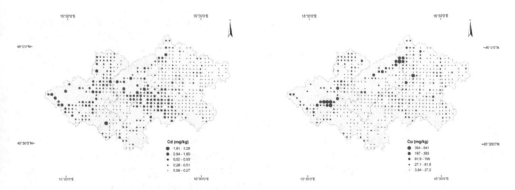

Fig. 4. Cadmium and copper contents at the point locations of the study area

4.2.2 Assessing spatial variability

Spatial variation of soil properties within a field, including trace element contents, is extremely complex, even in a small scale. Therefore it is important to apply methods that may describe this variability. First step in the data processing is the exploratory analysis in spatial data processing. Summary statistics calculation on the dataset (shown in Table 1) provided information about the frequency distribution on the concentrations of chemical elements in the topsoil; the results were compared with some reference values (Table 1).

Variable	Mean	Median	SD	Min	Max	Skewness	Kurtosis	Soils worldwide[a]	Upper continental crust[b]	MPC[c]
Ba (mg kg⁻¹)	224	220	68.2	12.7	421	0.19	3.09	500	584	-
Ca (g kg⁻¹)	18.4	5.20	37.3	0.67	214	3.25	13.4	14	29.5	-
Cd (mg kg⁻¹)	0.40	0.31	0.34	0.02	4.94	5.58	57.7	0.3	0.102	2
Co (mg kg⁻¹)	10.9	10.8	3.62	2.74	47.2	1.96	16.7	10	24	50
Cr (mg kg⁻¹)	54.6	51.2	21.1	11.5	400	5.86	91.2	80	126	100
Cu (mg kg⁻¹)	56.1	23.5	117	3.64	1335	4.56	29.3	25	25	100
Fe (g kg⁻¹)	29.7	28.9	7.77	5.85	59.1	0.26	2.91	35	30.9	-
Mg (g kg⁻¹)	8.04	6.51	5.43	0.77	36.6	2.98	12.7	9	13.5	-
Mn (mg kg⁻¹)	597	556	266	79.2	4537	4.28	56.0	530	527	-
Ni (mg kg⁻¹)	35.2	29.7	23.8	0.70	488	8.51	148	20	56	60
P (mg kg⁻¹)	722	675	300	213	3023	2.36	13.6	750	665	-
Pb (mg kg⁻¹)	23.2	19.6	14.4	1.00	216	5.24	54.2	17	14.8	150
S (mg kg⁻¹)	407	349	249	47.7	2683	2.52	16.7	800	-	-
Sr (mg kg⁻¹)	92.9	55.5	136	17.2	1846	6.17	56.5	240	333	-
Zn (mg kg⁻¹)	77.9	70.7	33.6	27.1	479	4.28	39.6	70	65	300

[a] Median, after Reimann and de Caritat (1998)).
[b] Mean, after Wedepohl (1995).
[c] Maximal permissible concentrations defined by the Croatian government

Table 1. Summary statistics of element concentrations in soils of the study area

The background concentrations of the elements in soils were mostly lower than the average element concentrations in the upper continental crust (Wedepohl, 1995) and were similar to the worldwide median values in soils (Reimann & de Caritat, 1998). Except for the cobalt, the concentrations of all other elements exceeded the maximum permissible concentration (MPC) defined by the Croatian government, with critical conditions recorded for copper, nickel and chromium. For these elements, the maximum concentrations exceed the MPC by up to ten times. Nevertheless, the mean values for all samples were lower than the critical thresholds, meaning that the high values are unevenly distributed in the region. Only a relatively small proportion of samples had concentrations of Ni, Cu and Cr exceeding the MPC (9%, 8% and 1.4% of all samples, respectively). As confirmed by the skewness values (Table 1), the concentrations of elements are characterized by large variability, with positively skewed frequency distributions. This is common for heavy metals, because they usually have low concentrations in the environment, so that the presence of a point source of contamination may cause a sharp increase of local concentration, so exceeding the thresholds. The concentrations of Ni and Cr higher than the MPC were mainly recorded near the urban area of Zagreb and on the northern edge of the city, whereas the Cu concentration exceeding MPC was found all over the piedmont belt surrounding the northern Paleozoic and Mesozoic relieves. Only the Fe and Ba data distributions are nearly normal, with small differences between the mean and median values and the skewness values close to 0. In contrast, the Mg and Ca concentrations in the topsoil showed a skewed distribution with high values where calcareous rocks outcrop.

As confirmed by the skewness values (Table 1), the concentrations of elements are characterized by large variability, with positively skewed frequency distributions. This is common for trace metals, because they usually have low concentrations in the environment, so that the presence of a point source of contamination may cause a sharp increase of local concentration, so exceeding the thresholds. Skewnes measures the asymmetry of the observations. Normal distribution is symmetrical and its mean, mode and median coincide at its centre. When the distribution is skewed than the mean does not represent the central data value that causes the unreliability of the statistics. The problem of processing geochemical and environmental data sets is elaborated in details elsewhere (Zang & Selinus, 1998; Reinman & Filmoser, 2000; Webster & Oliver, 2001). Data transformation enables approaching to the normal distribution, reduces the influence of high values, stabilizes variance and thereby enables the next data processing, as shown on Figure 5.

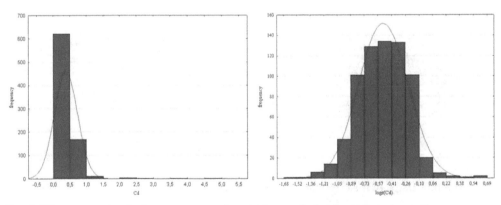

Fig. 5. Histogram of cadmium concentrations before and after logit transformation

Classic geostatistics methods based on univariate analysis can provide distribution maps for each pollutant and delineate the areas where its concentration exceeds the threshold values (Fig. 6). Although the concentration patterns of the elements could suggest probable sources of contamination (Atteia et al., 1995) each chemical element or pollutant might have its own distinctive spatial distribution making difficult to get an overall picture of the contamination.

To solve a problem of presenting overall polluted areas, Romic & Romic (2003) applied factor analysis prior to interpolation and then interpolated factor scores (Fig. 7). As a multivariate method, factor analysis (FA) facilitates the reduction, transformation and organization of the original data by the use of intricate mathematical techniques, which eventually results in a sample form of factor model. Factor analysis creates a new set of uncorrelated variables, which are the linear combination of the original ones with the same amount of information. Since the FA is conducted if the original variables have significant linear intercorrelations, the first few factors will include the largest part of the total variance. The interpretation of dominant factors was made by taking into account the highest factor loadings on chemical elements. The theoretical details of the FA are given by Johnson (1998).

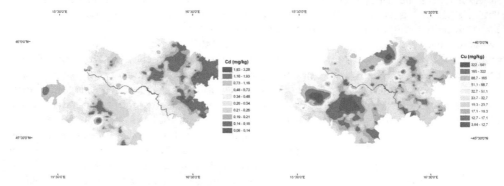

Fig. 6. Interpolated maps of cadmium and copper contents in soils of the study area

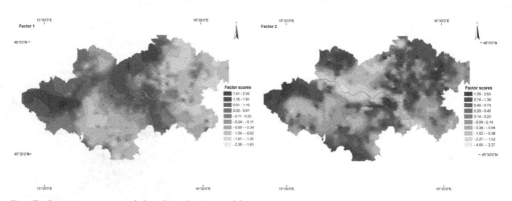

Fig. 7. Contour maps of the distribution of factor scores

Van der Gaast et al. (1998) used maps of background values of soil contaminants focusing on the 90-percentiles. Hanesch et al. (2001) tested fuzzy classification algorithms to distinguish different sources of pollution. Amini et al. (2004) classified HMCs using unsupervised fuzzy k-means to partition the values optimally. The final outputs are maps of memberships to each cluster, which commonly reflect the combination of most correlated heavy metals. In all these examples the procedures are statistically valid, but the meaning of such factors and continuous memberships is hard to interpret. In practice, decision makers usually only wish to see the areas that are polluted without any training in (geo)statistics. Legislative regulations on maximal permitted soil concentrations of potentially toxic elements from practical reasons are setting sharp boundaries. As additional criteria for the soil pollution assessment other soil properties that control metal behaviour in soil are usually set up. Therefore, geochemical data have to be integrated with the detailed soil characterization. Spatial variation of soil properties within a field is extremely complex, even in a small scale. Main questions interrogating in so doing are how we can measure and model this variability and how this variability affects land use decision-making and environmental quality.

For the study area of NW Croatia, it was assumed that the distribution of trace element contents is systematic, i.e. controlled by natural and anthropogenic factors. The problem appears when the complex interactions between these factors allows for local yield of high natural concentrations of potentially toxic metals in soils, which may exceed the threshold limits designated for contaminated areas (Myers & Thorbjornsen, 2004, Amorosi & Sammartino, 2007). Therefore, statistical and spatial analysis tools were utilised in order to combine the quantitative information obtained from the chemical analysis of the soil samples with the area-specific qualitative information. Romić et al. (2007) applied the regression block kriging for spatial interpolation of heavy metals. A list of potential predictors was used as auxiliary data in the RK model (geological and land cover map, NCVI, water table depth, slope, distance to urban area and roads and wind exposition. After the heavy metal concentrations had been interpolated, they were converted to limitation scores. This study was shown that in the Zagreb region, only 7.2% of the total area is critically polluted by one or more heavy metals.

The applied procedure for geostatistical analysis of heavy metal concentration data successfully identified a number of contamination hotspots in the studied region. The limitation of using the scores is that the high overall pollution can be due to very high values of single element, or due to a cumulative effect of large number of contaminants. But the methodological framework of cumulative limitation scores (CLS) opens several perspectives, i.e. to relate the CLS directly with the remediation costs (Broos et al., 1999), or to observe how heavy metal concentrations change at different scale. In addition, one might consider the methodology of error propagation (Heuvelink, 1998) to derive the composite uncertainty of the final soil pollution map. Geostatistical simulations would help us to get an idea about the propagated uncertainty, but can also be used as an input to a more complex environmental data modelling.

The maps of cumulative limitation scores might be advantageously used to target sampling and/or delineating contaminated zones with lower costs. However, the procedure is suitable in many other decision situations, especially for making decisions about soil remediation, health risk assessment or contaminated land management in general.

5. Conclusion

Risk assessment is the basic element of sustainable management of the agricultural environment because it provides answers to the question about how safe is a studied medium for a population in a given time period and under defined conditions of that medium and population. Risk assessment is the basis for selecting sustainable solutions for allocation of agricultural soils as well as for undertaking efficient measures for remediation of contaminated soils. In the case of trace metals, risk assessment is a demanding and complex task. Metals have a very complex chemical behaviour in the environment and, unlike organic compounds, are not subject to degradation processes. The problem of metals in agricultural environments is enhanced by their potential for mobilization under specific soil physical and chemical conditions. A direct effect of this mobilization is a faster translocation of metals into deeper soil layers, and thereby possible contamination of groundwater. Increased mobility of metals in soil is a frequent effect of fertilization or application of composts and other organic amendments in a solid form. Therefore, metals risk assessment is an important tool for establishing protective levels in different environmental media. The critical scientific issues that need addressing are the various properties of metals environmental chemistry, bioavailability and bioaccumulation. Bioavailability assessment methods need to be elaborated on as well.

Chemical speciation of metals, their bioavailability, bioaccumulation and toxicity are the key elements in assessing the potential harmful impacts of metals upon the environment and human health. Metal bioavailability is a complex issue that depends on a series of properties related to soil matrix, plant characteristics, and environmental conditions. An interdisciplinary approach is therefore required in risk assessment studies. Bioavailability can be used as an element of risk assessment only if all risk assessors apply the same approach. Such a uniform approach to bioavailability assessment will enable establishment of standard methods for soil analyses and contribute towards developing new legislation on soil protection, particularly those related to persistent immobile substances such as metals and polycyclic aromatic hydrocarbons.

Climate changes are currently the focus of numerous studies because their indirect consequences can affect the geochemical and biological cycling of metals as well. The main aim of these studies is to define the indicators that best describe bioavailability of a metal under given environmental conditions. Reliable information on soil contamination by trace metals is needed, as well as the application of geostatistical methods for determining trace metal variation in soil and identifying possible sources of contamination.

6. References

Adriano, D.C., Chlopecka, A., Kaplan, D.I., Clijsters, H., Vangronsveld, J. (1995). Soil contamination and remediation: philosophy, science and technology. In: Prost R., ed. *Contaminated Soils: Third International Conference on the Biogeochemistry of Trace Elements*, Paris, 15-19 May, 1995. (INRA editions, Paris, France, CD-room, full text)

Alloway, B.J. (1995). The mobilisation of trace elements in soils, In: *Prost R., ed. Contaminated Soils: Third International Conference on the Biogeochemistry of Trace Elements*, Paris, 15 - 19 May 1995. (INRA editions, Paris, France, CD-room, full text)

Alloway, B.J. & Steinnes, E. (1999). Anthropogenic Additions of Cadmium to Soils, In: *Cadmium in Soils and Plants, McLaughlin, M.J. and B.R. Singh (Eds.). Kluwer Academic Publishers*, Dordrecht, Netherlands, pp: 97- 123

Al-Wabel M.A, Heil D.M, Westfall D.G, Barbarick K.A. (2002). Solution chemistry Influence on metal mobility in biosolids-amended soils. *J. Environ. Qual.* 31: 1157-1165

Amini, M., Afyuni, M., Fathianpour, N., Khademi, H., Fluhler,H. (2004). Continuous soil pollution mapping using fuzzy logic and spatial interpolation. Geoderma 124 (3–4), 223–233

Amorosi, A., Sammartino, I. (2007). Influence of sediment provenance on background values of potentially toxic metals from near-surface sediments of Po coastal plain (Italy), *International journal of earth sciences*, 96, 2 , 389-396

Andrews, P., Town, R.M., Hedley, M.J., Loganathan, P. (1996). Measurement of plant-available cadmium in New Zealand soils. *Australian Journal of Soil Research*, 34 (3): 441-452.

Atteia, O., Thelin, P.H., Pfeifer, H.R., Dubois, J.P., Hunziker, J.C. (1995). A search for the origin of cadmium in the soil of the Swiss Jura. *Geoderma, 68,* 149–172

Belak, M., Tibljas, D. (1998). Discovery of blueschists in the Medvednica Mountain (northern Croatia) and their significance for the interpretation of the geotectonic evolution of the area, *Geologia Croatica* 51 (1), 27–32

Bross, M.J, Aarts, L., van Tooren, C.F., Stein, A. (1999). Continous soil pollution mapping using fuzzy logic and spatial interpolation, *Geoderma*, 124 (3-4), 223-233

Bross, M.J., Aarts, L., van Tooren, C.F., Stein, A. (1999). Quantification of the effects of spatially varying environmental contaminants into a cost model for soil remediation, *Journal of Environmental Management*, 56 (2), 133–145

Brun, L.A, Maillet J, Richarte J, Herrmann P, Remy J.C. (1998). Relationships between extractable copper, soil properties and copper uptake by wild plants in vineyard soils, *Environmental Pollution* , 102: 151-161

Chaignon V, Hinsinger P. (2003). Heavy Metals in the Environment: A Biotest for Evaluating Copper Bioavailability to Plants in a Contaminated Soil, *J. Environ. Qual.* ,32:824-833

Charlesworth, S., De Miguel, E., Ordóñez, A. (2011). A review of the distribution of particulate trace elements in urban terrestrial environments and its application to considerations of risk, *Environmental geochemistry and health*, 33, 2 , 103-123

Chen, W., Krage, N., Wu, L., Pan, G., Khosrivafard, M., Chang, A.C. (2008). Arsenic, cadmium, and lead in California cropland soils: role of phosphate and micronutrient fertilizers. Journal of Environmental Quality, 37(2), 689-95.

Cornu, J. Y., Staunton, S., Hinsinger, P. (2007). Copper concentration in plants and in the rhizosphere as influenced by the iron status of tomato (Lycopersicon esculentum L.). *Plant Soil* , 292:63–77

De Haan, F.A.M. (1996). Soil quality evaluation. In: *Soil pollution and soil protection*. Ed. F.A.M. de Haan, M.I. Visser-Reyneveld. Wageningen Agricultural University and International Training Centre (PHLO). Wageningen. pp. 1 -17

Fang J, Wen B, Shan X, Lin J, Owens G. (2007). Is an adjusted rhizosphere-based method valid for field assessment of metal phytoavailability? application of non-contaminated soils, *Environmental Pollution* 150: 209-217

FAO/ISRIC/ISSS (1998). World reference base for soil resources, *World Soil Resources Report*, 84, FAO, Rome

Feng M. H, Shan X. Q, Zhang S, Wen B. (2005). A comparison of the rhizosphere-based method with DTPA, EDTA, CaCl$_2$, and NaNO$_3$ extraction methods for prediction of bioavailability of metals in soil to barley, *Environmental Pollution* 137: 231-240

Ghallab, A., Usman, A.R.A (2007). Effect of Sodium Chloride-induced Salinity on Phyto-availability and Speciation of Cd in Soil Solution, *Water, Air, and Soil Pollution*, 185 (1-4). p.43

Gray, C.W., McLaren, R.G., Roberts, A.H.C., Condron, L.M. (1999). Cadmium phytoavailability in some New Zeland soils. *Australian Journal of Soil Research*, 37: 461-477

Goodchild, M.F., Parks, B.O., Steyaret, L.T. (Eds.) (1993). Environmental modeling with GIS, *Oxford University Press*, New York, USA

Gummuluru, S.R., Krishnamurti, G.S.R., Ravendra, N. (2002). Solid-Solution Speciation and Phytoavailability of Copper and Zinc in Soils. *Environmental Science & Tecnology*, 36 (12) 2645-2651

Hanesch, M., Scholger, R., Dekkers, M.J., (2001). The application of fuzzy c-means cluster analysis and non-linear mapping to a soil data set for the detection of polluted sites, *Physics and Chemistry of the Earth, Part A: Solid Earth and Geodesy*, 26 (11–12), 885–891

Harris, R.F.D., Karlen, D.L., Mulla, D.J. (1996). A conceptual framework for assessment and management of soil quality and health, *Methods for assessing soil quality*. Ed. Doran J.W., A.J. Jones, Soil Science Society of America, Special Publication, 49: 61-82

Hengl T., Heuvelink GBM., Stein A. (2004). A generic framework for spatial prediction of soil variables based on regression-kriging, *Geoderma*, 122(1-2), 75-93

Heuvelink, G. (1998). *Error Propagation in Environmental Modelling with GIS*. Taylor & Francis, London, UK, pp 127

Johnson, D.E. (1998). Applied multivariate methods for data analysts. Duxbury Press, Pacific Grove, CA, USA

Kaiser, K. & Guggenberger, G. (2003). Mineral surfaces and soil organic matter, *European Journal of Soil Science*, 54, 219-236.

Kelly, J., Thornton, I., Simpson, PR. (1996). Urban Geochemistry: A study of the influence of anthropogenic activity on the heavy metal content of soils in traditionally industrial and nonindustrial areas of Britain. *Applied geochemistry*, 11 , 1-2, 363-370

Manouchehri, N., Besancon, S., Bermond, A. (2006). Major and trace metal extraction from soil by EDTA: Equilibrium and kinetic studies. *Analytica Chimica Acta* 559: 105-112

McBride M.B. (1994). *Environmental chemistry of soils*, Oxford University Press, New York

McLean & Bledsoe, (1992). *EPA Ground Water Issue*, EPA 540-S-92-018:25 pp

Moolenaar, S.W., Lexmond, T.M. (1999). General aspects of cadmium, copper, zinc, and lead balance of agro-ecosystems, *Journal of Industrial Ecology*, 2 (4): 45-60

Myers, J; Thorbjornsen, K. (2004). Identifying metals contamination in soil: A geochemical approach, *Soil & sediment contamination*, 13 , 1 , 1-16

Nicholson, FA., Smith Sr., Alloway, BJ., Carlton-Smith, C., Chambers, BJ. (2003). An inventory of heavy metals inputs to agricultural soils in England and Wales, *Science of the total environment*, 311 , 1-3, 205-219

N'guessan, Y.M., Probst, J.L., Bur, T., Probst, A. (2009). Trace elements in stream bed sediments from agricultural catchments (Gascogne region, S-W France): Where do they come from? Science of the total environment 407, 2939– 2952

Odeh I, McBratney A, Chittleborough D. (1995). Further results on prediction of soil properties from terrain attributes: heterotopic cokriging and regression kriging, *Geoderma*, 67 (3-4) 215-226

Ondrašek, G., Romić, D., Rengel, Z., Romić, M., Zovko, M., (2009). Cadmium accumulation by muskmelon under salt stress in contaminated organic soil. *Science of the Total Environment*, 407(7), 2175-2182

O'Shea, L. (2002). An economic approach to reducing water pollution: point and diffuse sources, *Science of the Total Environment*, 282:49-63

Palumbo, B., Angelone, M., Bellanca, A., (2000). Influence of inheritance and pedogenesis on heavy metal distribution in soils of Sicily, Italy, *Geoderma*, 95 (3–4), 247–266

Peijnenburg, W.J.G.M, Jager, T. (2003). Monitoring approaches to assess bioaccessibility and bioavailability of metals: Matrix issues, *Ecotoxicology and Environmental Safety*, 56: 63–77

Peris, M., Recatalá L., Micó, C. , Sánchez, R., Sánchez, J. (2007). Increasing the Knowledge of Heavy Metal Contents and Sources in Agricultural Soils of the European Mediterranean Region, *Water Air Soil Pollut.*, 192: 25–37

Proctor, J., Baker, A.J.M. (1994). The importance of nickel for plant growth in ultramafic (serpentine) soils, In: *Toxic Metals in Soil-Plant System. Ed. S.M. Ross. John Wiley & Sons Ltd.* pp. 417-432

Reichman S.M (2002). The responses of Plants to Metal Toxicity: A review on Copper, Manganese and Zinc. *Australian Minerals & Energy Environment Foundation.* www.ameef.com.au

Reimann, C., de Caritat, P. (1998). *Chemical Elements in the Environment.* Springer.

Reimann, C., Filzmoser, P. (2000). Normal and lognormal data distribution in geochemistry: death of a myth. Consequences for statistical treatment of geochemical and environmental dana, *Environmental Geology*, 39 (9): 1001-1014

Romić M., Hengel T., Romić D., Husnjak S. (2007). Representing soil pollution by heavy metals using continuous limitation scores, *Computers & Geosciences*, 33(10): 1316-1326

Romić M. and Romić, D. (2003). Heavy metals distribution in agricultural topsoils in urban area. *Environmental Geology*, 43(7), 795-805

Salonen, V., Korkka-Niemi, K. (2007). Influence of parent sediments on the concentration of heavy metals in urban and suburban soils in Turku, Finland, *Applied Geochemistry*, 22, 906–918

Shuman, L.M. (1985). Fractionation method for soil microelements, *Soil science*, 140 (1): 11-22

Singh, B.R. (1994). Trace element availability to plants in agricultural soils, with special emphasis on fertilizer inputs, *Environmental Reviews*, 2: 133-146

Sollitto, D., Romić, M., Castrignano, A., Romić, D., Bakić, H. (2010). Assessing heavy metal contamination in soils of the Zagreb region (Northwest Croatia) using multivariate geostatistics. *Catena*, 80(3), 182-194

Tack, F.M.G., Singh, S.P., Verloo, M.G. (1998). Heavy metal concentrations in consecutive saturation extracts of dredged sediment derived surface soils. *Environmental Pollution*, 103(1), 109-115.

Vandecasteele, B., Quataert, P. and Tack, F.M.G. (2007). Uptake of Cd, Zn and Mn by willow increases during terrestrialisation of initially ponded polluted sediments. *Science of the Total Environment*, 380, 133-143.

Van der Gaast, N., Leenaers, H., Zegwaard, J. (1998). The grey areas in soil pollution risk mapping the distinction between cases of soil pollution and increased background levels, *Journal of Hazardous Materials*, 61 (1–3), 249–255

Vrsaljko, D., Pavelić, D., Bajraktarević, Z. (2005). Stratigraphy and palaeogeography of Miocene deposits from the marginal area of Žumberak Mt. and the Samoborsko Gorje Mts. (Northwestern Croatia), *Geologia Croatica* 58 (2), 133–150

Wang G., Su, M., Chen, Y., Lin, F., Luo, D., Gao, S. (2006). Transfer characteristics of cadmium and lead from soil to the edible parts of six vegetable species in southeastern China. *Environmental Pollution* , 144: 127–135

Wang, Z., Shan, X., Zhang S. (2002). Comparison between fractionation and bioavailability of trace elements in rhizosphere and bulk soils. *Chemosphere*, 46: 1163–1171

Webster, R., Oliver, M. (2001). *Geostatistics for Environmental Scientists*, John Wiley & Sons, LTD.

Wedepohl, K.H. (1995). The composition of the continental-crust, *Geochimica et Cosmochimica Acta*, 59(7): 1217-1232

Yanai, J., Yabutani, M., Kang, YM., Huang, B., Luo, GB., Kosaki, T. (1998). Heavy metal pollution of agricultural soils and sediments in Liaoning Province, China, *Soil science and plant nutrition*, 44 , 3, 367-375

Zhang, C., Selinus, O. (1998). Statistics and GIS in environmental geochemistry, some problems and solutions, *Journal of Geochemical Exploration*, 64: 339-354

Part 4

Remote Sensing

6

Integration of Satellite Imagery, Geology and Geophysical Data

Andreas Laake
WesternGeco Cairo
Egypt

1. Introduction

Satellite imagery is a large scale surface geological mapping too, which offers the unique opportunity to investigate the geological characteristics of remote areas of the earth surface without the need to access the area on the ground. The resolution of the technique is limited by the resolution of the imagery. This chapter explains how geological information can be extracted from satellite imagery and how this information can be merged with geological and geophysical data to build consistent geological models for the surface and subsurface. On the one hand, the interpretation of satellite imagery can generate start models prior to the start of geophysical surveys. On the other hand, geological and geophysical data can calibrate models derived from satellite imagery.

2. Methodology

When studying the shape of the earth surface in connection with the rock layers and their deformation by tectonic forces, we often notice a correlation between shapes and structures at the surface and in the subsurface (Short and Blair 1986). This opens the opportunity to map the characteristics of the surface and infer characteristics of the subsurface. We can describe the surface by its shape and by its structure. The surface shape depends on topography, terrain gradient and surface lithology, which we call **geomorphological properties**. The surface structure is determined by lithological boundaries and fracture zones outcropping at the surface, which we call **litho-structural properties**. Fracture zones can also be inferred from the characteristics of recent or paleo-drainage (Short and Blair 1986). Geomorphology and litho-structure allow building a **static geological model**. If information is available about the elevation change with time, then the statics model can be expanded into a **dynamic geological model**. Figure 1 gives an over of the building blocks for geological model building.

We illustrate the methodology at a simple layer-cake geological model which is deformed by a vertical fault (fig. 2). The surface is formed by a soft sandstone layer resting on a hard limestone layer. These two layers form the near-surface. We call the layers between the bottom of the near-surface and the top of the basement subsurface. Prior to the deformation by the fault only the soft sandstone was visible at the surface. The fault has lifted part of the layer package and exposed the near-surface sandstone and limestone layers at the fault plane and made them accessible for mapping by satellites.

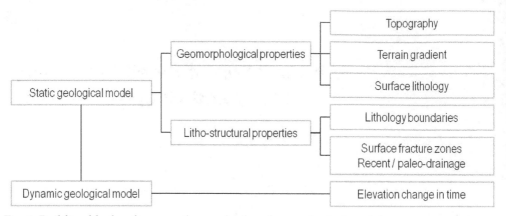

Fig. 1. Building blocks of near-surface and subsurface geological models

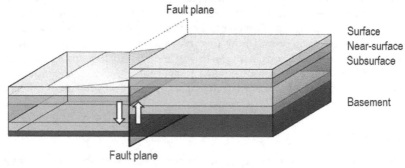

Fig. 2. Correlation of surface shape and subsurface geology

3. Satellite imagery

Earth observation satellites map the physical properties of the earth surface and near-surface. In the context of geological mapping we distinguish two types of electro-magnetic methods (see figure 3) :

- **passive optical methods**: use the sunlight as the source and measure the reflectance of the earth surface in the visible and infrared spectral bands. We used Landsat 7 ETM+ and the ASTER instrument from the Terra satellite.
- **active microwave radar methods**: use a microwave source onboard of the satellite and measure the back-scatter from the earth. We used Radarsat-1 and the radar sensor from the Shuttle Radar Tomographic Mission (SRTM).

Details about the acquisition and processing are provided among others by the USGS (2011) and Short (2010). For an introduction into the interpretation of satellite imagery see Sabins (1996). The **visible imagery (VIS)** covers the colors blue, green and red and provides information about water features, infrastructure and landuse as well as limited information about selected rock types. Infrared imagery is split into three classes : **very near infrared (VNIR)**, which detects specifically vegetation; **short wave infrared (SWIR)** which is the best option for the discrimination of sedimentary rocks; and finally **thermal infrared** (TIR). The thermal infrared radiation from the earth surface represents the property of the surface material to convert the solar spectrum into heat radiation. We distinguish between a warm response from dark materials such as non-sedimentary rocks and cool response from ground moisture or voids, where evaporation absorbs energy. In general optical imagery does not penetrate the earth surface.

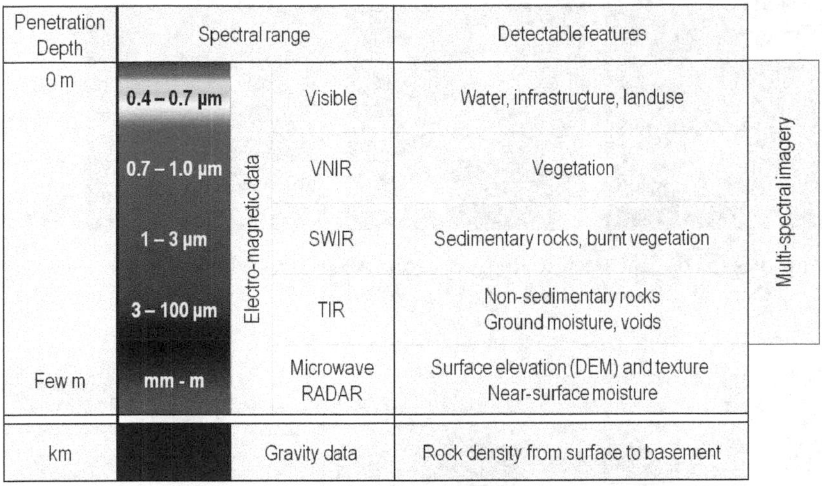

Penetration Depth	Spectral range		Detectable features	
0 m	0.4 – 0.7 μm	Visible	Water, infrastructure, landuse	Multi-spectral imagery
	0.7 – 1.0 μm	VNIR	Vegetation	
	1 – 3 μm	SWIR	Sedimentary rocks, burnt vegetation	
	3 – 100 μm	TIR	Non-sedimentary rocks Ground moisture, voids	
Few m	mm - m	Microwave RADAR	Surface elevation (DEM) and texture Near-surface moisture	
km		Gravity data	Rock density from surface to basement	

Fig. 3. Spectral overview of electro-magnetic satellite imagery

Microwave radar uses electro-magnetic waves in the mm to m range. At hard surfaces microwaves are almost completely back-scattered. Their travel time can be used to determine the distance between the satellite and the surface, which is used to generate digital elevation models (DEM). For soft, non-conductive surfaces the microwaves penetrate into the subsurface; the back-scattered signal is generated from volume scatter in the subsurface.

A special application of microwave radar is the estimation of **gravity anomalies**. Sandwell and Smith (2009) have studied anomalies in the orbits of radar satellites and inferred Bouguer gravity anomalies using a geoid model. These gravity anomalies can be interpreted for the thickness of the sedimentary cover above the crystalline basement.

The following examples illustrate the information obtainable from the various satellite imagery sets. Figure 4 shows an example of multi-spectral data from ASTER for each range : visible (fig. 4.a), short wave infrared (fig. 4.b) and thermal infrared bands (fig. 4.c) of the island of Bahrain with a colour scale from blue (low reflection intensities) via green and yellow to red (high intensities). The visible green band penetrates water for a few meters (blue shades). It is relatively insensitive to different rock types (red shades). The short wave infrared band does not penetrate water and therefore shows uniform blue color, whereas it allows the discrimination of the different rock types in the oval anticlinal structure covering most of the island (yellow and red shades). The thermal infrared band shows the thermal properties such as warm response from the built-up areas (dark red). Cool response (greenish areas in the coastal areas) is observed for wet coastal salt flats called sabkha.

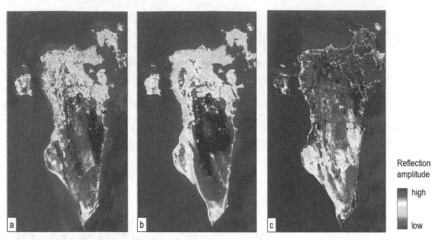

Fig. 4. Examples for optical satellite imagery from ASTER: a – visible blue, b – short wave infrared, c – thermal infrared [after Laake et al. 2006]

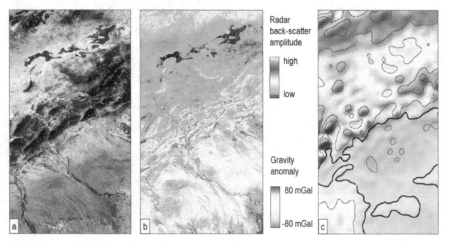

Fig. 5. Examples for optical and radar satellite imagery: a – optical data [Landsat 742 RGB] for comparison, b – radar scatter intensity [Radarsat-1], c – gravity anomaly derived from radar

The examples for microwave radar data show the Atlas area in Algeria. For orientation we have supplied a Landsat 742 RGB image (fig. 5.a), which shows mountain ranges (dark purple), gravel planes (yellow-gray), salt lakes (blue) and vegetation (green). The radar back-scatter intensity from Radarsat-1 (fig. 5.b) is displayed from blue (total absorption in conductive salt brine) via intermediate volume back-scatter (green and yellow) to strong surface back-scatter from hard rock (brown and white).

Figure 5.c shows the gravity anomaly inferred from radar satellite imagery. Negative anomalies correspond to low density rocks as from thick sedimentary cover at the foot of the

Atlas mountains whereas positive anomalies correlate with dense metamorphic and basement rocks in the Atlas ranges.

4. Extraction of information from satellite imagery

Satellite imagery is provided as sets of digital images, one image for each spectral band. Each image displays the measured values as intensity. The information contained in the satellite imagery can be extracted using either single bands or combinations of bands. **Single bands** are usually displayed as maps coding measured amplitude as colour.

Figure 6 shows radar data from the very dry desert in south-west Egypt : radar data from the shuttle radar tomographic mission (SRTM, Jarvis et al., 2008) interpreted for a DEM using topographic colour coding, i.e. green for low elevations, yellow to brown for high ground (fig. 6.a). Raw radar data from Radarsat-1 reveal high amplitudes from the back-scatter at hard sandstone (white and brown colors in fig. 6.b). Low radar back-scatter (fig. 6.c) correspond to volume back-scatter from microwaves penetrating sand sheets. They reveal buried paleo-rivers (blue colors in fig. 6.c) following the interpretation by El-Baz and Robinson (1997) and Robinson et al. (2007).

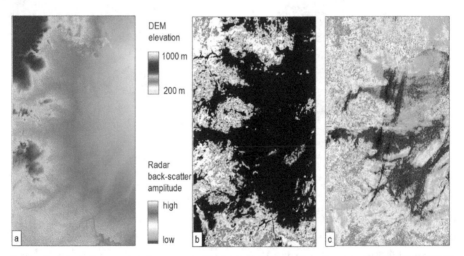

Fig. 6. Single band data examples : a – radar based digital elevation model [SRTM DEM], b – high intensity radar data [Radarsat-1], c – low intensity radar data [Radarsat-1]

Dual band images use combinations of two bands from Landsat for the north-western desert in Egypt (fig. 7) thereby enhancing subtle features in the data that would not be imaged by a single band alone. The ratio of the infrared bands 7 and 4 (fig. 7.a) highlights for example clay minerals which fill karst holes (red and cyan) in an otherwise homogeneous limestone plateau (dark blue). The band difference of the very near infrared band 4 and the green band 2 for the same area highlights difference in lithology between the pure limestone (yellow to red) and the more sandy cover (blue tones) towards the top of the image (fig. 7.b).

Multi-band images use three or more bands combined in continuous colour or red-green-blue (RGB) images. RGB images provide significantly more shades than single or dual band images: for 8 bit imagery an arithmetic combination of 3 bands provides 256 shades whereas an RGB image offers 16.8 million colors (Guo et al., 2008).

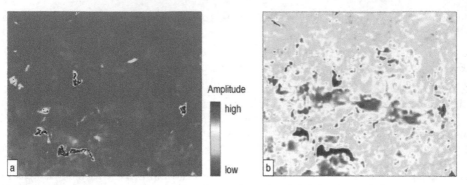

Fig. 7. Dual band data examples: a – band ratio [Landsat 7/4], b – band difference [Landsat 4-2] for the same area

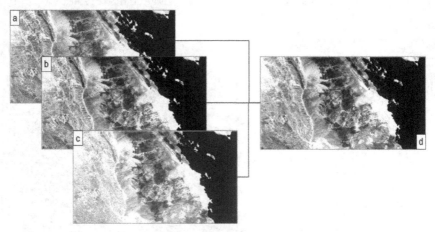

Fig. 8. Multi-band data example : merge of single band data (a to c) into RGB image (d) [Landsat 742 RGB]

To illustrate this we study an example from the southern Red Sea Mountains and the adjacent coastal area in south-east Egypt (fig. 8). The raw input bands (fig. 8 a-c) show only marginally different signatures for the very different rock types, whereas the multi-band RGB image (fig. 8.d)distinguishes clearly between the basement rocks (dark), the Mesozoic clastic sedimentary rocks (yellow tones), coastal carbonates (white) and the sea (black).

5. Integration of satellite imagery and geology

In this section we describe techniques to extract geological information from satellite imagery and how to integrate this information with geological data. Satellite imagery is interpreted for surface topography and lithology as well as for surface and subsurface structure with the goal of generating geological models for the near-surface and subsurface (Laake and Insley, 2007, and Laake et al., 2008). The techniques are illustrated through a series of case histories starting with simple layer-cake geology.

5.1 Layer-cake geology (Qattara Depression, Egypt)

The surface north of the Qattara depression in Egypt is dominated by flat layering of hard and soft rocks at the surface (fig. 9). The raw DEM (fig. 9.a) shows a platform (yellow), which is located between rough hills (brown tones) and a sharp escarpment towards the Qattara Depression (green to blue). The terrain classification map (fig. 10.c) allocates the different elevations to three classes. The lithological analysis of the multi-spectral image (fig. 9.b) allows a clear separation into rock types : two types of limestone (blue tones), two types of sandstone (yellow to orange) and evaporites of the sabkha at the bottom of the depression (cyan).

Ideally the surface lithology interpretation is validated in the field (see fig. 10) using GPS tracked lithological analysis. The combination of terrain classification and lithology suggests that the plane represents the top of the sandstone formation, which continues also below the limestone layers, which form the higher ground. We will use this geological model to estimate statics for seismic data processing in section 5.

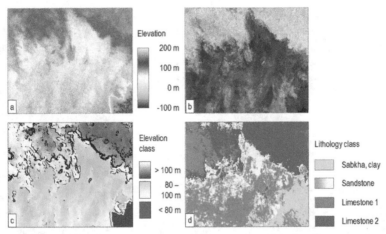

Fig. 9. Layer-cake geology near Qattara Depression, Egypt : a – digital elevation model [SRTM DEM], b – multi-band satellite image [ASTER 631 RGB], c – terrain classification map, d – lithology map [Cutts and Laake 2009]

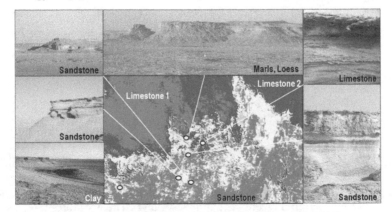

Fig. 10. Validation of lithology map in the field [after Coulson et al. 2009]

5.2 Anticline (Awali, Bahrain)

The outline of the island of Bahrain is determined by the topographical and lithological structure of the Awali anticline (fig. 11). We use satellite imagery from the ASTER sensor for the discrimination of clastic and carbonatic rocks (Laake et al., 2006).

The continuous colour image from short wave infrared and visible bands (ASTER 631 RGB in fig. 11.a) allows discriminating the coastal farmland in the north (green) from the carbonates of the anticline (purple tints) and coastal sabkha (cyan). The structure of the anticline can be delineated using the difference of visible and short wave infrared data (fig. 11.b), which can be traced even in the built-up area of Manama city in the north. The outer contour of the anticline appears highlighted in the west through the strong signature of the coastal sabkha (dark red). Draping the continuous colour image over the vertically exaggerated DEM generates a strong structural impression, which is useful for structural interpretation (fig. 11.c).

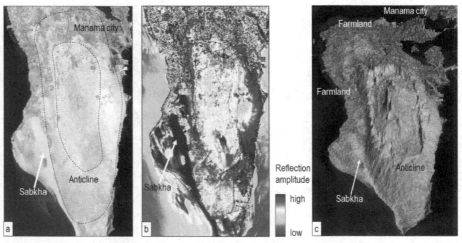

Fig. 11. Anticlinal structure of Awali, Bahrain : a – multi-spectral image [ASTER 631 RGB], geological structure from band difference image [ASTER visible minus short wave infrared], c – rendering of multi-spectral image on DEM [after Laake et al. 2006]

5.3 Mapping of basins from gravity and radar data (Illizi Basin, Algeria)

The mapping of basins from satellite imagery is the only remote sensing technique which infers deep geological structures. Bouguer gravity anomalies inferred from satellite radar data give an indication of the thickness of the sedimentary cover above the basement. However, this interpretation requires additional information for example from magnetic data to constrain the model. Following the concept of geomorphology, which correlates surface and subsurface geology and litho-structure, surface structural lineaments can be used to infer subsurface structures.

In figure 12 we show structural maps from satellite imagery for the Illizi Basin, which is indicated by the black continuous line in each figure. The lithology image (Landsat 742 RGB, fig. mapping 12.a) distinguishes dark paleozoic limestone (blue tones) and Mesozoic clastics (brown tones), which surround recent sand dunes (yellow). The geomorphological map (fig. 12.b) highlights the litho-structural boundaries, which are co-located with topographic ridges. Satellite gravity data (fig. 12.c) reveal a basin (blue). The combination of surface geomorphology and gravity (fig. 12.d) gives an indication of the basin outline.

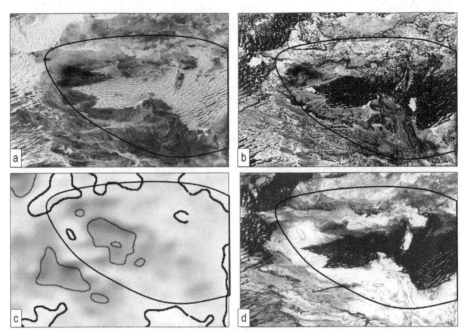

Fig. 12. Basin mapping of Illizi basin, Algeria: a – multi-spectral image (Landsat 742 RGB], b – geomorphological map [SRTM DEM, Landsat 742 RGB], c – gravity anomaly derived from radar, d – overlay of geomorphology on gravity data

5.4 Mapping of fracture zones below sand cover (Gilf Kebir, Egypt)

When sand dunes cover fracture zones in the near-surface, satellite radar data can map ancient river courses, from which fracture related weak zones in the near-surface can be derived (fig. 13). We start with the outcrops of faults in the hard rock surrounding the sand dune field mapping straight valleys in the hard sand stone. The topography of the study area (fig. 13.a) is composed of the flat Gilf Kebir plateau (brown tones) with gentle slopes (yellow) and a large plane (green to white). This corresponds to a surface lithology (fig. 13.b) of hard sandstones (greenish-brown) and basement rocks (dark brown) as well as belts of sand dunes (white), which are locally discoloured by hematitic iron (purple). The shape of the escarpment and the valleys intersecting the Gilf Kebir plateau reveal fairly straight fault lines (black lines). Below sand neither topographical nor optical satellite imagery can reveal buried fracture zones. Therefore we use satellite radar which can penetrate dry sand for up to 20 m to map the clay contained in buried paleo-river beds. The radar data (fig. 13.c) do not only map the fracture zones in the sandstone of the plateau (brown lineaments), it also shows W-E and SW-NE trends in the paleo-river courses (low radar intensities in blue) which continue from the lineaments in the rock outcrops. The overlay of the paleo-river courses from radar data on the geomorphology delineates the outcropping fracture zones across all terrains (fig. 13.d).

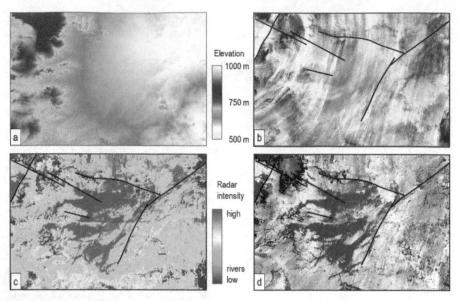

Fig. 13. Structural delineation from radar data close to Gilf Kebir, Egypt: a – DEM, b – multi-spectral image [Landsat 742 RGB], c – low intensity radar data [Radarsat1], d – merge of radar data with geomorphology. Faults indicated by lines.

5.5 Mapping of glacial moraine structures (Pechora Basin, Russia)

The last case study concerns the mapping of post-glacial structures in the Pechora basin in northern Russia using vegetation and water features as indicators; the lithology is not exposed in the study area. The structures studied comprise different types of moraines as well as drainage features in front of the glaciers (Laake 2009, Astakhov et al. 1999). We distinguish terminal and lateral moraines, which are composed of gravel and glacial till, from ground moraines, which are characterized by undulating terrain. Moraine ridges are often drier than the surrounding terrain because the elevated gravels cannot support a shallow water table. In contrast to this, ground moraines deposit more glacial till, which provides the basis for a shallow water table with very wet terrain and lakes. Significant differences in the ground moisture attract different plant species, which can be distinguished by their different response in the very near infrared and visible bands of satellite imagery. The topography allows delineating of the lateral and terminal moraines as long as they are still elevated above the surrounding plane (fig. 14.a).

Rendering the geomorphological map on the DEM in 3D improves the detection of these moraines. The location of lakes, which is indicative of ground moraines, is obtained from the landuse analysis of short wave infrared and visible green data. (fig. 14.c). The combination of all maps yields a clear delineation of the glacial moraines particularly when rendered in 3D (fig. 14.d). In this case four glacial stages can be interpreted (numbers in fig. 14) : an initial stage where the entire area was covered by a thick ice shield (1), which deposited ground moraines over the entire study area. The second (2) and the third (3) phase comprise two distinguishable glaciers, where the glacier from phase two created the partly broken moraine wall in the west. The glacier in phase three covered the centre of the study area thereby partly levelling the lateral moraines from phase two. After melting, this glacier left the huge Lake Komi between the lateral and terminal moraines, which forms an extensive swamp today. The final glacier advance (4) covered only the northern third of the study area, leaving an irregular line of terminal moraines and extensive planes of ground moraines behind.

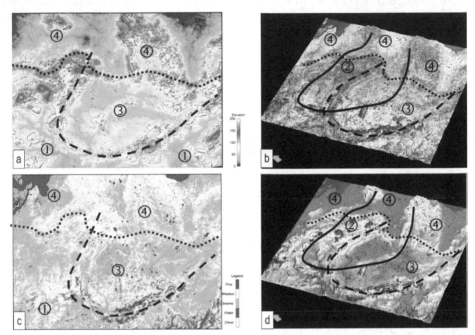

Fig. 14. Identification of glacial structures in Pechora basin, Russia: a – DEM [SRTM DEM], b – rendering of geomorphological map in 3D, c – landuse classification from Landsat, d – interpretation of geomorphology for glaciers and moraines [after Laake 2009]

6. Integration of satellite imagery and geophysics

The geological information extracted from satellite imagery can be used to build geological models from basin scale (several 10^5 sqkm) to survey scale (few 10^3 sqkm) well before any geophysical data are acquired on the ground. These models can be used for the estimation of logistic risks for personnel and vehicles as well as for the estimation of the quality of the acquired geophysical data (Laake and Insley (2004a and b, Laake and Cutts 2007, Coulson et al. 2009). In turn, geophysical data can be employed to calibrate geophysical characteristics of near-surface layers inferred from satellite imagery.

We distinguish the following types of geophysical surveys :

1. **Frontier exploration** aiming at identification of new basins in large unexplored areas. Satellite imagery can assist defining the outline of potential basins and the definition of scouting surveys.

2. **Structural imaging** focuses on potential structures once the outline and character of a basin has been identified. Satellite imagery can support the design and logistic planning

of geophysical surveys on the ground and can provide estimates for the quality of the acquired data before the data are acquired.

3. **Reservoir characterization** targets the most comprehensive study of the subsurface geological structure and fluids for individual reservoirs and therefore requires the best data quality. Satellite imagery can provide detailed models of the surface and near-surface which provide input to data quality estimation before and during acquisition. For data processing satellite imagery can supply input to processes that correct for noise related to near-surface properties.

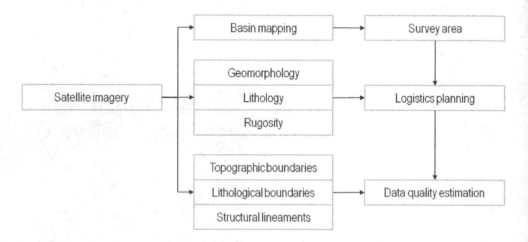

Fig. 15. Workflow for application of satellite imagery to geophysical methods

An outline of the information satellite imagery can provide for geophysical methods and the parameters studied is shown in figure 15. This section focuses on the aspects of survey design and data quality estimation.

The design of geophysical surveys requires an understanding of the logistics and data quality aspects of the target area to correctly estimate the effort required to provide the desired quality of the final subsurface geological product. Satellite imagery can provide the information about the surface and near-surface before the start of the data acquisition (Cutts and Laake, 2009a and b). The method uses satellite imagery to generate geomorphological and litho-structural models of the surface and near-surface as described in the previous section to derive logistic planning and data quality estimation maps.

Figure 16 shows the result of the method for a survey in the Western Desert of Egypt. DEM and multi-spectral data provide topography (fig. 16.a) and lithology class maps (fig. 16.b). The planning of the logistics requires information about limitations for access and maneuver for personnel and vehicles. The terrain classification provides the locations and steepness of escarpments as well as the surface roughness related to hard limestone. The combination of these terrain attributes defines the logistics risk (fig. 16.c). A major obstacle for data quality for this survey is the scattering of seismic waves at the numerous escarpments resulting from the different weathering of limestone and sandstone. Fig. 16.d shows an estimate map for the scatter risk associated with escarpments.

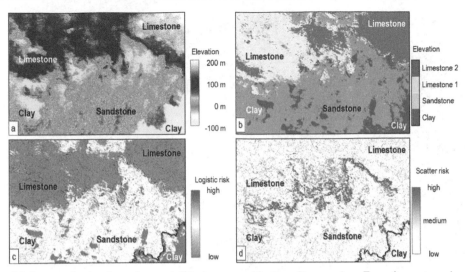

Fig. 16. Survey design based on satellite imagery (Qattara Depression, Egypt) : a – terrain class map based on DEM, b – lithology map, c – logistic risk estimate, d – data quality risk estimate [after Cutts and Laake 2008]

6.1 Logistic planning in volcanic terrain

Vegetation may obstruct the assessment of the logistic risk (Laake 2005b). The basalt plateau close to the Payun Volcano in the Andean foothills of Argentina exposes the risk of big basalt blocks obstructing the access to large areas whereas in other parts of the survey basalt grit exposes no maneuver risk at all (fig. 17.a to c). The analysis of the multi-spectral imagery from ASTER for big basalt blocks is challenging because in both cases – big basalt blocks and bush as well as basalt grit and grass – the ratio of basalt and vegetation is very similar. The resolution of the satellite image of 15 m does not allow the direct mapping of the basalt blocks (fig. 17.d). However, the thermal properties of big basalt blocks are sufficiently different from basalt grit, which allows the discrimination of the basalt block size using the thermal infrared bands from ASTER imagery (fig. 17.e), which is supported by the intermediate back scatter intensity recorded by radar data from Radarsat-1 (fig. 17.f).

6.2 Planning of safe operation in sand dunes

In high sand dunes, both logistics for wheeled vehicles and high absorption for seismic waves may impact the survey design severely (Laake and Insley, 2004a). The analysis of the topography can assist in outlining the dunes and characterizing their shape (fig. 18). Digital elevation models derived from radar or stereo optical data (ASTER in our case) provide the slope index, which can be represented as an index for safe operation. In seismic operation surface gradients above 15 degrees are considered unsafe for operation, whereas in areas with 10 to 15 degrees slope an inspection of the terrain would be advised. Figure 18 shows a composite image of seismic line crossing a dune field. The visible band ASTER image (fig. 18.b) gives a photographic impression of the surface, whereas the DEM (fig. 18.c) provides an impression of the elevations. The slope risk map (fig. 18.d) is obtained from the gradient of the DEM.

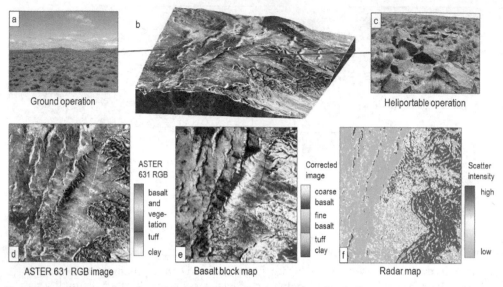

Fig. 17. Satellite imagery based logistic planning in Andean foothills, Argentina : a – basalt gravel plane, b – rendering of ASTER image, c – basalt blocks, d – landuse image [(ASTER 631 RGB], e - basalt block map, f – surface rugosity from Radarsat-1 [after Laake 2005b]

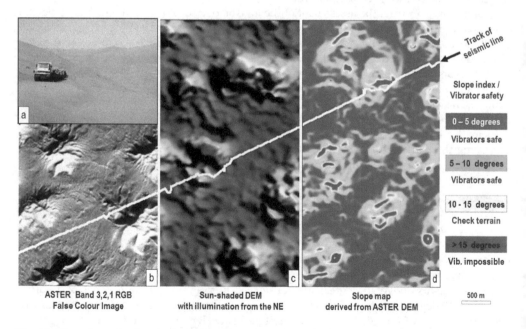

Fig. 18. Planning of safe vibroseis operation in high sand dunes, Berkine basin, Algeria: a – vibrator array in high dunes, b – multi-spectral satellite image [ASTER 321 RGB], c – sun-shaded DEM, d – Slope map derived from DEM [ASTER DEM] [Laake and Insley, 2004a]

6.3 Prediction of accessibility and data quality in sabkha

In arid coastal areas often sabkha called salt flats pose a severe threat to personnel and equipment (Cutts and Laake 2009b). The salt flats may be subject to seasonal inundation, which may change their accessibility significantly. Analysis of the multi-spectral satellite Landsat imagery can allow the detection of halite minerals found in the surface crusts of sabkha. Thermal infrared imagery provides information about the different thermal properties of wet and dry sabkha. The visible band Landsat image in figure 19.b gives an indication of the location of the wet sabkha. The risk that the sabkha surface would not support heavy vehicles (see fig. 19.a) is directly correlated with presence of sabkha and can be derived from a sabkha detection map (fig. 19.c). In addition to its impact on the logistics sabkha also affects the vibrator data quality. High distortion of the vibrator signal, an undesirable data property, is directly related to the presence of sabkha (fig. 19.d).

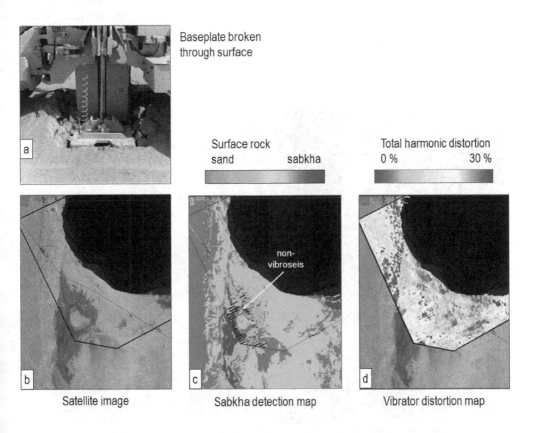

Fig. 19. Prediction of accessibility and data quality of vibroseis operations in sabkha, Sabkha Matti, UAE: a – vibrator baseplate broken through surface crust, b – satellite image [Landsat 321 RGB], c - sabkha detection map, d – map of vibrator total harmonic distortion [after Cutts and Laake 2008]

6.4 Estimation of the vibrator baseplate-to-ground coupling

Hard rock terrain affects the coupling of the vibrator baseplate to the ground (Laake and Tewekesbury 2005). Good ground coupling is achieved when the entire baseplate is in contact with the ground for example on gravel (fig. 20.c). If the baseplate rests on a big boulder, the coupling surface from the baseplate to the ground may be reduced to a very small area resulting in substantial distortion of the transmitted signal and in severe damage to the baseplate (fig. 20.a).

In this case the so-called point loading risk is correlated with hard limestone. The limestone prediction map draped over the DEM is shown in figure 20.b. But even for gravel the ground coupling may be compromised if the force level of the vibrator shaker and the frequency exceed the levels at which the ground supports the weight of the vibrator. The time signal of the baseplate amplitude (fig. 20 bottom) shows the increasing frequency of the shaker force with time along with measured signal distortion. At three moments of the sweep photos of the shaker were taken. As long as the coupling of the baseplate to the ground is perfect the signal distortion is low (fig. 20.d and e). When the ground surface breaks dust is whirled up by the baseplate motion and the distortion goes up (fig. 20.f). Usually frequencies higher than the frequency at which the baseplate breaks through the surface show high phase distortion and deteriorate the data quality. The high distortion for very low frequencies (corresponding to fig. 20.d) is associated with motions of the entire vibrator and are not considered here.

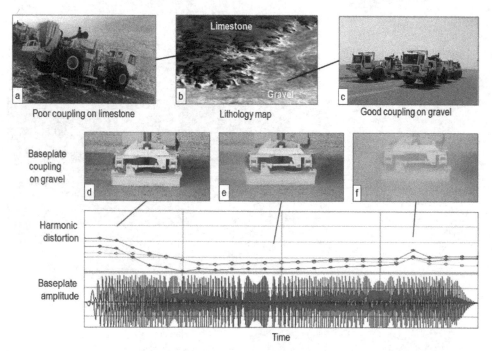

Fig. 20. Prediction of accessibility and data quality for vibroseis operations in hard rocky terrain, Tademait, Algeria: a, c – operations photos, b – virtual 3D lithology map, d - f – photos from baseplate coupling, bottom – baseplate distortion and amplitude time signal [after Laake and Tewkesbury 2005]

Soft terrain may have an impact on both the source coupling as well as on the receiver signal output (Laake 2005a). Due to higher absorption in soft sand, the signal level on sandy surface (fig. 21.b and d) is generally lower than on hard gravel plain (fig. 21.a and c). Satellite imagery can discriminate gravel with shallow evaporite pans on the higher ground from sandy terrain on the lower ground (fig. 21.e). The quality of the source signal also correlates with the terrain character : signals from the sandy terrain show higher distortion (fig. 21.f) and lower ground stiffness (fig. 21.g) than signals from the gravel plain. Interpretation of satellite imagery can provide estimates for the quality of seismic signals obtained from the terrain characterization.

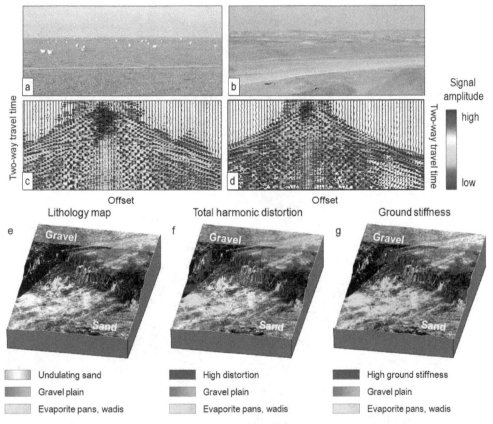

Fig. 21. Correlation of source and receiver data quality with terrain, Kuwait: a – gravel plain, b – undulating sandy surface, c, d – shot record displays, e – surface lithology, f – total harmonic distortion, g – ground stiffness [after Laake 2005a]

6.5 Improvement of static corrections
Near-surface geological models can be generated from the interpretation of satellite imagery for geomorphology and lithology (Laake et al. 2008). Using standardized seismic velocities the geological model is converted into an elastic model, from which estimates for statics corrections can be computed (Laake and Zaghloul 2009). The analysis of the DEM delivers

the topographic classification into three layers (see fig. 22): the platform (sandstone), which occupies most of the study area, as well as the lower platform of the Qattara Depression and the rough higher layer (limestone). The interpretation of the shortwave infrared data reveals that the platform is composed of sandstone, whereas the higher ground consists of limestone (fig. 22 top). When choosing standardized P-wave velocities of 2200 m/s for sandstone and 3300 m/s for limestone we can compute estimated statics values (fig. 22 bottom). The estimated values are compared against refraction static corrections from a 3D seismic survey (fig. 23).

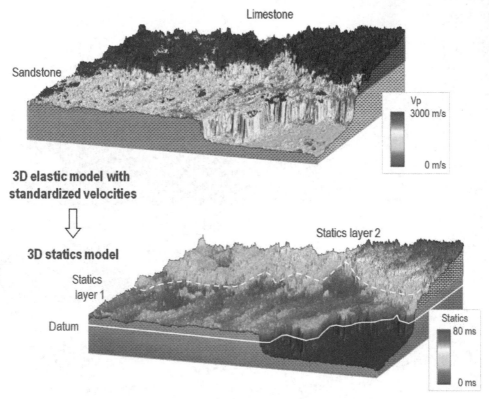

Fig. 22. Estimation of statics corrections from shallow geological model, near Qattara depression, Egypt : top – near-surface geological model from satellite imagery, bottom – 3D statics model [after Laake and Zaghloul 2009]

The static corrections for model 1 (fig. 23.a) use the geological model generated from satellite imagery. The statics contain the lithological details of the sandstone plateau and retain the sharp velocity contrast at the edge of the rough limestone plateau (see fig. 23.c for comparison with the lithology). The statics estimates for model 2 are based on the first break picks from a 3D seismic survey, which are remarkably smoother than the statics from model 1 (fig. 23.d). First break picking involves velocity smoothing over offsets up to 1500 m, which results in a spatial low pass filter. This may result in attenuation of sharp structural lineaments such as fracture zones outcropping at the surface. The difference between the two models (fig. 23.d) reveals another effect of the spatial smoothing: local velocity anomalies may not be corrected properly and might remain in the data as an artefact, which may lead to the generation of structural artefacts in the deeper seismic data.

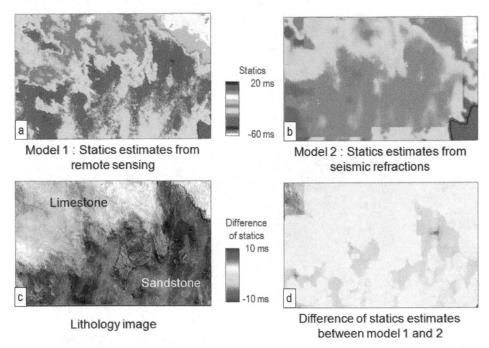

Model 1 : Statics estimates from
remote sensing

Model 2 : Statics estimates from
seismic refractions

Lithology image

Difference of statics estimates
between model 1 and 2

Fig. 23. Comparison of statics from near-surface geological model and from seismic refractions, Qattara depression, Egypt : a – statics corrections from satellite imagery, b – statics from seismic refractions, c – lithology image, d – difference between statics from both models [after Laake and Zaghloul 2009]

7. Consistent deep geological models from data integration

In the final section we will show the benefits of merging interpreted satellite imagery with geological and geophysical data on the western shore of the Gulf of Suez in Egypt by demonstrating the correlation of the geomorphology with the subsurface litho-structure . The processing of multi-spectral imagery from Landsat for lithology and drainage reveals fault structures which are masked by recent wadi deposits (Laake 2010, Laake et al. 2011). The approach follows the idea that tectonic movements deviate the courses of wadis in a characteristic way.

7.1 Mapping of fault outcrops at surface from satellite imagery

The geological setting of the study area is determined by rift faulting in NW – SE direction and transform faults in the perpendicular direction (Darwish and El-Araby 1993, Alsharhan and Salah 1993). The natural colour Landsat 321 RGB image (fig. 24.a) shows the almost featureless beige gravel plain between the Red Sea Mountains in the west and the Gulf of

Suez in the East. The only visible features are SW – NE running wadis delineated through their light sediments. The inverted natural colour image directs the eye to the wadi courses, which reveal several anomalously straight sections (fig. 24.b). When we use all data from Landsat and spectrally enhance the resulting image boundaries along the main tectonic directions are imaged (fig. 24.c): parallel to the coast NW – SE trending linear anomalies point to outcropping parallel fracture zones (dashed lines). The dominating wadi in the southern part of the study area is confined between two straight SW – NE trending lineaments which point to faults running perpendicular to the rift orientation (dotted lines). This map also reveals differences in the mineral composition of the bedrock of the Red Sea Mountain granites. The erosion fans from these two granites may actually be used as tracers to highlight the structural boundaries on the gravel plain. The lineaments are also highlighted in the drainage map, which results from the ratio of the thermal and pan-chromatic bands (fig. 24.d).

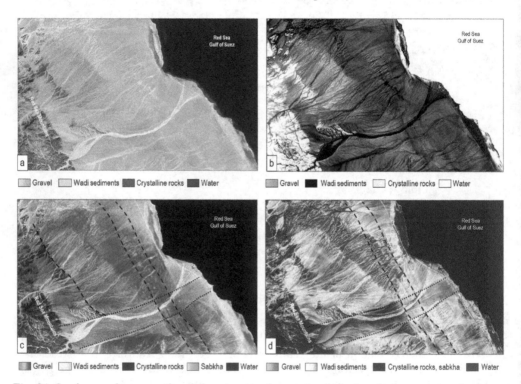

Fig. 24. Surface and near-surface litho-structural mapping from Landsat : a – natural colour image [Landsat 321 RGB], b – inverted natural colour image, c – high-discrimination lithology image, d – drainage map [Landsat bands 6 – 8] [after Laake et al. 2011]

7.2 Mapping of faults in near-surface from shallow geophysical data

In the study area, the fault outcrop pattern mapped from satellite imagery is calibrated with surface features and shallow seismic data from a 3D seismic survey. The only tree in the study area has been found at the intersection of two fault outcrops (fig. 25.a) indicated by the arrow in the lithology map (fig. 25.b). The surface structure becomes evident when draping the litho-structural map over the vertically exaggerated DEM (fig. 25.c). Arrows indicate the outcrop of faults at the surface. The correlation with faults in the near-surface is achieved by extracting weak zones of low surface wave velocity from seismic 3D data (fig. 25.d). Surface wave velocity analysis provides an iso-velocity horizon which corresponds to a shallow lithologic horizon. This horizon shows very low velocities along the rift parallel fault outcrop zones mapped by remote sensing. In the direction orthogonal to the rift faults a high-velocity structure below the wadi is revealed but no sharp boundaries. This may be due to the transform character of these faults which does not provide an impedance contrast that could be detected from the seismic.

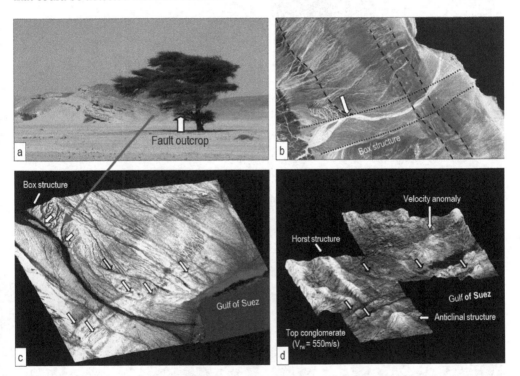

Fig. 25. Calibration of satellite imagery results with near-surface seismic data: a – tree located at a fault intersection, b – high-discrimination lithology map, c – virtual 3D rendering of the litho-structural map – arrows indicate fault outcrops, d – near-surface formation surface obtained from the velocity analysis of seismic surface waves [after Laake et al. 2011]

7.3 Mapping of faults in the subsurface using surface tamplates

The characteristics of drainage patterns delineated from satellite imagery may also be used to generate shape templates for the detection and extraction of similar structures in subsurface seismic data (fig. 26). We use the drainage map to define the template for a wadi (fig. 26.a) consisting of the braided stream and the fan delta. In our case the braided stream is confined by the SW – NE trending faults perpendicular to the rift orientation, whereas the fan delta part shows so little elevation change that the fan delta crosses the boundary faults (fig. 26.b).

The shape template is used in the processing of the shallow seismic data to search for the correct seismic attribute suited to map the faults which are hidden in most other seismic attributes. In our case the instantaneous frequency attributes were identified for mapping the SE – NE faults because this attribute highlights locally high spectral amplitudes which were interpreted as resonances within the palao-wadi structures (fig. 26.c). Once the shallow paleo-wadi has been identified the same methodology is applied to the deeper data; a total of four paleo-wadis spanning the entire rift history were mapped (fig. 26.d).

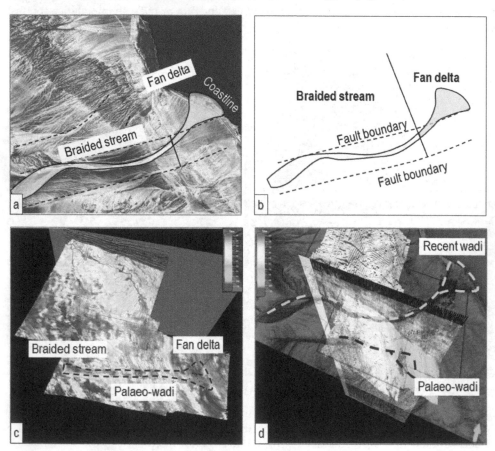

Fig. 26. Satellite based generation of templates for subsurface geobody extraction: a – drainage map, b – extracted wadi pattern, c – detection of similar pattern in shallow seismic data, d – delineation of geobodies in deep seismic data [after Laake et al. 2011]

7.4 Mapping of shallow drilling risk related to faults

Finally the surface and shallow subsurface mapping can be merged into a shallow structural map with indications of shallow drilling risks related to outcropping fault zones (fig. 27). The map basis is a landuse map (Landsat 742 RGB) onto which the near-surface topography is projected. The fault lines delineate areas of shallow drilling risk related to fault induced weak zones, which might lead to collapsing boreholes and/or loss of circulation.

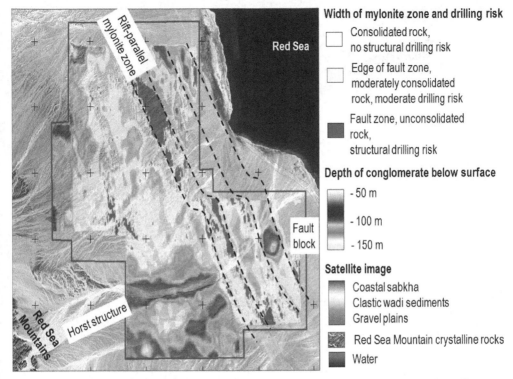

Fig. 27. Shallow structural and drilling risk mapping guided by interpretation of satellite imagery [after Laake et al. 2011]

8. Conclusion

Satellite imagery enables the investigation of the properties of the earth surface in remote areas and over large areas through the mapping of physical properties from satellite based sensors. Optical imagery delivers information about land-use, water features and surface rocks. Microwave radar imagery maps surface roughness when back-scattered at hard surfaces and paleo-drainage when penetrating dry sand cover. Microwave radar distance measurements can be converted into surface topography maps or digital elevation models. The joint interpretation of radar elevation measurements and the geoid shape delivers satellite based gravity anomaly maps, which reflect the thickness of the sedimentary layers above the crystalline basement.

The interpretation of satellite imagery can assist in the planning of surface seismic surveys through assessment of logistic and data quality risks early in the planning specifically when exploring in frontier areas. Interpretation of satellite imagery can provide estimates of the source and receiver data quality and static corrections. Consistent geological models can be generated from the interpretation of satellite imagery for geomorphological and litho-structural properties integrated with geological and geophysical data.

Fig. 28 and table 1 give an overview of the geological features detectable from satellite imagery and their impact on seismic data quality.

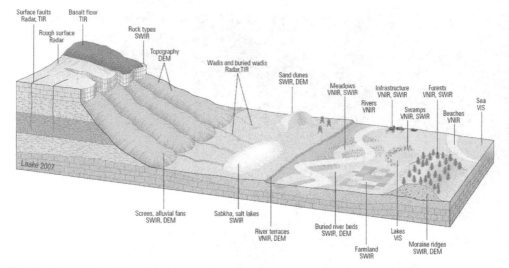

Fig. 28. Characterization of surface and near-surface properties from satellite imagery

Surface Class	Surface Feature	Satellite Imagery	Impact on Data Quality	Impact on Logistics
Topography and texture	Escarpments, river terraces	DEM, radar	Scattering noise	Severe risk for 15 - 25% slope, no access above
Topography and texture	Rough surfaces	Radar	Poor source / receiver coupling	Severe risk of tire damage for vehicles
Topography and texture	Surface faults	TIR, radar	Scattering noise	Only on escarpments
Land use	Swamps, marshes	VIS–NIR	Resonance; velocity statics	If wet, no access for vibrators and vehicles
Land use	Water features	VIS	Special equipment	No vehicle access
Lithology	Basalt flows	TIR	Poor coupling; strong scattering	Often risk for vibrator and vehicle tires
Lithology	Claypans	DEM, NIR	Resonance	No access if wet
Lithology	Hard rock outcrops	NIR, SWIR, radar	Poor source / receiver coupling	Limited risk of access for vibrators
Lithology	Sabkhas, salt lakes	DEM, SWIR	Resonance; velocity statics	Severe risk for vibrator and vehicle access
Lithology	Buried rivers	TIR, radar	Velocity statics	No risk
Geomorphology	Moraine ridges	SWIR, DEM	Low velocity, high attenuation	No risk
Geomorphology	Sand dunes	SWIR, DEM	Elevation statics; strong attenuation	Access for vibrators severely limited

Table 1. Detection of surface geological features from satellite imagery and impact on seismic data. For satellite imagery codes see section 3.

9. Acknowledgements

The author thanks ADNOC, Anadarko, Apache Oil Egypt, Bahrain Oil Company, Dara Petroleum Company, Egypt General Petroleum Corporation, Gaz de France, Kuwait Ministry of Oil, Kuwait Oil Company Ltd., Repsol YPF, Sonatrach, TransGlobe Energy, and WesternGeco for the permission to publish the data.
The original data for Landsat 7 ETM+ and MrSID is provided by NASA. The original data for the SRTM DEM is provided by NASA through the CIAT-CSI SRTM website (http://srtm.csi.cgiar.org), the Canadian Space Agency provided Radarsat microwave data. ASTER original data and ASTER GDEM are products of METI and NASA, the original data is property of METI and NASA. Radar based satellite gravity data are provided by the University of California, La Jolla Satellite.
The author also thanks Elisabeth DeTemple, Charles Woodward and Nick Moldoveanu for their reviews, Andrew Cutts for the support with GIS and Claudio Strobbia, Mohamed Sheneshen and Larry Velasco for their contributions to the data integration.

10. References

Alsharhan, A.S. and Salah, M.G. (1995). Geology and hydrocarbon habitat in rift setting: Northern and central Gulf of Suez, Egypt, Bulletin of Canadian Petroleum Geology, Vol. 43, 156–176.

Astakhov, V. I., Svendsen, J. I, Matiouchkov, A., Mangerud, J., Maslenikova, O. and Tveranger, J. (1999), Marginal formations of the last Kara and Barents ice sheets in northern European Russia. Boreas, Vol. 28, pp. 23–45. Oslo.

Coulson, S., Grabak, O., Cutts, A., Sweeney, D., Hinsch, R., Schachinger, M., Laake, A., Monk, D. and Towart, J. (2009). Satellite sensing : risk mapping for seismic surveys, Schlumberger Oilfield Review, Winter 2008/2009, pp. 40-51.

Cutts, A. and Laake, A. (2009). An Analysis of the Near Surface Using Remote Sensing for the Prediction of Logistics and Data Quality Risk, Tunis 2009 – 4th North African/Mediterranean Petroleum and Geosciences Conference & Exhibition, Tunis, March 2009, paper S30.

Darwish, M. and El-Araby, A.M. (1993). Petrography and diagenetic aspects of some siliciclastic hydrocarbon reservoirs in relation to the rifting of the Gulf of Suez. In: Philobbos, E.R. & Purser, B.H. (eds) Geodynamics and Sedimentation of the Red Sea–Gulf of Aden Rift System. Geological Society of Egypt, Special Publication, 1, pp. 155–187.

El-Baz, F. and Robinson, C. (1997), Paleo-channels revealed by SIR-C data in the Western Desert of Egypt : implications to sand dune accumulations, 12th International Conference and Workshops on Applied Geologic Remote Sensing, Denver November 1997.

Guo, H., Lewis, S. and Marfurt, K.J. (2008). Mapping multiple attributes to three- and four-component colour models – A tutorial. Geophysics, Vol. 73, pp. W7–W19.

Jarvis A., H.I. Reuter, A. Nelson, E. Guevara, 2008, Hole-filled seamless SRTM data V4, International Centre for Tropical Agriculture (CIAT), available from: http://srtm.csi.cgiar.org.

Laake, A. (2005a). Application of Landsat data to seismic exploration – Case study from Kuwait, Kuwait First International Remote Sensing Conference and Exhibition, Kuwait September 2005.

Laake, A. (2005b). Remote sensing application for vibroseis data quality estimation in the Neuquen Basin, Argentina, VI Congreso de Exploración y Desarrollo de Hidrocarburos, Mar del Plata, November 2005.

Laake, A. (2009). Hybrid near-surface modeling for seismic property estimation in arctic areas, 71st EAGE Conference & Exhibition, Amsterdam, June 2009, paper T003.

Laake, A. (2010). Enhancing the value of remote sensing data through integration with ground based data in 3D, ESA workshop on satellite earth observation for the oil and gas sector, Frascati, September 2010.

Laake, A. and Cutts., A. (2007). The role of remote sensing data in near-surface seismic characterization, First Break, Vol. 25, pp. 51 – 55.

Laake, A. and Insley, M. (2007). Near-surface characterization from remote sensing data, ENVISAT Symposium 2007, Montreux, April 2007, A457029.

Laake, A. and Tewkesbury, A. (2005). Vibroseis data quality estimation from remote sensing data, Proceedings of the 67th Conference and Exhibition, Madrid, June 2005, Expanded abstracts, paper G017.

Laake, A. and Zaghloul, A. (2009). Estimation of static corrections from geologic and remote-sensing data, The Leading Edge, February 2009, pp. 192-196.

Laake, A., Al-Alawi, H. and Gras, R. (2006). Integration of remote sensing data with geology and geophysics – Case study from Bahrain, GEO 2006, Manama, March 2006.

Laake, A., and Insley, M. (2004a). Applications of satellite imagery to seismic survey design, The Leading Edge, Vol. 23, No. 10, 1062-1064.

Laake, A., and Insley, M. (2004b). Satellite-based seismic technology, World Oil, Vol. 225, No. 9, pp. 27-33.

Laake, A., Sheneshen, M., Strobbia, C., Velasco, L. and Cutts, A. (2011). Integration of 4surface/subsurface techniques reveals faults in Gulf of Suez oilfields, Petroleum Geoscience, Vol. 17, 2011, pp. 165–179.

Laake, A., Strobbia, C. and Cutts, A. (2008). Integrated approach to 3D near-surface characterization, First Break, Vol. 26, pp. 109-112.

Robinson, C.A., Werwer, A., El-Baz, F., El-Shazly, M., Fritch, T. and Kusky, T. (2007), The Nubian aquifer in Southwest Egypt, Hydrogeology Journal, Vol. 15, 2007, pp. 33-45.

Sabins, F. (1996). Remote sensing, principle and interpretation (3rd ed.), Freeman, ISBN 0716724421, New York.

Sandwell, D.T. and Smith, W.H. (2009). Global marine gravity from retracked Geosat and ERS-1 altimetry: Ridge segmentation versus spreading rate, Journal of Geophysical, Vol. 114, B01411, pp. 1-18.

Short, N., (2010), The remote sensing tutorial, NASA 2010, Date of access : 15/06/2011, Available from : http://landsat.gsfc.nasa.gov/about/L7_td.html

Short, N.M.Sr., and Blair, R.W.Jr. (eds.) (1986), Geomorphology from Space, NASA 1986, Date of access : 19/09/2007, Available from : http://geoinfo.amu.edu.pl/wpk/geos/GEO_COMPLETE_TOC.html

USGS (2011), Landsat 7 science data users handbook, Date of access : 24/06/2011, Available from: http://landsathandbook.gsfc.nasa.gov/pdfs/Landsat7_Handbook.pdf

Part 5

Environmental Sciences

7

Multivariate Analysis Techniques in Environmental Science

Mohammad Ali Zare Chahouki

Department of Rehabilitation of Arid and Mountainous Regions, University of Tehran, Iran

1. Introduction

One of the characteristics of environmental data, many of them and the complex relationships between them. To reduce the number variables, different statistical methods exist. Multivariate statistics is used extensively in environmental science. It helps ecologists discover structure and previous relatively objective summary of the primary features of the data for easier comprehension. However, it is complicated in theorical structure and in operational methodology.

In this chapter some important statistical methods such as Principal component analysis (PCA), Canonical correspondence analysis (CCA), Redundancy analysis (RDA), Cluster analysis, and Discriminate function analysis will be explained briefly.

This chapter too cover the statistical analysis of assemblage data (species by samples matrices of abundance, area cover etc) and/or multi variable environmental data which arise in a wide range of applications in ecology and environmental science, from basic ecological studies (e.g. of dietary composition or population size-structure), through community-based field studies, environmental impact assessments and monitoring of large-scale biodiversity change, to purely physical or chemical analyses.

The use of multivariate analysis has been extended much more widely over the past 20 years. Much more is included on techniques such as Canonical Correspondence Analysis (CCA) and Non-metric Multidimensional Scaling (NMS) and another technique to include organisms and organism-environment relationships other than vegetation. Spatially constrained data analysis will be introduced and the importance of accounting for spatial autocorrelation will be emphasized. Use of the methods within ecology and in environmental reconstruction will also be covered. A study and review of the application of multivariate analysis in biogeography and ecology is provided in: Kent, M. (2006).

2. Landscape ecology

Landscape is simply an area of land (at any scale) containing an interesting pattern that affects and is affected by an ecological process of interest. Landscape ecology, then, involves the study of these landscape patterns, the interactions among the elements of this pattern, and how these patterns and interactions change over time. In addition, landscape ecology involves the application of these principles in the formulation and solving of real-world problems (Turner et al, 2001).

Landscape ecology is perhaps best distinguished by its focus on: 1) spatial heterogeneity, 2) broader spatial extents than those traditionally studied in ecology, and 3) the role of humans in creating and affecting landscape patterns and process (Turner et al, 2001).

In effect the role of ecology, and especially that of vegetation science, has been mainly restricted to the evaluation of the landscape with respect to particular demands: either evaluation as an assessment of the qualities of the ecosystem or evaluation as a socio-economic procedure intended to estimate the functions the natural environment fulfills for human societ (Van der Ploeg and Vlijm, 1978).

Landscape ecology theory stresses the role of human impacts on landscape structures and functions. It also proposes ways for restoring degraded landscapes. Landscape ecology explicitly includes humans as entities that cause functional changes on the landscape (Mielke& Berry, 2001).

Landscape ecology theory includes the landscape stability principle, which emphasizes the importance of landscape structural heterogeneity in developing resistance to disturbances, recovery from disturbances, and promoting total system stability (Mielke& Berry, 2001). This principle is a major contribution to general ecological theories which highlight the importance of relationships among the various components of the landscape. Integrity of landscape components helps maintain resistance to external threats, including development and land transformation by human activity. Analysis of land use change has included a strongly geographical approach which has led to the acceptance of the idea of multifunctional properties of landscapes (Mielke et al, 1976). There are still calls for a more unified theory of landscape ecology due to differences in professional opinion among ecologists and its interdisciplinary approach (Bastian 2001).

Landscape ecology is distinguished by its focus on broader spatial extents than those traditionally studied in ecology. This stems from the anthropocentric origins of the discipline.

Despite early attention to the effects of sample area on measurements, such as species-area relationships, the importance of scale was not widely recognized until the 1980's. Recognition that pattern-process relationships vary with scale demanded that ecologist give explicit consideration to scale in designing experiments and interpreting results.

It became evident that different problems require different scales of study, and that most problems require multiple scales of study. The theory of scale and hierarchy emerged as a framework for dealing with scale. The emergence of scale and hierarchy theory provided a partial theoretical framework for understanding pattern-process relationships, which became the basis for the emergence of landscape ecology as a discipline (Turner MG. 1989).

3. Sampling methods

Sampling design varies considerably with habitat type and specific taxonomic groups. Sampling design begins with a clear statement of the question(s) being asked. This may be the most difficult part of the procedure because the quality of the results is dependent on the nature of the original design.

If the sampling is for densities of organisms then at least five replicate samples per sampling site are needed because many statistical tests require that minimal number. Better yet, consider 20 replicates per sampling site and in some cases 50 or more. If sample replicates are less than five then bootstrapping techniques can be used to analyze the data (name). Some type of random sampling should be attempted (e.g., stratified random sampling) or a line intercept method used to estimate densities (e.g., Strong Method).

Measurements of important physical–chemical variables should be made (e.g.,temperature, salinity, sediment grain size, etc...). Field experiments need to be carried out with carefully designed controls. The correct spatial scale needs to be considered when planning experiments (Stiling, 2002).

If vegetation is correlated with geomorphologic landforms, for this reason stratified random sampling method is better and this method employ (Greig-Smith, 1983; Ludwig and Reynold, 1988)

Environmental impact assessments ideally attempt to compare before and after studies. There were differences in sampling sites, sampling dates, effort, replication, taxonomic categories, and recovery data. (Ferson et al., 1986).

Gotelli and Ellison (2004) and Odum and Barrett (2005) studied on sampling design. Diserud and Aagaard (2002) present a method that tests for changes in community structure based on repeated sampling. This may be the plot method of sampling generally consists of three major types:

(1) Simple random or random sampling without replacement, (2) stratified random, and (3) systematic (Cochran, 1977)

Simple random sampling with replacement is inherently less efficient than simple random sampling without replacement (Thompson, 2002). It is important not to have to determine whether any unit in the data is included more than once. Simple random sampling consists of using a grid or a series of coordinate lines (transects) and a table of random numbers to select several plots (quadrats), the size depending on the dimensions and densities of the organisms present.

4. Classification methods

4.1 Measures of similarity and difference (similarity and dissimilarity)

Dissimilarity or distance measures can be categorized as metric, semi-metric, or nonmetric (McCune et al., 2002). Symmetric distances are extremely useful in community ecology. There are different Similarity coefficients based on binary data. Two methods that are simple and give good results are Jaccard and Czekanowski (Sorenson) coefficients range 0 to 1.o.

Following equations are for Jaccard (1901) and Sorenson similarity coefficients

$$J = \frac{c}{a+b+c}$$

Where J= Jcacard coeffiecient, a= number of occurrences of species a alone, b= number of occurrences of b alone, c= number of co-occurences of two species (a and b).

$$C = \frac{2c}{2a+b+c}$$

Where: (same as in Jaccard)

There are different Dissimilarity coefficients based on meristic or metric data. Coefficient that are widely used in ecology studies are Bray-Curtis (1975) dissimilarity coefficient and Morisita' s Index.

Bray-Curtis is recommended by Clarke and Warwick (2001) and others as the most appropriate dissimilarity coefficient for community studies.

$$BC = \frac{\sum_{j=1}^{n} |X_{1j} - X_{2j}|}{(X_{1j} + X_{2j})}$$

Where
Σ= sum (from 1 to n)
X1j= # organisms of species j (attribute) collected at site 1 (entity).
X2j= # organisms of species j collected at site 2.
BC = Bray – Curtis coefficient of distance
| |= absolute value
J= 1 to n
N= number of species
Krebs (1999) considers Morisita' s index as the best similarity index, as follows:

$$C_\lambda = \frac{2\sum X_{ij}X_{ik}}{(\lambda_1 + \lambda_2)N_j N_k}$$

$$\lambda_1 = \frac{\sum \left[X_{ij}(X_{ij} - 1) \right]}{N_i(N_i - 1)}$$

$$\lambda_2 = \frac{\sum \left[X_{ik}(X_{ik} - 1) \right]}{N_k(N_k - 1)}$$

Where
Σ= sum
Cm= Morisita' s Index of Similarity
Xij= No. individuals of species I in sample j
Xik= No. individuals of species I in sample k
Nj= Total No. individuals in sample j
Nk= Total No. individuals in sample k
And Coefficients of Association are two types, ranging from -1 to +1 (e.g., Pearson Correlation Coefficient) and from 0 to X (e.g., χ^2) and applicable to binary and continuous data. Pearson coefficient moment correlation uses conjoint absences, the use of which is inappropriate for comparing sites and appropriate for comparing species (Clarke and Warwick, 2001)
Euclidian Distance Euclidean distance is another measure of distance that can be applied to a site by species matrix. It has been widely used in the past because it is compatible with virtually all cluster techniques.

4.2 Cluster analysis
Clustering is a straightforward method to show association data, however, the confidence of the nodes are highly dependent on data quality, and levels of similarity for cluster nodes is dependent on the similarity index used. Krebs (1989) shows that mean linkage is superior to single and complete linkage methods for ecological purposes because the other two are extremes, either producing long or tight, compact clusters respectively. There are, however, no guidelines as to which mean-linkage method is the best (Swan, 1970).

The purpose of two-way clustering (also known as biclustering) is to graphically expose the relationship between cluster analyses and your individual data points. The resulting graph makes it easy to see similarities and differences between rows in the same group, rows in different groups, columns in the same group, and columns in different groups. You can see graphically how groups of rows and columns relate to each other. Two-way clustering refers to doing a cluster analysis on both the rows and columns of your matrix, followed by graphing the two dendrograms simultaneously, adjacent to a representation of your main matrix. Rows and columns of your main matrix are re-ordered to match the order of items in your dendrogram.

Group Linkage Methods

1.	Nearest Neighbor	5.	Centroid
2.	Farthest Neighbor	6.	Ward's Method
3.	Median	7.	Flexible Beta
4.	Group Average	8.	McQuitty's Method

Ward's is also know as Orloci's and Minimum Variance Method

Table 1. Major types of hierarchical, agglomerative, polythetic clustering strategies

Cluster analysis can be performed using either presence–absence or quantitative data. Each pair of sites is evaluated on the degree of similarity, and then combined sequentially into clusters to form a dendrogram with the branching point representing the measure of similarity.

4.3 TWINSPAN

The TWINSPAN method (from Two Way Indicator Species Analysis, Hill 1979; Hill et al. 1975) is a very popular method among community ecologists and it was partially inspired by the classificatory methods of classical phytosociology (use of indicators for the definition of vegetation types). Two popular agglomerative polythetic techniques are Group Average and Flexible. McCune et al. (2002) recommend Ward' s method in addition. Gauch (1982) preferred to use divisive polythetic techniques such as TWINSPAN.

This method works with qualitative data only. In order not to lose the information about the species abundances, the concepts of pseudo-species and pseudo-species cut levels were introduced. Each species can be represented by several pseudo-species, depending on its quantity in the sample. A pseudo-species is present if the species quantity exceeds the corresponding cut level.

TWINSPAN is a program for classifying species and samples, producing an ordered two-way table of their occurrence. The process of classification is hierarchical; samples are successively divided into categories, and species are then divided into categories on the basis of the sample classification. TWINSPAN, like DECORANA, has been widely used by ecologists.

4.4 Indicator species analysis

Indicator species analysis is a divisive polythetic method of numerical classification applicable to large sets of qualitative or quantitative data.

This method provides a simple, intuitive solution to the problem of evaluating species associated with groups of sample units Dufrêne and Legendre's (1997). It combines information on the concentration of species abundance in a particular group and the faithfulness of occurrence of a species in a particular group. This method produces indicator values for each species in each group. These are tested for statistical significance using a Monte Carlo technique. It requires data from two or more sample units. Indicator values (range 0 for no indication to 100 for perfect indication) are presented for each species. The statistical significance of the maximum indicator value recorded for a given species is generated by a Monte Carlo test. There are many types of indicator species ranging from individual species (Dufrêne & Legendre's ,1997).

The identification of characteristic or indicator species is traditional in ecology and biogeography. Field studies describing sites or habitats usually mention one or several species that characterize each habitat. There is clearly a need for the identification of characteristic or indicator species in the fields of monitoring, conservation, and management. Indicator species add ecological meaning to groups of sites discovered by clustering, they can compare typologies derived from data analysis, identify where to stop dividing clusters into subsets, and point out the main levels in a hierarchical classification of sites.

Indicator species differ from species associations in that they are indicative of particular groups of sites. Good indicator species should be found mostly in a single group of a typology and be present at most of the sites belonging to that group (Hill, 1975).

With Classical problem in community ecology and biogeography, species are the best indicators we have for particular environmental conditions. And in long-term environmental follow-up, conservation, ecological management, researchers are looking for bioindicators of habitat types to preserve or rehabilitate.

McGeoch & Chown (1998) found the indicator value method important to conservation ecosystem because it is conceptually straightforward and allows researchers to identify bioindicators for any combination of habitat types or areas of interest, e.g. existing conservation areas, or groups of sites based on the outcome of a classification procedure.

Indicator species are species that, due to their niche preferences, can be used as ecological indicators of community types, habitat conditions, or environmental changes (McGeoch 1998, Carignan and Villard 2002, Niemi and McDonald 2004).

They are usually determined using an analysis of the relationship between the observed species presence–absence or abundance values in a set of sampled sites and a classification of the same sites (Dufrêne and Legendre 1997).

Finally this method may represent the groups of sites in the classification, in different qualitative characteristics of the ecosystem, such as habitat or community types, environmental or succession states, or the levels of controlled experimental designs.

Since indicator species analysis relates two elements, the species and the groups of sites, it can be used for gaining information on either or both. Indeed, indicator species analysis allows the characterization of the qualitative environmental preferences of the target species (for instance, when the groups are habitat types), and identifyes indicators of particular groups of sites, which can be used in further surveys.

The applications of indicator species analysis are many, including conservation, land management, landscape mapping, or design of natural reserves. Indicator species are commonly referred to as 'diagnostic species' in vegetation studies (Chytr et al. 2002).

4.5 Multi-response permutation procedures (MRPP)

Multi-response Permutation Procedure (MRPP) was introduced by Mielke, Berry, and Johnson (1976) as a technique for detecting the difference between a priori classified groups.

It turned out to be an extremely versatile data-analytic framework from which a number of applications fall out, such as the measurement of agreement, multivariate correlation and association coefcients, and the detection of autocorrelation (see Mielke & Berry, 2001 for a complete coverage of applications of the MRPP framework).

MRPP is a non-parametric method for testing the hypothesis of no difference between two or more groups of entities. The groups must be a priori. For example, one could compare species composition between burned and unburned plots to test the hypothesis of no treatment effect.

Discriminant analysis is a parametric method that can be used on the same general class of questions. However, MRPP has the advantage of not requiring assumptions (such as multivariate normality and homogeneity of variances) that are seldom met with ecological community data (Bakus, 2007).

Multiple Response Permutation Procedure (MRPP) provides a test of whether there is a significant difference between two or more groups of sampling units. This difference may be one of location (differences in mean) or one of spread (differences in within-group distance). Function MRPP operates on a data frame matrix where rows are observations and responses data matrix. The response(s) may be uni or multivariate. The method is mathematically allied with analysis of variance, in that it compares dissimilarities within and among groups. If two groups of sampling units are really different (e.g. in their species composition), then average of the within-group compositional dissimilarities ought to be less than the average of the dissimilarities between two random collection of sampling units drawn from the entire population.

The MRPP method is simply the overall weighted mean of within-group means of the pairwise dissimilarities among sampling units.

The MRPP algorithm first calculates all pairwise distances in the entire dataset, then calculates delta. It then permutes the sampling units and their associated pairwise distances, and recalculates a delta based on the permuted data. It repeats the permutation step permutations times.

The function also calculates the change-corrected within-group agreement. And it also calculates classification strength which is defined as the difference between average between group dissimilarities and within group dissimilarities (Van Sickle 1997).

If the first argument data can be interpreted as dissimilarities, they will be used directly. In other cases the function treats datas observations, and uses vegdist to find the dissimilarities. The default distance is Euclidean as in the traditional use of the method, but other dissimilarities in vegdist also are available.

Function meandist calculates a matrix of mean within-cluster dissimilarities (diagonal) and between-cluster dissimilarities (off-diagonal elements), and an attribute n of grouping counts. Function summary finds the within-class, between-class and overall means of these dissimilarities, and the MRPP statistics with all weight type options and the classification strength.

The MRPP a robust alternative to the traditional normal theory based parametric tests, such as the t test and the analysis of variance. However, MRPP is not widely known to researchers, and part of the reason is that it has not been incorporated into major statistical packages.

4.6 Mantel test

The Mantel test evaluates the null hypothesis of no relationship between two dissimilarity (distance) or similarity matrices. The Mantel test is an alternative to regressing distance matrices that circumvents the problem of partial dependence in these matrices.

For example evaluating the correspondence between two groups of organisms from the same set of sample units or comparing community structure before and after a disturbance is evaluate by mantel test.

In Consequency, Two methods are available in PC-ORD: Mantel's asymptotic approximation and a randomization (Monte Carlo) method.

The Mantel Test (Mantel, 1967) compares two dissimilarity (distance) matrices using Pearson correlation. The matrices that use in this test must be of the same size (i.e., same number of rows). It would appear that the Mantel Test could be used to evaluate the internal structure in two sets of samples by comparing the two (dis)similarity matrices.

McCune and Mefford (1999) have a computer program for performing the Mantel test. After the Mantel statistic has been calculated, the statistical significance of the relationship is tested by a permutation test or by using an asymptotic t-approximation test.

The minimum number of permutations (randomizations) recommended by Manly (1997) is 1000. The Mantel and ANOSIM procedures produce similar probabilities (Legendre and Legendre, 1998).

Some of the problems associated with the Mantel test include (1) weakness in detecting spatial autocorrelation where the spatial pattern is complex and not easily modeled with distance matrices, (2) a larger number of data points may be needed for field experiments than is usually obtained, and (3) multivariate data are summarized into a single distance or dissimilarity and it is not possible to identify which variable(s) contributed the most (Fortin and Gurevitch, 2001).

The use of Partial Mantel tests can distinguish the relative contributions of the factors of a third matrix considered as covariables.

Mantel and partial Mantel tests can produce complementary information that other methods such ANOVA cannot provide and may do a better job than other method, in detecting block effects (Fortin and Gurevitch, 2001).

Thus, it is possible to distinguish the effects of spatial pattern from those of experimentally imposed treatment effects with these Mantel tests (Bakus, 2007).

4.7 The best clustering strategies

The similarity or dissimilarity measures, forming the backbone of most multivariate clustering and ordination techniques, which are favored by many ecologists, are Jaccard and Bray–Curtis.

The most popular clustering techniques in ecological studies are Group Average and Flexible group average clustering. For many of these techniques, it is impossible to choose a "best method" because of the heuristic nature of the methods (Jongman et al., 1995; Anderson, 2001).

The choice of an index must be made based on the investigator's experience, the type of data collected, and the ecological question to be answered. When comparisons are made, the Jaccard index is among the least sensitive of the similarity (or dissimilarity) indices (Jongman et al. 1995). Similarly, hierarchical agglomerative techniques (e.g., particularly Group Average) have proven to be very useful to ecologists in the construction of dendrograms.

As part of a study is needed information on plant communities and their distribution. Then the objective of best method of multivariate analysis is to:

1. Identify and describe plant communities on the ecosystem
2. Map these communities to provide a tool for ecological studies and monitoring of ecological community;

3. Characterize the interrelationships between plant communities, soil particle size, moisture availability, grazing pressure and elevation.

5. Ordination methods

Ordination is a collective term for multivariate techniques which adapt a multi-dimensional swarm of data points in such a way that when it is projected onto a two dimensional space any intrinsic pattern the data may possess becomes apparent upon visual inspection (Pielou, 1984). As mentioned in the introduction, ordination is the arrangement of samples along gradients. Indeed, ordination can be considered a synonym for multivariate gradient analysis.

Basically, ordination serves to summarize community data by producing a low-dimensional ordination space in which similar species and samples are plotted close together, and dissimilar species and samples are placed far apart.

Indirect gradient analysis
   Distance-based approaches
   Polar ordination, PO (Bray-Curtis ordination)
   Principal Coordinates Analysis, PCoA (Metric multidimensional scaling)
   Nonmetric Multidimensional Scaling, NMDS
Eigenanalysis-based approaches
   Linear model
   Principal Components Analysis, PCA
   Unimodal model
   Correspondence Analysis, CA (Reciprocal Averaging)
   Detrended Correspondence Analysis, DCA
Direct gradient analysis
   Linear model
   Redundancy Analysis, RDA
   Unimodal model
   Canonical Correspondence Analysis, CCA
   Detrended Canonical Correspondence Analysis, DCCA

5.1 Polar ordination (Bray- Curtis)

(Polar Ordination) arranges samples with respect to "poles" (also termed end points or reference points) according to a distance matrix (Bray and Curtis 1957).. These endpoints are two samples with the highest ecological distance between them (objective approach), OR two samples suspected of being at opposite ends of an important gradient (subjective approach). This procedure is especially useful for investigating ecological change (e.g., succession, recovery).

Advantages of this method is Ideal for evaluating problems with discrete endpoints.

Polar Ordination ideal for testing specific hypotheses (e.g., reference condition or experimental design) by subjectively selecting the end points Disadvantages:

This technique does not provide a general-purpose description of the community (perspective is biased) Very sensitive to outliers (by definition – "end points") Select a distance measure (usually Sorensen Index) and calculate matrix of distances (D) between all pairs of points (Beals, 1984).

In the earliest versions of PO, these endpoints were the two samples with the highest ecological distance between them, or two samples which are suspected of being at opposite ends of an important gradient (thus introducing a degree of subjectivity).

Beals (1984) extended Bray-Curtis ordination and discussed its variants, and is thus a useful reference. The polar ordination, simplest method is to choose the pair of samples, not including the previous endpoints, with the maximum distance of separation.

5.2 Principal component aanalysis

PCA was invented in 1901 by Karl Pearson (Dunn, et al, 1987) Now it is mostly used as a tool in exploratory data analysis and for making predictive models.

Principal Components Analysis is a method that reduces data dimensionality by performing a covariance analysis between factors.

It can be done by eigenvalue decomposition of a data covariance matrix or singular value decomposition of a data matrix, usually after mean centering the data for each attribute.

The results of a PCA are usually discussed in terms of component scores (the transformed variable values corresponding to a particular case in the data) and loadings (the weight by which each standarized original variable should be multiplied to get the component score (Feoli and Orl¢ci. 1992).

Principal components analysis is the basic eigenanalysis technique. It maximizes the variance explained by each successive axis.

PCA was one of the earliest ordination techniques applied to ecological data. PCA uses a rigid rotation to derive orthogonal axes, which maximize the variance in the data set.

Both species and sample ordinations result from a single analysis. Computationally, PCA is basically an eigenanalysis. The sum of the eigenvalues will equal the sum of the variance of all variables in the data set. PCA is relatively objective and provides a reasonable but crude indication of relationships.

This method is a mathematical procedure that uses an orthogonal transformation to convert a set of observations of possibly correlated variables into a set of values of uncorrelated variables called principal components.

The number of principal components is less than or equal to the number of original variables. This transformation is defined in such a way that the first principal component has as high a variance as possible (that is, accounts for as much of the variability in the data as possible), and each succeeding component in turn has the highest variance possible under the constraint that it be orthogonal to (uncorrelated with) the preceding components.

Principal components are guaranteed to be independent only if the data set is jointly normally distributed. It is sensitive to the relative scaling of the original variables.

Depending on the field of application, it is also named the discrete Karhunen–Loève transform (KLT), the Hotelling transform or proper orthogonal decomposition (POD).

Broken-stick eigenvalues are provided to help evaluating statistical significance. Principal component analysis (PCA) (ter Braak and S˘ milauer, 1998) was used to determine the association between plant communities and environmental variables, i.e. in an indirect non-canonical way (ter Braak and Loomans, 1987).

While PCA finds the mathematically optimal method (as in minimizing the squared error), it is sensitive to outliers in the data that produce large errors PCA tries to avoid. It therefore is common practice to remove outliers before computing PCA.

However, in some contexts, outliers can be difficult to identify. For example in data mining algorithms like correlation clustering, the assignment of points to clusters and outliers is not known beforehand.

A recently proposed generalization of PCA based on Weighted PCA increases robustness by assigning different weights to data objects based on their estimated relevancy.

Although it has severe faults with many community data sets, it is probably the best technique to use when a data set approximates multivariate normality. PCA is usually a poor method for community data, but it is the best method for many other kinds of multivariate (Bakus, 2007).

5.3 Principal coordinate analysis (PCoA)

Principal Coordinate Analysis (PCoA) is a method to represent on a 2 or 3 dimensional chart objects described by a square matrix containing that contains resemblance indices between these objects.

This method is due to Gower (1966). It is sometimes called metric MDS (MDS: Mutidimensional scaling) as opposed to the MDS (or non-metric MDS). Both methods have the same objective and produce similar results if the similarity matrix f square distances are metric and if the dimensionality is sufficient.

Principle coordinates are similar to principal components in concept. The advantage of PCoA is that it may be used with all types of variables (Legendre and Legendre, 1998). Because of this, PCoA is an ordination method of considerable interest to ecologists.

Most metric (PCoA) and nonmetric MDS plots are very similar or even identical, provided that a similar distance measure is used. The occurrence of negative eigenvalues, lack of emphasis on distance preservation, and other problems are discussed in detail by Legendre and Legendre (1998) and Clarke and Warwick (2001).

One of the biggest differences between PCA and PCoA is that the variables (i.e. species) representing the original axes are projected as biplot arrows. In the bryophyte communities, these biplot arrows greatly aid in interpretation (Bakus, 2007).

In most applications of PCA (e.g. as a factor analysis technique), variables are often measured in different units. For example, PCA of taxonomic data may include measures of size, shape, color, age, numbers, and chemical concentrations. For such data, the data must be standardized to zero mean and unit variance (the typical default for most computer programs).

For ordination of ecological communities, however, all species are measured in the same units, and data should not be standardized. In matrix algebra terms, most PCAs are eigenanalyses of the correlation matrix, but for ordination they should be PCAs of the covariance matrix.

In contrast to Correspondence Analysis and related methods, species are represented by arrows. This implies that the abundance of the species is continuously increasing in the direction of the arrow, and decreasing in the opposite direction. Thus PCA is a 'linear method'.

Although the discussion above implies that PCA is distinctly different from PCoA, the two techniques end up being identical, if the distance metric is Euclidean.

Unfortunately, this linear assumption causes PCA to suffer from a serious problem, the horseshoe effect, which makes it unsuitable for most ecological data sets (Gauch 1982).

Principal Coordinates Analysis (PCoA, = Multidimensional scaling, MDS) is a method to explore and to visualize imilarities or dissimilarities of data. It starts with a similarity matrix or dissimilarity matrix (= distance matrix) and assigns for each item a location in a low-dimensional space, e.g. as a 3D graphics.

PCOA tries to find the main axes through a matrix. It is a kind of eigenanalysis (sometimes referred as "singular value decomposition") and calculates a series of eigenvalues and eigenvectors. Each eigenvalue has an eigenvector, and there are as many eigenvectors and eigenvalues as there are rows in the initial matrix.

Eigenvalues are usually ranked from the greatest to the least. The first eigenvalue is often called the "dominant" or leading" eigenvalue. Using the eigenvectors we can visualize the main axes through the initial distance matrix. Eigenvalues are also often called "latent values".

The result is a rotation of the data matrix: it does not change the positions of points relative to each other but it just changes the coordinate systems!

By using PCoA we can visualize individual and/or group differences. Individual differences can be used to show outliers.

There is also a method called 'Principal Component Analysis' (PCA, sometimes also misleadingly abbreviated as 'PCoA') which is different from PCOA.

4.   Principal Coordinates Analysis (Principal Coordinates Analysis (PCoA PCoA or PCO) or PCO)
5.   Maximizes the linear correlation between distance measures and distance in the ordination.
6.   useful if one has only a distance (or similarity) matrix
7.   The underlying model is that there a fixed number of explanatoryoriginal variables. In contrast, PCA, RA, and DCA assume that there are potentially many variables, but of declining importance.
8.   One cannot easily put new points in a PCoA.
9.   For Euclidean distance, PCoA= PCA
10. PCoA can be expressed as an eigenanalysis.

5.4 Factor analysis (FA)

FA and PCA (principal components analysis) are methods of data reduction. Take many variables and explain them with a few "factors" or "components".

Correlated variables are grouped together and separated from other variables with low or no correlation. Patterns of correlations are identified and either used as descriptives (PCA) or as indicative of underlying theory (FA). It process of providing an operational definition for latent construct (through regression equation).

FA and PCA are not much different than canonical correlation in terms of generating canonical variates from linear combinations of variables. Although there are now no "sides" of the equation.

For calcuting this method we should:

1.   Selecting and Measuring a set of variables in a given domain
2.   Data screening in order to prepare the correlation matrix
3.   Factor Extraction
4.   Factor Rotation to increase interpretability
5.   Interpretation

Factor analysis is seldom used in ecology. Several statisticians state that it should not be used because it is based on a special statistical model.

Estimating the unique variance is the most difficult and ambiguous task in FA (McGarigal et al., 2000).

5.5 Redundancy analysis

RDA is a linear method and since it is a linear method, species as well as environmental variables are represented by arrows. In most cases, it is best to represent the two sets of arrows in two figures for ease of display. Thus, if have a gradient along which all species are positively correlated, RDA will detect such a gradient while CCA will not. RDA can use 'species' that are measured in different units. If so, the data must be centered and standardized. But in fact, as an ordination technique, the species should not be standardized.

Redundancy analysis is the linear method of direct ordination. It also goes under the name of principal components analysis with respect to instrumental variables, which in our case are environmental variables (Sabatier et al., 1989). It is also called least-squares reduced rank regression so as to emphasize its link with multivariate regression (ter Braak & Prentice, 1988; ter Braak & Looman, 1994). In statistical textbooks on multivariate analysis, redundancy analysis is usually neglected.

RDA is useful when gradients are short. In particular, RDA may be the method of choice in a short-term experimental study. In such cases, the treatments are the explanatory variables (and are usually dummy variables). The sample ID or block might be a covariable in a partial RDA, if one wishes to factor out local effects (Bakus, 2007).

'variance explained' is actually a variance explained, and not merely inertia. Thus, variance partitioning, and interpretation of eigenvalues, are more straightforward than for CCA (Lepš & Šmilauer, 1999).

The relatively short space devoted to RDA should not be given as an indication that it is less valuable than, or inferior to, other method. It is simply used for different purposes.

Consequently, the linear method of direct ordination, redundancy analysis, can often efficiently display the interesting effects, and there is no need for methods that also work when the range of community variation is larger (such as canonical correspondence analysis) nor for unconstrained nonmetric multidimensional scaling.

5.6 Correspondence analysis (CA) or reciprocal averaging (RA)

Reciprocal averaging is also known as correspondence analysis (CA) because one algorithm for finding the solution involves the repeated averaging of sample scores and species scores (citations). It is a graphical display ordination technique which simultaneously displays the rows (sites) and columns (species) of a data matrix in low dimensional space (Gittins, 1985). Row identifiers (species) plotted close together are similar in their relative profiles, and column identifiers plotted close together are correlated, enabling one to interpret not only which of the taxa are clustered, but also why they are clustered (Bakus,2007).

Reciprocal averaging (RA) yields both normal and transpose ordinations automatically. Like DCA, RA ordinates both species and samples simultaneously. Instead of maximizing 'variance explained', CA maximizes the correspondence between species scores and sample scores.

If species scores are standardized to zero mean and unit variance, the eigenvalues also represent the variance in the sample scores (but not, as is often misunderstood, the variance in species abundance).

Since CA is a unimodal model, species are represented by a point rather than an arrow. This is (under some choices of scaling; see ter Braak and Šmilauer 1998) the weighted average of the samples in which that species occurs. With some simplifying assumptions (ter Braak and Looman 1986), the species score can be considered an estimate of the location of the peak of the species response curve.

The CA distortion is called the arch effect, which is not as serious as the horseshoe effect of PCA because the ends of the gradients are not incurved. Nevertheless, the distortion is prominent enough to seriously impair ecological interpretation (Bakus, 2007).

In other words, the spacing of samples along an axis may not affect true differences in species composition.

Gradient compression can be quite blatant in simulated data sets. The problems of gradient compression and the arch effect led to the development of Detrended Correspondence Analysis (Bakus, 2007).

5.7 Detrended correspondence analysis (DCA)

Detrended Correspondence Analysis (DCA) eliminates the arch effect by detrending (Hill and Gauch 1982). It's a series of rules that are used to reshape data to make it friendlier for analysis. Once again, primarily used for ecological data, but can be extended to anything (data simply can't contain negative values).

The reason that this technique is used is to over come the arch effect (the horseshoe effect).

Found in data whenever "PCA or other distance conserving ordination techniques are applied to data which follow a continuous gradient, along which there is a progressive turnover of dominant variables." Such as in ecological succession

After ordination by a distance conserving technique and the first two axes are plotted against each other, one would find an arch shape.

DCA is another eigenanalysis ordination technique that based on reciprocal averaging (RA; Hill 1979). DCA is geared to ecological sets, is based on samples and species. DCA ordinates both species and samples simultaneously.

There are two basic approaches to detrending: by polynomials and by segments (ter Braak and Šmilauer 1998). Detrending by polynomials is the more elegant of the two: a regression is performed in which the second axis is a polynomial function of the first axis, after which the second axis is replaced by the residuals from this regression. Similar procedures are followed for the third and higher axes.

The compression of the ends of the gradients is corrected by nonlinear rescaling. Rescaling shifts sample scores along each axis such that the average width is equal to 1.

Rescaling has a beneficial consequence: the axes are scaled in units of beta diversity (SD units, or units of species standard deviations). Thus if the underlying gradient is important well known, it is possible to plot the DCA scores as a function of the gradient, and there by determine whether the species 'perceive' the gradient differently than we measure it.

The shape of the species response curves may change if axes are rescaled. Thus, skewness and kurtosis are largely artifacts of the units of measurement for which we choose to measure the environment.

5.8 Nonmetric multimentional scaling (MDS, NMDS, NMS, NMMDS)

Nonmetric Multidimensional Scaling (NMDS) rectifies this by maximizing the rank order correlation. For this proceeds at first the user selects the number of dimensions (N) for the solution, and chooses an appropriate distance metric and then The distance matrix is calculated. And initial configuration of samples in N dimensions is selected. This configuration can be random, though the chances of reaching the correct solution are enhanced if the configuration is derived from another ordination method.

And finally, the final configuration of points represents your ordination solution. The configuration is dependent on the number of dimensions selected; e.g. the first two axes of a 3-dimensional solution does not necessarily resemble a 2-dimensional solution. The stress will typically decrease as a function of the number of dimensions chosen; this function can aid in the selection of the results (Bakus, 2007).

This is why it is sometimes useful to rotate the solution (such as by the Varimax method) – although there is no theory that states that the final solution will represent a 'gradient' Other problems and advantages of NMDS will be discussed later, when comparing it to Detrended Correspondence Analysis (Bakus, 2007).

5.9 MANOVA and MANCOVA

A factorial MANOVA may be used to determine whether or not two or more categorical grouping variables (and their interactions) significantly affect optimally weighted linear combinations of two or more normally distributed outcome variables.

These parametric multivariate techniques (Multivariate Analysis of Variance and Multivariate Analysis of Covariance) are similar to ANOVA and ANCOVA MANOVA (Wilks' Lambda) and ANCOVA are advantageous in that performing multiple univariate tests can inflate the a value, leading to false conclusions (Scheiner, 2001).

MANOVA seeks differences in the dependent variables among the groups (McCune et al, 2002). Assumptions of MANOVA include multivariate normality (error effects included), independent observations, and equality of variance-covariance matrices (Paukert and Wittig, 2002). Because of these assumptions, among others, MANOVA is not often used in ecology although its use in increasing.

The power of traditional MANOVA declines with an increase in the number of response variables (Scheiner, 2001). Unequal sample sizes are not a large problem for MANOVA, but may bias the results for factorial or nested designs. Before ANCOVA is run, tests of the assumption of homogeneity of slopes need to be performed (Petratis et al., 2001).

Early attempts to develop nonparametric multivariate analysis include those of Mantel and Valand (1970). They are more complex as they handle three or more variables simultaneously. They are frequently used with the analysis of experimental studies, especially laboratory experiments.

More recently, the Analysis of Similarities was developed to compare communities or changes in communities because of pollution (Clarke, 1993). For typical species abundance matrices, an Analysis of Similarities (ANOSIM) permutation procedure is recommended over MANOVA.

Multiple analysis of variance (MANOVA) is used to see the main and interaction effects of categorical variables on multiple dependent interval variables. MANOVA uses one or more categorical independents as predictors, like ANOVA, but unlike ANOVA, there is more than one dependent variable. Where ANOVA tests the differences in means of the interval dependent for various categories of the independent(s), MANOVA tests the differences in the centroid (vector) of means of the multiple interval dependents, for various categories of the independent(s). One may also perform planned comparison or post hoc comparisons to see which values of a factor contribute most to the explanation of the dependents.

There are multiple potential purposes for MANOVA.

To compare groups formed by categorical independent variables on group differences in a set of interval dependent variables.

To use lack of difference for a set of dependent variables as a criterion for reducing a set of independent variables to a smaller, more easily modeled number of variables.

Multiple analysis of covariance (MANCOVA) is similar to MANOVA, but interval independents may be added as "covariates." These covariates serve as control variables for the independent factors, serving to reduce the error term in the model. Like other control procedures, MANCOVA can be seen as a form of "what if" analysis, asking what would happen if all cases scored equally on the covariates, so that the effect of the factors over and beyond the covariates can be isolated. The discussion of concepts in the ANOVA section also applies, including the discussion of assumptions.

5.10 Discriminate analysis

Discriminate Analysis (DA) is a powerful tool that can be used with both clusters of species data and environmental variables. It determines which variables discriminate between two or more groups, that is, independent variables are used as predictors of group membership (McCune et al., 2002). It is very similar to MANOVA and multiple regression analysis (Statsoft, Inc., 1995; McGarigal et al., 2000). Clusters can be identified by several methods using raw data: (1) constructing a dendrogram, (2) using PCA (even if you have field data) for initial visual identification of clusters, and (3) point rotation in space by rotating ordinations (i.e., rotating axes – see McCune and Mefford, 1999). If any method indicates groups or clusters of data then DA can be used. However, the number of groups is set before the DA analysis. DA finds a transform for the minimum ratio of difference between pairs of multivariate means and variances in which two clusters are separated the most and inflated the least.

DA produces two functions: (1) classification function consisting of 2 groups or clusters of points (this information can be used for prediction with probabilities) and (2) discriminate function containing environmental variables that can be used to discriminate differences among the groups.

DA differs from PCA and Factor Analysis in that no standardization of data is needed (PCA and FA need standardization because of scaling problems) and the position of the axes distinguishes the maximum distance between clusters. (Davis, 1986).

DA assumes a multivariate normal distribution, homogeneity of variances, and independent samples (Paukert and Wittig, 2002). Violations of normality are usually not fatal (i.e., somewhat non-normal data can be used). A description of the procedure to use DA with Statistica is given in the Appendix. Multiple Discriminate Analysis (MDA) is the term often used when three or more clusters of data are processed simultaneously. MDA is particularly susceptible to rounding error. Calculations in double precision for at least the eigenvalue-eigenvector routines are advisable (Green, 1979). Limitations of Discriminate Analysis are discussed by McGarigal et al. (2000) Dytham (1999).

5.11 Canonical correspondence analysis (CCA)

In ecology studies, the ordination of samples and species is constrained by their relationships to environmental variables. When species responses are unimodal (hump-shaped), and by measuring the important underlying environmental variables, CCA is most likely to be useful.

It was used to examine the relationships between the measured variables and the distribution of plant communities (Ter Braak, 1986). CCA expresses species relationships as linear combinations of environmental variables and combines the features of CA with canonical correlation analysis (Green, 1989). This provides a graphical representation of the relationships between species and environmental factors.

Canonical Correlation Analysis is presented as the standard method to relate two sets of variables (Gittins, 1985). However, the latter method is useless if there are many species compared to sites, as in many ecological studies, because its ordination axes are very unstable in such cases.

The best weight for CCA describes environment variables with the first axis shows. Species information structure using a reply CCA Nonlinear with the linear combination of variables will consider environmental characteristics of acceptable behavior characteristics of species with environment shows. CCA analysis combined with non-linear species and environmental factors shows the most important environmental variable in connection with the axes shows.

In Canonical Correspondence Analysis, the sample scores are constrained to be linear combinations of explanatory variables. CCA focuses more on species composition, i.e. relative abundance.

When a combination of environmental variables is highly related to species composition, this method, will create an axis from these variables that makes the species response curves most distinct. The second and higher axes will also maximize the dispersion of species, subject to the constraints that these higher axes are linear combinations of the explanatory variables, and that they are orthogonal to all previous axis.

Monte Carlo permutation tests were subsequently used within canonical correspondence analysis (CCA) to determine the significance of relations between species composition and environmental variables (ter Braak, 1987)

The outcome of CCA is highly dependent on the scaling of the explanatory variables. Unfortunately, we cannot know a priori what the best transformation of the data will be, and it would be arrogant to assume that our measurement scale is the same scale used by plants and animals. Nevertheless, we must make intelligent guesses (Bakus, 2007).

In CCA possible that patterns result from the combination of several explanatory variables. And many extensions of multiple regression (e.g. stepwise analysis and partial analysis) also apply to CCA.

It is possible to test hypotheses (though in CCA, hypothesis testing is based on randomization procedures rather than distributional assumptions). Explanatory variables can be of many types (e.g. continuous, ratio scale, nominal) and do not need to meet distributional assumptions.

Another advantage of CCA lies in the intuitive nature of its ordination diagram, or triplot. It is called a triplot because it simultaneously displays three pieces of information: samples as points, species as points, and environmental variables as arrows (or points).

If data sets are few, CCA triplots can get very crowded then should be separate the parts of the triplot into biplots or scatterplots (e.g. plotting the arrows in a different panel of the same figure) or rescaling the arrows so that the species and sample scores are more spread out. And we can only plotting the most abundant species (but by all means, keep the rare species in the analysis).

Noise in the species abundance data set is not much of a problem for CCA (Palmer, 1988). However, it has been argued that noise in the environmental data can be a problem (McCune 1999). It is not at all surprising that noise in the predictor variables will cause noise in the sample scores, since the latter are linear combinations of the former.

It is probably obvious that the choice of variables in CCA is crucial for the output. Meaningless variables will produce meaningless results. However, a meaningful variable that is not necessarily related to the most important gradient may still yield meaningful results (Palmer, 1988).

If many variables are included in an analysis, much of the inertia becomes 'explained'. Any linear transformation of variables (e.g. kilograms to grams, meters to inches, Fahrenheit to Centigrade) will not affect the outcome of CCA whatsoever.

There are as many constrained axes as there are explanatory variables. The total 'explained inertia' is the sum of the eigenvalues of the constrained axes. The remaining axes are unconstrained, and can be considered 'residual'. The total inertia in the species data is the sum of eigenvalues of the constrained and the unconstrained axes, and is equivalent to the sum of eigenvalues, or total inertia, of CA. Thus, explained inertia, compared to total inertia, can be used as a measure of how well species composition is explained by the variables. Unfortunately, a strict measure of 'goodness of fit' for CCA is elusive, because the arch effect itself has some inertia associated with it (Bakus, 2007).

The ordination diagrams of canonical correlation analysis and redundancy analysis display the same data tables; the difference lies in the precise weighing of the species (ter Braak & Looman, 1994; Van der Myer, 1991)

One of limitations to CCA is that correlation does not imply causation, and a variable that appears to be strong may merely be related to an unmeasured but 'true' gradient. As with any technique, results should be interpreted in light of these limitations (McCune, 1999).

5.12 Multiple regression (MR) (multiple linear regression)

The mechanics of testing the "significance" of a multiple regression model is basically the same as testing the significance of a simple regression model, we will consider an F-test, a t-test (multiple t's) and R-sqrd. However, unlike simple regression where the F & t tests tested the same hypothesis, in multiple regression these two tests have different purposes. R-sqrd is still the percent of variance explained but is no longer the correlation squared (as it was with in simple linear regression) and we will also introduce adjusted R-sqrd. When considering a multiple regression (MR) model the most common order to interpret things consists of first looking at the R-sqrd, then testing the entire model by looking at the F-test, and finally looking at each individual coefficient individually using the t-tests. NOTE: The term "significance" is a nice convenience but is very ambiguous in definition if not properly specified. Thus when taking this class you should avoid simply saying something is significant without explaining (1) how you made that determination, and (2) what that specifically means in this case. You will see from the examples that those two things are always done. If you cannot do that then any time you use the word "significant" you are potentially hurting yourself in two ways; (1) you won't do well on the quizzes or exams where you have to be able to be more explicit than simply throwing out the word "significant", and (2) you will look like a fool in the business world when somebody asks you to explain what you mean by "significant" and you are stumped. Remember if you can't explain your results in managerial terms than you do not really understand what you are doing.

In order to show the relationship between biotic (principal component axes) and abiotic factors (environmental factors), a multiple regression type analysis is used. Multiple regression solves simultaneously normal equations and produces partial regression coefficients. Partial regression coefficients each give the rate of change or slope in the dependent variable for a unit of change in a particular independent variable, assuming all other independent variables are held constant. MR is not considered by some statisticians as a multivariate procedure because it includes only one dependent variable (Paukert and Wittig, 2002).

The objective of multiple regression is to determine the influence of independent variables on a dependent variable, for example, the effect of depth, sediment grain size, salinity, temperature, and predator density on the population density of species.
 The parameters are estimated by the least-squares method, that is, minimizing the sum of squares of the differences between the observed and expected response (Jongman et al., 1995)
Normally component loadings (e.g., scores in PCA) suggest which variables are most important. However, with species abundances the only variable in some ordinations, one must use other techniques to attempt to suggest what may have produced the gradients for Axis 1, 2, 3, and so forth.
Univariate analyses such as the Spearman rank correlation coefficient are not as ecologically realistic as multivariate analyses such as multiple regression because some variables are correlated and there are interaction effects between variables (Jongman et al., 1995).
The highest standardized partial regression coefficients (positive or negative) suggest the most important factors (e.g., sediment size, predator density, etc.) in controlling the population density of species X. Significance tests and standard errors then can be calculated from the data.
Multiple regression has many potential problems such as the type of response curve, error distribution, and outliers that may unduly influence the results (Jongman et al., 1995). Multiple regression variables may be highly correlated, therefore, examine the correlation coefficients first (i.e., run a multiple correlation analysis between variables) to exclude some of them before doing multiple regressions.
Multiple regression generally should employ a maximum of 6 variables. Legendre and Legendre (1998) suggest a stepwise procedure for reducing numerous variables. This involves a process of alternating between forward selection and backward elimination (Kutner et al., 1996).
Lee and Sampson (2000) took ordination scores, representing a gradient in fish communities, and regressed them against a group of environmental variables and time. Many of scientist study on multiple regression and computer program (Such as Davis, 1986; and Sokal and Rohlf, 1995)
In multiple regression, it is typical to include quadratic terms for explanatory variables. For example, if you expect a response variable to reach a maximum at an intermediate value of an explanatory variable, including this explanatory variable AND the square of the explanatory variable may allow a concave-down parabola to provide a reasonable fit.
This is an analogous situation to multiple regression: the multiple r^2 or 'variance explained' increases as a function of the number of variables included.
Both multiple regression and CCA find the best linear combination of explanatory variables, they are not guaranteed to find the true underlying gradient (which may be related to unmeasured or unmeasurable factors), nor are they guaranteed to explain a large portion of variation in the data.

5.13 Path analysis
In statistics, path analysis is used to describe the directed dependencies among a set of variables. This includes models equivalent to any form of multiple regression analysis, factor analysis, canonical correlation analysis, discriminate analysis, as well as more general families of models in the multivariate analysis of variance and covariance analyses (MANOVA, ANOVA, ANCOVA).

Path analysis is a straightforward extension of multiple regression. Its aim is to provide estimates of the magnitude and significance of hypothesised causal connections between sets of variables. This is best explained by considering a path diagram.

Path analysis is an extension of multiple linear regression, allowing interpretation of linear relationships among a small number of descriptors (Legendre and Legendre, 1998).

This method was originally developed by Sewall Wright in which he introduced the concept of a path diagram. It handles more than one dependent variable and the effects of dependent variables on one another (Mitchell, 2001).

Path analysis assumes a normal distribution of residuals, additive and linear effects, inclusion of all important variables, that residual errors are uncorrelated, and that there is no measurement error.

Path analysis was developed as a method of decomposing correlations into different pieces for interpretation of effects (e.g., how does parental education influence children's income 40 years later?). Path analysis is closely related to multiple regression; you might say that regression is a special case of path analysis.

Some people call this stuff (path analysis and related techniques) causal modeling. The reason for this name is that the techniques allow us to test theoretical propositions about cause and effect without manipulating variables. However, the causal in causal modeling refers to an assumption of the model rather than a property of the output or consequence of the technique. That is, people assume some variables are causally related, and test propositions about them using the techniques.

5.14 Canonical correlation analysis (CVA)

Canonical variate analysis (CVA) is a widely used method for analyzing group structure in multivariate data. It is mathematically equivalent to a one-way multivariate analysis of variance and often goes by the name of canonical discriminate analysis. Change over time is a central feature of many phenomena of interest to researchers. This dissertation extends CVA to longitudinal data. It develops models whose purpose is to determine what is changing and what is not changing in the group structure. Three approaches are taken: a maximum likelihood approach, a least squares approach, and a covariance structure analysis approach. All methods have in common that they hypothesize canonical variates which are stable over time.

The maximum likelihood approach models the positions of the group means in the subspace of the canonical variates. It also requires modeling the structure of the within-groups covariance matrix, which is assumed to be constant or proportional over time.

In addition to hypothesizing stable variates over time, one can also hypothesize canonical variates that change over time. Hypothesis tests and confidence intervals are developed. The least squares methods are exploratory. They are based on three-mode PCA methods such as the Tucker2 and parallel factor analysis. Graphical methods are developed to display the relationships between the variables over time.

Stable variates over time imply a particular structure for the between-groups covariance matrix. This structure is modeled using covariance structure analysis, which is available in the SAS package Proc Calis.

Canonical Variate Analysis is a special case of Canonical Correlation Analysis (Jongman et al., 1995). It is also described as a type of linear discriminate analysis (McGarigal et al., 2000). The set of environmental variables consists of a single nominal variable defining the classes.

CVA is usable only if the number of sites is much greater than the number of species and the number of classes. Many ecological data sets cannot be analyzed by CVA without dropping many species, thus CVA is not used much in ecology.

Instead, they usually give the impression that there is only one such hypothesis, and therefore only one statistical technique is needed-Hotelling's canonical variate analysis (CVA).

Most discussions of CVA are restricted almost entirely to a description of the underlying mathematical theory, computing directions, and perhaps an example of the computations. Very little is usually said about the logic of the method, so that the reader is unable to judge for himself whether the method described can actually be used to test the hypothesis of interest to him.

Even when CVA is appropriate, several prominent sources have recommended misleading interpretations of the statistics computed in CVA.

Perhaps because of these deficiencies, many behavioral scientists have concluded incorrectly that CVA has few or no valid and important uses in the behavioral sciences. The use originally proposed by Hotelling has been rejected by most behavioral scientists.

6. Ordination and classification methods in various ecology studies

The more applied an ecological study is, the more the emphasis is on the effects on ecological communities of particular environmental factors, for example pollutants, management regimes, and other human-induced changes in the environments. (ter Braak, 1994).

Correspondingly, the statistical analysis should not 'just' show the major variation in the species assemblage, but focus on the effects on the variables of prime interest. Applied studies thus call for direct methods of ordination, typically with a very limited number of (qualitative or quantitative) environmental variables (ter Braak, 1994)

The range of community variation in an applied study tends to be quite small compared to that in the early ordination studies (e.g. Whittaker, 1965; Hill & Gauch, 1980).

Determining which factors control the distribution patterns of plant communities remains a central goal in ecology Classification assumes from the outset that the species assemblages fall into discontinuous groups, whereas ordination starts from the idea that such assemblages vary gradually.

Ordination compares sites on their degree of similarity, and then plots them in Euclidian space, with the distance between points representing their degree of Similarity. Ordination techniques include principal components analysis (PCA), detrended correspondence analyses (DCA), and nonmetric multidimensional scaling (NMS)

Ordination (or inertia) methods, like principal component and correspondence analysis, and clustering and classification methods are currently used in many ecological studies (Zare Chahouki et al., 2009, Anderson, 2002; Gauch et al., 1977; Orloci, 1975; Whittaker (ed.), 1967; Legendre & Legendre, 1988).

Ordination methods can be divided in two main groups, direct and indirect methods. Direct methods use species and environment data in a single, integrated analysis. Indirect methods use the species data only (Jongman et al., 1987). In contrast, if a unimodal response model is assumed, the relationships are unimodal. Unimodal relations are usefully summarized by their modes or - more conveniently - weighted averages (ter Braak & Looman, 1986), so that a sensible coefficient for the species $\times$ environment table is the weighted average. The other way round, if a correlation coefficient is chosen, the implied response model is linear or approximately linear and if the chosen coefficient is the weighted average than the implied response model is unimodal (i.e. if the true model is bimodal, the ordination will fail, and if the true model is linear, the ordination will be inefficient). Assumptions about the response model and the choice of coefficients to use in secondary tables are thus interrelated.

New methods of exploring differences among groups include the nonparametric, recursive classification, and regression tree (CART). It is used to classify habitats or vegetation types and their environmental variables (McCune et al., 2002). It produces a top-to-bottom visual classification tree that undergoes a "pruning" or optimization process. CART is used to generate community maps, wildlife habitats, and land cover types in conjunction with a GIS (see p. 195 in Chapter 3). Another multivariate technique is Structural Equation Modeling (unfortunately termed SEM), a merger of factor analysis and path analysis (McCune et al. (2002). It is a method of evaluating complex hypotheses (e.g., effects of abiotic factors on plant species richness).

With multiple causal pathways among variables. It requires the initial development of a path diagram. It is an analysis of covariance relationships, effectively limited to about 10 variables. See Shipley (2000), McCune et al. (2002), Pugesek et al. (2002) and Bakus (2007).

7. References

Anderson, M.J. (2001). A new method for non-parametric multivariate analysis of variance. Austral. Ecol. 26:32-46.

Alisauskas, R.T. (1998). Winter range expansion and relationships between landscape and morphometrics of midcontinent Lesser Snow Geese. Auk 115(4):851-862.

Anderson, C.W., Barnett, V., Chatwin, P.C. and El-Shaarawi A.H. (2002). Quantitative Methods for Current EnvironmentalIssues. Springer-Verlag, New York.

C.T. Bastian, S.R. Koontz and Menkhaus D.J. (2001). The Impact of Forward Contract Information on the Fed Cattle Market: An Experimental Investigation into Mandatory Price Reporting, UW Agricultural and Applied Economics Seminar. August 31, 2001. (Presented by Bastian).

Bakus Gerald J. (2007). Quantitative Analysis of Marine Biological Communities Field Biology and Environment. WILEY-INTERSCIENCE, A John Wiley & Sons, Inc., Publication, 453p.

Beals, E.W. (1973). Ordination: Mathematical elegance and ecological naivete. J. Ecol. 61:23–35.

Bray, J.R. and Curtis J.T. (1957). An ordination of the upland forest communities of southern Wisconsin.Ecol. Monogr. 27:325-349.

Carignan, V. and Villard M. (2002). Selecting indicator species to monitor ecological integrity: a review. Environ. Monitor. Assess. 78: 45-61.

Chytr, M. (2002). Determination of diagnostic species with statistical fidelity measures. J. Veg. Sci. 13: 79–90.

Clarke, K.R. (1993). Non-parametric multivariate analyses of changes in community structure. Aust. J. Ecol.18:117-143.

Clarke, K.R. and Warwick R.M. (2001). Change in Marine Communities: An Approach to Statistical Analysis and Interpretation. 2nd edition. PRIMER- E, Plymouth Marine Laboratory, Plymouth, U.K.

Clifford, H.T. and Stephenson W. (1975). An introduction to numerical classification. Academic Press, New York, pp. 229.

Cochran, W.G. (1977). Sampling Techniques. Wiley, New York.

Davis, J.C. (1986). Statistics and Data Analysis in Geology. Wiley, New York.

Diserud, O.H. and Aagaard K. (2002). Testing for changes in community structure based on repeated sampling. Ecology, 83(8): 2271-2277.

Dufrene, M. and Legendre P. (1997). Species assemblages and indicator species: the need for a flexible asymmetrical approach. Ecol. Monogr. 67:345-366.

Dunn, C. P., and Stearns F. (1987). Relationship of vegetation layers to soils in southeastern Wisconsin forested wetlands. Am. Midl. Nat. 118:366-74.

Dytham, C. (1999). Choosing and Using Statistics: A Biologist's Guide. Blackwell Publishing, Williston, VT.

Ferson, S., Downey P., Klerks P., Weissburg M., Kroot S.I., S. Jacquez O., Ssemakula J., Malenky R. and Anderson K. (1986). Competing reviews, or why do Connell and Schoener disagree? Am. Nat., 127: 571-576.

Feoli, E., and Orl¢ci L. (1992). Thre properties and interpretation of observations in vegetation study. Coenoses 6:61-70.

Fortin, M-J. and Gurevitch J. (2001). Mantel tests: Spatial structure in field experiments. pp. 308-326 in: Scheiner S.M. and J. Gurevitch (eds.). Design and Analysis of Ecological Experiments. Oxford University Press, Oxford.

Gauch, H.G. (1977). ORDIFLIX — A flexible computer program for four ordination techniques: weighted averages, polar ordination, principal components analysis, and reciprocal averaging. In: Ecology and Systematics, Cornell University, Ithaca, N.Y.

Gauch, H.G., J. (1982). Multivariate Analysis in Community Ecology. Cambridge University Press, New York.

Gittins, R. (1985). Canonical analysis: a review with applications in ecology. Springer-Verlag, Berlin.

Gotelli, N.J. and Ellison A.M. (2004). A Primer of Ecological Statistics. Sinauer Associates, Sunderland, Maine

Gower, J. C. (1966). Some Distance Properties of Latent Root and Vector Methods used in Multivariate Analysis. Biometrika 53, 325-338.

Green, R.H. (1979). Sampling Design and Statistical Methods for Environmental Biologists. John Wiley and Sons, New York.

Greig-Smith, P. (1983). Quantitative Plant Ecology, 3rd Edition. Blackwell Scientific Publications, London, 359 pp.

Green, R.H. 1989. Power analysis and practical strategies for environmental monitoring. Environ. Res. 50:195-205.

Hill, M.O., Bunce R.G.H. & Shaw M.V. (1975). Indicator species analysis, a divisive polythetic method of classification, and its application to survey of native pinewoods in Scotland. Journal of Ecology, 63: 597-613

Hill, M.O. (1979). TWINSPAN – a FORTRAN Program for Arranging Multivariate Data in an Ordered Two-way Table by Classification of the Individuals and Attributes. Ithaca: Section of Ecology and Systematic, Cornell University.

Hill M.O. & Gauch H.G. (1980). Detrended correspondence analysiss, an improved ordination technique. Vegetatio, 42: 47-58

Jaccard, P. (1901): Etude comparative de la distribution florale dans une portion des Alpes et du Jura. Bulletin de la Socie´ te´ Vaudoisedes SciencesNaturelles, 37: 547–579.

Jongman, R.H.G., ter Braak C.J.F. and Van Tongeren O.F.R. (1987). Data analysis in community and landscape ecology. Cambridge University Press, Cambridge, UK.

Jongman, R.H.G., Ter Braak, C.J.F. and van Tongeren O.F.R. (eds.) (1995). Data Analysis in Community and Landscape Ecology. Cambridge University Press, Cambridge.

Kent, M. (2006). Numerical classification and ordination methods in biogeography. Progress in Physical Geography, 30: 399-408

Krebs, C.J. (1989). Ecol. Method. Harper Collins, New York.

Krebs, C.J. (1999). Ecological Methodology. Harper & Row, New York.

Kutner, M.H., Nachtscheim, C.J., Wasserman, W. and Neter J. (1996). Applied linear statistical models. WCB/McGraw-Hill, New York.

Lee, Y.W. and Sampson D.B. (2000). Spatial and temporal stability of commercial ground fish assemblages off Oregon andWashington as inferred from Oregon travel logbooks. Canadian J. Fish. Aqua. Sci., 57:2443-2454.

Legendre, P. & Gallagher E.D. (2001). Ecologically meaningful transformations for ordination of species data. Oecologia, 129: 271–280

Legendre, P. and Legendre L. (1998). Numerical Ecology. 2nd Edition.Elsevier, Amsterdam.

Lepš, Jan & Šmilauer P. (1999). Multivariate Analysis of Ecological Data Faculty of Biological Sciences, University of South Bohemia Ceské Budejovice,110pp

Ludwig, J.A., Reynold, J.F. (1988). Statistical Ecology. Wiley, New York, 337pp

Niemi, G.J. and McDonald M.E. (2004). Application of ecological indicators. – Annu. Rev. Ecol. Evol. Syst. 35: 89–111.

Manly, B.F.G. (1997). Randomization, Bootstrap and Monte Carlo Methods in Biology. Chapman and Hall, London.

Mantel, N. 1967. The detection of disease clustering and generalized regression approach. Cancer Res. 27:209-220.

Mantel, N. and Valand R.S. (1970). A technique of nonparametric multivariate analysis. Biometrics 26:547–558.

McCune, B. and Mefford M.J. (1999). PCORD, Multivariate Analysis of Ecological Data, Version 4. MjM Software Design, Gleneden Beach, Oregon, USA.

McCune, B., Grace J.B. and Urban D.L. (2002). Analysis of Ecological Communities. MjM Software Design, Gleneden Beach, Oregon.

McGarigal, K., Cushman, S. and Stafford S. (2000). Multivariate Statistics for Wildlife and Ecology Research. Springer-Verlag, New York

McGeogh, M.A. (1998). The selection, testing and application of terrestrial insects as bioindicators. Biol. Rev. 73: 181–201.

McGeoch, M.A. and Chown. S.L. (1998). Scaling up the value of bioindicators. Trends Ecol. Evol. 13: 46-47.

Mielke, P.W. & Berry K.J. (2001). Permutation methods: A distance function approach. New York: Springer-Verlag.

Mielke, P. W., Berry K.J. & Johnson E. S. (1976). Multi-response permutation procedures for a priori classications. Communications in Statistics- Theory and Methods, 5: 1409-1424

Mitchell, R.J. (2001). Path analysis: Pollination. pp. 217–234 in: Scheiner S.M. and J. Gurevitch (eds.). Design and Analysis of Ecological Experiments. Oxford University Press.

Petratis, P.S., Beaupre S.J. and Dunham A.E. (2001). ANCOVA: Nonparametric and Randomization Approaches. pp. 116–133 in: Scheiner S.M. and Gurevitch (eds.). Design and Analysis of Ecological Experiments. Oxford University Press, Oxford.

Pugesek, B., Tomer A. and von Eye A. (eds.) (2002). Structural Equations Modeling: Applications in Ecological and Evolutionary Biology Research. Cambridge University Press, Cambridge, U.K.

Paukert, C.P. and Wittig T.A. (2002). Applications of multivariate statistical methods in fi sheries. Fisheries 27(9):16-22

Palmer, M.W. (1988). Fractal geometry: a tool for describing spatial patterns of plant communities. Vegetatio 75:91–102.

Odum, E.P. and Barrett G.W. (2005). Fundamentals of Ecology. Thomas Brooks/Cole, Belmont, CA.

Orloci, L. (1975). Multivariate Analysis inVegetation Research. Junk, The Hague.

Rohlf, F.J. (1995). BIOM: A Package of Statistical Programs to Accompany the Text Biometry. Exeter Software, Setauket, New York.

Scheiner, S.M. (2001). Theories, hypotheses, and statistics. pp. 3-13 in: Scheiner S.M. and J. Gurevitch (eds.). Design and Analysis of Ecological Experiments. Oxford University Press, Oxford.

Shipley, B. (2000). Cause and Correlation in Biology. Cambridge University Press, Cambridge, U.K.

Statsoft, Inc. 1995. STATISTICA for Windows. 2nd edition. Tulsa, OK.

Stiling, P. (2002). Ecology: Theories and Applications. Prentice Hall, Upper Saddle River, N.J.

Swan, J.M.A. (1970). An examination of some ordination problems by use of simulatedvegetational data. Ecology 51: 89–102.

Ter Braak C.J.F. & Looman C.W.N. (1986). Weighted averaging, logistic regression and the Gaussian response model. Vegetatio, 65: 3-11

Ter Braak, C.J.F. (1987). The analysis of vegetation-environment relationships by canonical correspondence analysis. Vegetatio, 69:69-77.

Ter Braak, C. J. F. (1994). Canonical community ordination. Part I: Basic theory and linear methods. Ecoscience 1 (2), 127-140.

Ter Braak C.J.F. & Šmilauer P. (1998). CANOCO Reference Manual and User's Guide to Canoco for Windows. Microcomputer Power, Ithaca, USA. 352 pp.

Thompson, S.K. (2002). Sampling. John Wiley & Sons, New York. Second Edition.

Turner MG. (1989). Landscape ecology: the effect of pattern on process. Ann. Rev. Ecol. Syst. 20:171-197.

Turner MG, RH Gardner and RV O'Neill (2001). Landscape Ecology in Theory and Practice: Pattern and Process. Springer, New York.

van der Meer, J. Heip C.H., Herman P.J.M., Moens T. and van Oevelen D. (2005). Measuring the Flow of Energy and Matter in Marine Benthic Animal Populations. pp. 326–408 in: Eleftheriou. A. and A. McIntyre (eds.). 2005. Methods for Study of Marine Benthos. Blackwell Science Ltd., Oxford, UK.

Van der Plogeg, S.W.F. & Vlijm L. (1978). Ecological evaluation, nature conservation and land use planning with particular reference to the methods used in the Netherlands.Biol.Consero.14:197-221.

Van Sickle, J. (1997). Using mean similarity dendrograms to evaluate classifications. Journal of Agricultural, Biological, and Environmental Statistics, 2:370-388.

Whittaker R.H. (1965). Dominance and diversity in land plant communities. Science 147: 250–260.

Whittaker, R.H. (1967). Gradient analysis of vegetation. Biol. Rev. 42:207-264.

Zare Chahouki, M. A. Azarnivand H., Jafari M. & Tavili A. (2009). Multivariate Statistical Methods as a Tool for Model_Based Prediction of Vegetation Types, Russian Journal of Ecology, 41(1): 84-94.

Zare Chahouki, M.A. (2006). Modeling the spatial distribution of plant species in arid and semi-arid rangelands. PhD Thesis in Range management, Faculty of Natural Resources, University of Tehran, 180 p. (In Persian).

Schluter. D. and Grant P.R. (1982). The distribution of *Geospiza difficilis* on Galapagos islands: test of three hypotheses. Evolution 36:1213-1226

8

Climate History and Early Peopling of Siberia

Jiří Chlachula

Laboratory for Palaeoecology, Tomas Bata University in Zlín,
Czech Republic

1. Introduction

Siberia is an extensive territory of 13.1 mil km² encompassing the northern part of Asia east of the Ural Mountains to the Pacific coast. The geographic diversity with vegetation zonality including the southern steppes and semi-deserts, vast boreal taiga forests and the northern Arctic tundra illustrates the variety of the present as well as past environments, with the most extreme seasonal temperature deviations in the World ranging from +45°C to -80°C. The major physiographical units – the continental basins of the Western Siberian Lowland, the Lena and Kolyma Basin; the southern depressions (the Kuznetsk, Minusinsk, Irkutsk and Transbaikal Basin); the Central Siberian Plateau; the mountain ranges in the South (Altai, Sayan, Baikal and Yablonovyy Range) and in the NE (Stavonoy, Verkhoyanskyy, Suntar-Hajata, Cherskego, Kolymskyy Range) constitute the relief of Siberia. The World-major rivers (the Ob, Yenisei, Lena, Kolyma River) drain the territory into the Arctic Ocean.

Siberia has major significance for understanding the evolutionary processes of past climates and climate change in the boreal and (circum-)polar regions of the Northern Hemisphere. Particularly the central continental areas in the transitional sub-Arctic zone between the northern Siberian lowlands south of the Arctic Ocean and the southern Siberian mountain system north of the Gobi Desert characterized by a strongly continental climate regime have have been in the focus of most intensive multidisciplinary Quaternary (palaeoclimate, environmental and geoarchaeological) investigations during the last decades. Siberia is also the principal area for trans-continental correlations of climate proxy records across Eurasia following the East-West and South-North geographic transects (Fig. 1). Among the terrestrial geological archives, loess (fine aeolian dust) represents the most significant source of palaeoclimatic and palaeoenvironmental data, together with Lake Baikal limnological records, with bearing for reconstruction of the past global climate history. The Siberian loess, being a part of the Eurasian loess-belt, has provided chronologically the most complete evidence of past climate change in the north-central Asia (Chlachula, 2003).

The Cenozoic neo-tectonic activity with the Pleistocene glaciations and interglacial geomorphic processes modeled the configuration of the present relief of Siberia and the adjoining Ural Mountains. During the cold Pleistocene periods, the vast extra-glacial regions of West Siberia south of the NW Arctic ice-sheet were transformed into a large periglacial super zone (Arkhipov, 1998), which became a major sedimentation area of aeolian (silty) deposits cyclically derived by winds from the continental ice-front ablation surfaces. Main palaeoenvironmental records, spanning over several hundred thousand years, have been preserved in deeply stratified sections within the major basins (Ob, Yenisei, Angara and

Lena River) exposed after progressive erosion triggered by constructions of large dams (Drozdov et al., 1990; Medvedev et al., 1990). Equally important sources of the Quaternary (geological and biotic) palaeoclimate proxy data originate from open-coalmines and other modern industrial surface disturbances (Zudin et al., 1982; Foronova, 1999).

Studies of Quaternary climates in Siberia, encompassing the last 2.5 million years, have advanced considerably in recent years, mainly because of increased awareness of the value of reconstructing geological and natural histories for understanding of the present-day ecosystems, and applied as a means of predicting the probable extent and consequences of future climatic changes. Because of the multi-factorial nature of a long-term climatic and environmental evolution, Siberian palaeoclimate-oriented studies have become increasingly interdisciplinary, integrating Quaternary geology and palaeogeography, palaeopedology, palaeontology, palaeobotany, Palaeolithic archaeology and other fields (Chlachula, 2001a-c, 2010a-b). Reconstructions of past environments in specific regions and time periods have been used to assess the effects of orbital variations on seasonal and latitudinal distribution of solar radiation and atmospheric circulation patterns, and the consequential changes in regional temperatures, precipitation and moisture balance. Some long-term models provide means of predicting future climatic evolution in the context of the global climate history and help in the assessment of the modern human factor in environmental change. Because of the pronounced climatic continentality of the territory, even minor variations in atmospheric humidity and temperature led to major transformations in local ecosystems, particularly in the open southern Siberian continental sedimentary basins and the upland depressions.

Principal information on past climates and climate change comes from the southern Siberian loess regions that have been intensively studied since the 1990's, following the initial field investigations and chronological interpretations of long-term sub-aerial sequences. The most recent studies have gradually shifted to more detailed and high-resolution Late Quaternary records and refinement of the regional loess-palaeosol chronostratigraphies, illustrating the landscape development and biota evolution particularly for the last 130 000 years by using advanced chronometric, geo-chemical and biostratigraphic marker analyses. The most complete Late Pleistocene palaeoclimate archives include four main stages (the Kazantsevo Interglacial, the Ermakovo /early Zyriansk/ Glacial, the Karginsk Interpleniglacial, and the Sartan /late Zyriansk/ Glacial) correlated with the Marine Isotope Stages (MIS) 5-2.

The broad Siberia also has the key relevance for elucidation of timing and conditions of environmental adaptation of the prehistoric and early historic people to high latitudes of Eurasia, as well as the initial colonization of the Pleistocene Beringia, including the north-western part of the American continent. The particular geographic location and the diversity in topographic configuration of regional landscape reliefs together with changing Quaternary environments governed by the past global climate change played the key role in this long and complex process. The spatial and contextual distribution of the documented archaeological sites reflects a climatic instability and a timely discontinuous inhabitability of particular geographical areas of Siberia delimited by the Central Asian mountain system in the south and the Arctic Ocean in the north. The cyclic nature of the glacial and interglacial stages led to periodic geomorphic transformations and generation of specific ecosystems adjusted to particular topographic settings and responding to acting atmospheric variations. Diversity of the present relief and environments (Fig. 1B), reflecting the past climate change, played the key role in the process of the initial peopling of the immense Siberian territory. Palaeoenvironmental databases (palynological, palaeontological as well as early cultural records) provide unique evidence of strongly fluctuating Pleistocene glacial and interglacial climates, corroborating the geological stratigraphic archives.

The human occupation of Siberia used to be traditionally associated with the Late Palaeolithic cultures. Systematic geoarchaeology investigations during the last 20 years across the entire Siberia (with the key research loci in the Tran-Ural region of West Siberia, the Altai region, the Upper Yenisei, Angara and Lena Basins, as well as at the easternmost margins of the Russian Far East in Primoriye and on the Sakhalin Island) revealed several hundred of Palaeolithic and Mesolithic sites (Serikov, 2007; Chlachula et al., 2003, 2004b; Derevianko & Markin,1999; Derevianko & Shunkov, 2009; Medvedev et al., 1990; Mochanov, 1992; Vasilevsky, 2008; Zenin, 2002). Particularly the discoveries of numerous Palaeolithic sites, some of potentially great antiquity (> 0.5 Ma), located in large-scale surface exposures (river erosions and open-pit mines) followed by systematic archaeological investigations within the major river basins of south and central Siberia between the Irtysh River in the west and the Lena River / Lake Baikal in the east (Fig. 1A), have provided overwhelming evidence of a much greater antiquity of human presence in broader Siberia and capability of early people to adjust to changing Pleistocene environments. Cultural remains are located in diverse geomorphic settings (i.e., lowland plains, mountain valleys, upland plateaus) and geological contexts (aeolian, fluvial, lacustrine, palustrine, alluvial, glacial and karstic), with the highest concentrations in the Pleistocene periglacial parkland steppe and the boreal tundra-forest foothill zone. Particularly the geographically extensive and deeply stratified loess-palaeosol sections in the southern Siberian open parklands have revealed a long and chronologically well-documented cultural sequence of human occupation. The variety of cultural finds provides witness to several principal stages of inhabitation of the Pleistocene Siberia, possibly encompassing the time interval close to 1 Ma with the earliest (Early and Middle Pleistocene stages) represented by typical "pebble tool" industries, followed by the Middle Palaeolithic complexes, including the (Neanderthal) traditions with the Levallois prepared-core stone-flaking technology, and the regionally diverse Late/Final Palaeolithic blade complexes eventually replaced by the microlithic Mesolithic cultural facies that developed in response to major natural transformations during the final Pleistocene.

A further northern geographic expansion of humans into the Arctic regions reflects a progressive cultural adaptation to extreme climatic conditions of (sub)polar Pleistocene environments (e.g., Mochanov & Fedoseeva, 1996, 2001). The occupation sites in the Polar Urals and North Siberia (Svendsen & Pavlov, 2003; Pitulko et al., 2004) provide eloquent evidence that people reached the Arctic coast already before the Last Glacial (>24 000 years ago). All these discoveries logically lead to revision of the traditional perceptions on a late peopling of northern Asia as well as the "late chronology" models of the initial human migrations across the exposed land-bridge of Beringia to the North American continent (Chlachula, 2003b). Geoarchaeology studies, particularly in the poorly explored and marginal geographic regions of northern and eastern Siberia (Pitulko et al., 2004; Vasilevsky, 2008), are of utmost importance for reconstruction of past climate change as well as the early human history in north Eurasia. Evolutionary processes in natural environments, and specific behavioral Palaeolithic adaptation patterns and material-technological conditions, as well as documentation of sequenced climatic events stored in geological records are the principal objectives of the current multidisciplinary Quaternary investigations in Siberia.

This contribution summarizes in a general overview the present evidence on the Quaternary climates and climate development in western, southern and eastern Siberia in respect to associated environmental transformation and the stages of early peopling of particular geographic areas of this extensive and fascinating territory.

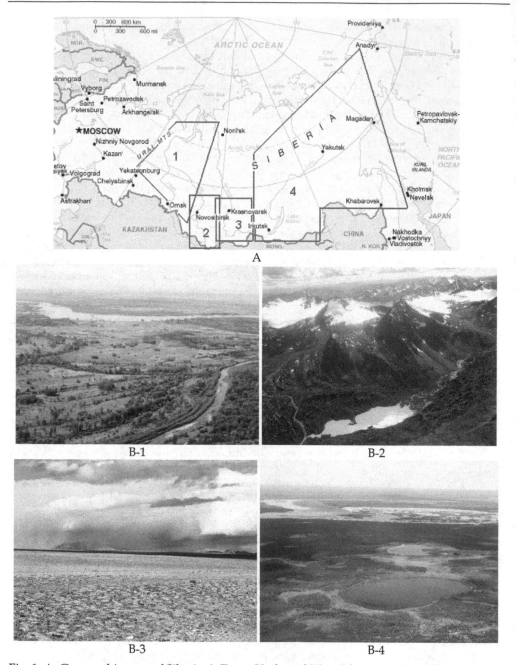

Fig. 1. A: Geographic map of Siberia. 1. Trans-Urals and West Siberia, 2. Altai, SW Siberia,
3. upper Yenisei Basin, 4. East Siberia (Lake Baikal, Lena Basin); B: Landscapes & ecosystems
1. parkland-steppe (West Siberian Lowland), 2. the Altai Mountain tundra; 3. semi-desert
(Chuya Basin, southern Altai); 4. northern taiga with thermo-karst lakes (central Yakutia).

2. Pleistocene climates and natural environments of Siberia

Because of its vast territory and the geographical isolation between the major Central Asian (Himalayan) mountain systems and the Arctic Ocean, Siberia represents a unique place that is of major potential for mapping the climatic history of northern Eurasia, but also general pathways and rates of global change in the Northern Hemisphere. Tectonic uplift in the Miocene, continuing through to the late Pliocene, initiated formation of the present relief. Breaking up the original pre-Cenozoic continental Siberian Platform significantly influenced the Pleistocene atmospheric circulation by blocking a free influx of warm and humid air masses from the southeast. The dominance of the arctic atmospheric streams gradually led to the present strongly continental climatic regime, with dry winters with little snow cover, and warm to hot summers, of mean annual temperatures -0.5 to -3°C (with the recorded minimum -71.2°C at Oymyakon in central Yakutia). Permafrost underlies most of north and central Siberia (Fig. 1B-4), while perennial mountain ground ice is locally distributed on the upland plateaus north of Mongolia. Glaciers occur above 3000 m altitude in the Altai and Sayan Mountains. The NW-SE oriented southern Siberian mountain ranges (Fig. 1B-2) form a barrier between the northern sub-arctic uplands and Arctic lowlands, and the southern steppes and rocky deserts of Mongolia and north-west China. The regional geography and relief shape the pronounced latitudinal and altitudinal vegetation zonation, with (parkland)-steppe, taiga forest, alpine tundra, polar tundra-steppe and mountain semi-deserts (Fig. 1B). Steppe chernozems with thick humic (Ah) horizons, cryogenically distorted by frost wedges, are distributed in the southern loess zone, while podzols with brown forest soils and gleyed tundra soils prevail in the mountains and the northern regions, respectively.

The pre-Quaternary (>2.5 Ma) geology of the southern and eastern Siberian areas was controlled by a series of tectonic events. Both relief and the geological structure contributed to intensive geomorphic processes in the past. The Middle and Late Palaeozoic orogenesis formed uplands separated by isolated continental depressions. Later Hercynian and Oligocene tectonics modified the original configuration of the upper Lena, Angara, Kuznetsk, Minusinsk and other basins north of the Altai-Sayan Mountains and the Northern Baikal-Yablonovyy Mountain Range. Accumulation of extensive proluvia near the mountain fronts and lacustrine /alluvial formations in the principal sedimentary basins continued throughout the Miocene and Pliocene. During the Late Pliocene and Early Pleistocene, early fluvial systems were established, accompanied by a progressive regional uplift. In response to climate change in conjunction with orogenic activity, a series of river terraces gradually developed in the climatically most favorable and first inhabited basins of southern Siberia, with the earliest, Early and Middle Pleistocene (70-90 m, 110-130 m, 150-170 m-high) terraces preserved mostly in relics and buried by 10-50 m loess deposits. The Siberian river valleys (Fig. 1A) served as main migrations corridors of both Pleistocene fauna and early humans.

During glacial periods, the northward drainage of the Ob and Yenisei Rivers was diverted by the ice barrier of the northern inland ice-sheet southwest into the Aral and Caspian Sea (Arkhipov, 1998). The relief of southern Siberia with deep intermountain depressions preconditioned formation of major river dams blocked by mountain glaciers that periodically caused high-magnitude floods (Rudoy & Baker, 1993). These large-scale catastrophic events undoubtedly had a dramatic impact on the early human habitat. The geographical position close to the geographical centre of Asia also contributed to extreme continental climatic conditions with widespread permafrost developed during cold stadials. Siberia is thus of bearing significance for mapping Pleistocene changes in Arctic air-mass circulation above the central and eastern Eurasia, as reflected in textural and compositional variations in the aeolian deposits. Among the continental sediments used as climatically significant palaeo-archives, loess (fine aeolian dust) has attracted most attention because of its high environmental sensitivity and the long-term stratigraphic records it often provides. Increased rates of loess accumulation correspond to cold and dry (glacial) stages followed by soil development during warm (interglacial and interstadial) intervals. The Siberian loess and the associated palaeoclimate proxies contribute to the trans-Eurasian palaeoclimatic correlation, linking the European, Central Asian and Chinese loess provinces (Rutter et al., 2003; Bábek et al., 2011). Ultimately, the Siberian loess-palaeosol sections, rich in fossil micro- and macro-fauna and early cultural remains, provide a contextual and chronological framework for documentation of timing, natural conditions and processes acting during the early human peopling of northern Asia and the Pleistocene Beringia.

The principal distribution zone of loess and loess-like deposits lies in the southern part of the Siberian territory (50°-60° N and 66°-104° E) west of the Ural Mountains within a broad belt 500-1500 km wide (N to S), encompassing ca. 800 000 km² between the Ob and Angara River basins along the Altai, Salair and Sayan Mountains (Fig, 2A). The Siberian loess provides a detailed, high-resolution record of climatic shifts that may not be detectable in more uniform loess-palaeosol formations elsewhere on the Eurasian continent. Thickness of loess sections is from a few meters in the Lake Baikal area to 40 m in the Yenisei River valley and up to 150 m on the Ob River (Priobie) loess plateau and the North Altai Plains. The regional loess-palaeosol record spans throughout the Quaternary, yet it may be locally fragmentary, partly re-deposited, or may be completely absent for the earlier stages. The geomorphological setting of southern Siberia with the extensive open lowlands in the west and depressions towards the east, combined with dominantly westerly (NW-SW) winds, points to location of the major deflation surfaces (loess source areas) on the eastern/steppe Altai Plains between the Irtysh and Ob Rivers with accumulations of fine aeolian dust eastwards on the plains north of Altai-Sayan Mountains. Ice-sheet marginal areas, large alluvial floodplains, exposed valley floors and margins of (glacio-) lacustrine basins were important sources of the silty sediment. Local geological sources in the southern intramontane basins (Kuznetsk, Minusinsk, Irkutsk Basin), with aeolian dust discharge from the glaciated Altai and Sayan Mountains, the Kuznetskiy Alatau and the Baikal Range, played an important role in the regional input of wind-blown sediment in the adjacent river basins. The present westerly (SW/NW) winds likely prevailed throughout the Quaternary Period with several Late Pleistocene stages of sub-aerial deflation and sediment deposition.

The most continuous and high-resolution sections of aeolian aerosol dust deposits, interbedded with variously developed fossil soils from the Priobie Loess Plateau, the North Altai Plains, the upper Yenisei and Angara River basins, produced a consistent and unique evidence of the Late Quaternary climate evolution and the corresponding landscape development in parkland-steppe of southern Siberia. The temporarily preceding Early and Middle Pleistocene loess records are less complete with regional climatostratigraphic hiatuses. Climate-indicative proxy data (magnetic susceptibility, grain-size, % $CaCO_3$, % TOC mineral color parameters) supplemented by pedological, including thin-section studies from the most continuous loess-palaeosol reference sites (i.e., Iskitim, Biysk, Kurtak and Krasnogorskoye) (Fig. 2A, 2D) show marked and cyclic climatic variations spanning the last ca. 250 000 years (250 ka) corresponding to Marine Oxygen Isotope Stages (MIS) 7-1, most pronounced during the last interglacial-glacial cycle in congruence with the Lake Baikal detrital records and the globally indicative deep-sea isotope records (Fig. 2B) (Chlachula, 2003a; Evans et al., 2003; Bábek et al., 2011; Grygar et al., 2006; Prokopenko et al., 2006).

The loess stratigraphy shows aeolian dust deposition during dry and cold stages, and soil formation during warm intervals with surface stabilization and a subsequent cryogenic distortion by frost actions followed by colluviation due to climatic warming. The massive (late) Middle Pleistocene loess accumulation in the southern periglacial tundra-steppe zone is linked with a major Pleistocene glaciation of the Altai and Sayan Mountains during the penultimate glacial (MIS 6, 170-130 ka BP). An intensified pedogenic activity indicates a rapid onset of the last interglacial (MIS 5, 130-74 ka BP) (Fig. 2D) with a strongly continental warm climate during the interglacial climate optimum (MIS 5e, 125 ka BP). The palaeo-temperature data indicate an increase of temperature during the last interglacial peak in average by ca +3°C in the southern part of Siberia and up to +6°C in the arctic regions. This was succeeded by a dramatic interglacial cooling (MIS 5d) with a temperature minimum at ca. 115 ka BP evidenced by a major cryogenic event (with deformation of the MIS 5e chernozem by 1-3 m deep frost-wedges blanked by a thin loess cover) in congruence with the Lake Baikal stratigraphic record. Permafrost remained widely distributed in east Siberia. Cooler conditions during the following interglacial sub-stages (MIS 5c and 5a) correspond to the zonal vegetation shifts, with a gradual replacement of parkland-steppe and mixed southern taiga by boreal forests, interspersed by periglacial tundra-steppe (MIS 5d and 5b). The progressing cooling climatic trend of the early Last Glacial (MIS 4, 74-59 ka BP) culminated in several hyper-arid and cold stadials (MAT by ca. 10°C under present values) interrupted by intervals of climatic amelioration with (gleyed) periglacial forest-tundra soils formation and isolated brush and tree communities *(Betula nana, Salix polaris, Pinus sibirica)*. The following cycle with relatively short warm as well as very cold climate variations characterizes the mid-last glacial interstadial stage (MIS 3, 59-24 ka BP) with cryogenically distorted chernozemic soils, pointing to the existence of a high permafrost table for most of this (Karginsk) interstadial. This was ensued by a new stage of a massive loess accumulation during the late Last Glacial (MIS 2, 24-12 ka BP) with incipient (forest/steppe)-tundra gleysols formed in response to warming oscillations prior to the present Holocene surface stabilization (MIS 1, 12-0 ka BP). A marked drop in MAT by about 8°C (to -9/-10°C) and a decline of precipitation by 250 mm is assumed for the last glacial maximum (22-19 ka BP) (Velichko, 1993). The highest loess accumulation rates recorded during the Late Pleistocene glacial stages (Ermakovo / MIS 4 and Sartan /MIS 2) on the Altai Plains and in the upper Yenisei Basin indicate the most intensive aeolian dust deposition after the glacial maxima, with the most recent interval dated to ca. 19-15 ka BP (Chlachula, 2003a; Evans et al., 2003).

The interglacial and interstadial palaeosol markers in the loess series, characterized by an intensive syndepositional pedogesis, attest to a strongly fluctuating climate evolution and the corresponding landscape development with major biotic transformations in north-central Asia during the Late Quaternary. The past continental atmospheric shifts reflected by changes in the main vegetation zones display a cyclic pattern of interglacial/interstadial parkland-steppe and mixed taiga ecosystems replaced by boreal tundra-forest and arid periglacial tundra-steppe during cold stadial stages. The northern expansion of parkland-steppes during warm periods as well as periglacial steppes undoubtedly promoted a geographical enlargement of the Palaeolithic oikumene. Cold climatic stages furthermore stimulated human biological and cultural adaptation in the process of peopling of Siberia.

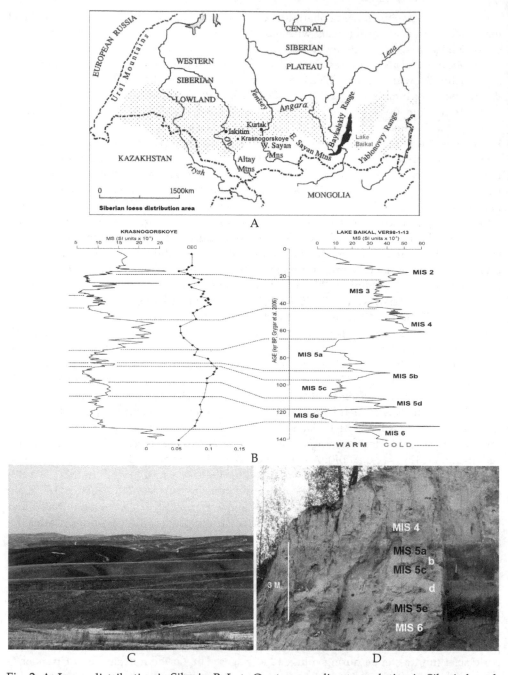

Fig. 2. A: Loess distribution in Siberia; B: Late Quaternary climate evolution in Siberia based on loess-palaeosol and Lake Baikal proxy records; C: the Altai loess landscape; D: the last interglacial (MIS 5) pedocomplex (130-74 ka BP) at Iskitim, with thee parkland chernozems.

3. Early human occupation and environmental adaptations

The vast territory of Siberia and the adjacent Trans-Ural region has principal bearing for elucidating the historical processes and environmental contexts of the Pleistocene expansion of people from the south-eastern parts of Europe into West-Central Siberia and the northern Russian Arctic areas. Interactions of past climate change and the regional relief modeling triggered by the neotectonic activity and reflected by natural transformations of ecosystems attest to the complexity of the Quaternary landscape development, ultimately affecting timing, intensity and adaptations of the earliest human occupation of north-central Asia.

The contextual geology, palaeoecology and paleontology records from the investigated archaeological sites and stratified geological sections provide evidence of pronounced (palaeo)environmental and biotic shifts triggered by the global climate evolution as well as the associated glacial and interglacial geomorphic processes. Quaternary climatic cycles regulated spatial and temporal movements of prehistoric people into the high latitudes of Eurasia. Integrated ecology multi-proxy databases document trajectories of the complex and long occupation history of this extensive, but still marginally known part of the World.

3.1 The Urals and the Trans-Ural regions of West Siberia

The Urals with the adjoining regions of the West Siberian Plain (*Zauralye* – the Trans-Urals) are of key relevance for multi-disciplinary Quaternary studies focusing on initial peopling of northern Eurasia in respect to the strategic geographical location of this mountain chain, forming the natural borderline between Europe and North Asia. Except for documenting the Pleistocene and Holocene climate evolution and related environmental transformations, the territory of the Urals has a major bearing for mapping the main time intervals of migrations of Pleistocene hominids / early humans into the high latitudes / sub-polar and polar regions of Eurasia, and reconstruction of adaptation strategies to past climate variations.

3.1.1 Geography and natural setting

The Urals, the principal mountain range separating the East European Plains and the West Siberian Lowland as the geographical milestone between Europe and Asia, extends for over 2000 km from the Arctic coast in the north to the arid steppes of western Kazakhstan in the south (Fig. 1A). It is characterized by a pronounced vegetation zonation with polar tundra-forest, taiga and open steppe being the dominant biotopes. The archaeologically most productive Central Urals (56-59°N) is one of the five physiogeographic zones of the Ural mountain range (1895 m asl.) , connecting the Southern Urals with the Northern, Sub-polar and Polar Urals (Fig. 3A). Most of the regional hilly relief (700-1500 m asl.) is transected by mountain valleys with the principal E-W oriented fluvial drainage discharge following the continental topo-gradient (Fig. 3C). The adjoining geographical area in the west *(Priuralye /* Fore-Urals) includes the loess plateau of the upper Kama River basin. The Trans-Ural region is delimited by the southern limits of the Sverdlovsk District, and the Tobol and Ob River basins from the (South-)East with the tributary valleys (Severnaya Sosva, Sosva, Tura, Tagil', Neiva, Pyshma) draining the eastern Ural foothills (Fig. 3D). Its western part is formed by small hills (300-400 m asl.) separated by shallow river valleys, lakes and bogs, extending east into the West Siberian Lowland. The present soil cover is characterized by well-developed parkland-steppe chernozems, mixed / coniferous taiga forest brunisols / gleysols and forest-tundra regosols, all with fossil analogues in the Quaternary pedostratigraphic record.

3.1.2 Palaeogeography and climate history

The topographically and biotically diverse territory of the broader Ural area attests to the complex palaeogeography and palaeoecology history reflecting large-scale landscape restructurings affected by orogenesis, and the Arctic and northern mountain glaciations (Arkhipov et al., 1986; Astakhov, 1997; 2001; Velichko, 1993; Mangerud et al., 2002), The regional Quaternary stratigraphy is based on geological, palaeontological, bio-stratigraphic and palaeomagnetic records related to the continental basin transgressions and regressions, and stages of loess deposition (Arkhipov et al., 2000; Vereshagina, 2001; Stefanovskiy, 2006). The pre-Quaternary history of the broader Urals is closely linked with formation of the main hydrological systems, particularly in the Fore-Urals area, connected to the Pechora and Caspian Basins. This process was triggered by past global climate change in conjunction with the re-activated orogenesis of the Eurasian Plate, with dynamic sea-transgressions and prolonged periods of stands and regressions, shaping the continental relief and resulting in formation of the present northern (Pechora) and southern (Kama-Volga) drainage systems. In warm climatic stages, the basins were filled by alluvial and lake deposits superimposed by periglacial alluvial formations (Yakhimovich et al., 1987). The original N-S oriented drainage of the Oligocene-Miocene basins persisted until Pliocene. A new tectonic phase contributed to a palaeo-relief restructuring with shallow river basins. The Early Pleistocene interglacial biotic records document a forest-steppe habitat with presence of arboreal taxa absent in the area today *(Fagus, Acer, Fraxinus, Ulmus, Tilia, Castanea)* and brackish settings inhabited by *Archidiscodon meridionalis, Hipparion* sp., *Coelodonta* sp. (Stefanovskiy, 2006).

Climatic cooling at the end of the Early Pleistocene (ca. 740 ka BP) is manifested by glacial deposits distributed in the Pechora and Upper Kama basins, with regressions of the Caspian and Arctic seas. The reactivated tectonics at the beginning of the Middle Pleistocene in the broader Urals-Caucasus area accelerated restructuring of the former drainage system. Increased aridity put up to an intensified weathering in the mountain regions of the Urals. A reduced fluvial activity in the eastern (Trans-Ural) West Siberian lowlands contributed to genesis of closed lake and boggy basins. Floral and faunal communities indicate dry steppe environments replacing former woodlands with the Mediterranean vegetation *(Pterocarya fraxinifolia, Rhododendron* sp., *Taxus baccata, Castanea sativa)* (Stefanovskiy, 2006). The new hydrological network of the Tobol (MIS 9) Interglacial (390-270 ka) correlates with the 60 m-terrace, with the Tiraspol Fauna Complex of large herbivorous species *(Paleoloxodont* sp., *Mammuthus primigenius* Blum., *Bison priscus gigas, Alces alces, Megaloceros gigantheus, Cervus elaphus, Equus hemionus, Equus caballus, Camelus* sp. and *Ovis amon* sp.) indicative of warm semi-arid parklands. The following Samarovo (MIS 8) Glacial (270-244 ka) correlates with the maximum expansion of the Arctic ice to ca. the present line Perm-Nizhny Tagil, blocking the northern Ob / Irtysh River drainage of the West Siberian glacial waters diverted south into the Aral-Caspian Basin. Temperate parklands characterized the following the Shirta (MIS 7) Interglacial (244-170 ka). Recession of the ice-masses is assumed for the final Middle Pleistocene Tazov (MIS 6) Glacial (160-130 ka) with mountain glaciers confined to the Northern - Polar Urals (Arkhipov et al., 1986). The late Middle Pleistocene in the south-central Ural basins is associated with 20-30 m thick fluvial deposits with fossil fauna of the Khozarian Complex *(Mammuthus primigenius, Coelodonta antiquitatis, Bison priscus longicornis, Megaloceros gigantheus* and *Equus caballus chosaricus)* indicative of a moderately cold climate. A final Middle Pleistocene neotectotic stage accentuated the regional geomorphological relief restructuring triggered by continuing uplifts of the Central and Southern Ural ranges.

The beginning of the Late Pleistocene brought up a reduced regional topographic differentiation associated with lateral planation processes in the Trans-Ural area and the initial last interglacial (MIS 5e, 130-120 ka BP) transgression. The Ural foothills were covered by a mixed spruce and pine-dominated taiga; open parklands occupied lowlands inhabited by the "Mammoth Fauna Complex" (Bolikhovskaya & Molodkov, 2006).

The Late Pleistocene stratigraphy and glacial geomorphology reflects reducing geographical limits of the continental ice-masses during the Zyriansk (MIS 4) and the Sartan (MIS 2) glacial stages. Predisposed by the territorial topography, most of the present West Siberian Lowland was inundated by a large ice-dammed lake formed between the northern Arctic ice-sheet and the southern continental water divides during the early Last Glacial (ca. 90 ka BP) (Mangerud et al., 2001). Moraines in the Pechora Basin on the NE European Plains point to the expanding Arctic glaciations from the Kara and Barents Sea ice lobes, presumably reaching their Late Pleistocene maximum extent (Mangerud et al., 1999) in corroboration with the West Siberian glacial records (Astakhov, 1997, 2001). Dynamic climatic fluctuations of the second half of the glacial stage are linked with a gradual recession of the continental ice limits in the Northern Urals and activation of intensive periglacial and gravity-slope processes resulting in massive accumulations of up to 80 m thick polygenic colluvia in the Trans-Ural foothills. An intensified loess deposition formed a 20-25 m thick cover on the Fore-Ural plateaus, overlying periglacial alluvia of the Upper Kama, reflects the prevailing dry and cold climatic conditions of periglacial tundra-steppe.

The subsequent mid-last glacial (MIS 3) interstadial warming (59-24 ka) promoted expansion of spruce- and pine-dominated taiga. The new stage of glacial advance during the late Last Glacial /Sartan (MIS 2), probably confined to the Polar and Northern Urals with localized ice-caps, led to expansion of tundra-steppe and tundra-forests in the southern lowland and mountain areas, respectively (Velichko et al., 1997). Most of north-central Siberia presumably remained ice-free (Zemtsov, 1976, Mangerud et al., 2008), opposing the model of a more extensive ice-cover in the Russian Arctic (Grosswald & Hughes, 2002). The geographic distribution of glacial landforms and glacigenic deposits provides evidence of the Arctic ice expanding from the Kara Sea basin along the western slopes of the Urals, likely merging with the piedmont glaciers (Mangerud et al., 2002). Most of the eastern slopes of the Urals likely hosted ice-free environment possibly due to a rapid disintegration of the early Last Glacial (MIS 4) ice and dry climate. Accelerated loess accumulation in the Fore-Ural area marks onset of the full last glacial conditions associated with remains of large Pleistocene fauna (*Mammuthus primigenius, Coelodonta antiquitatis, Equus* sp., *Bison priscus*) and small species typical of periglacial tundra-steppe (Yakhimovich et al., 1987).

The late Pleistocene ecosystems in the Trans-Ural region are linked to alluvial formations of the fluvial basins draining the eastern slopes of the Urals, and incorporating the typical large and small periglacial fauna of periglacial tundra-steppe (mammoth, woolly rhinoceros, horse, bison, reindeer, elk, saiga) with smaller open-steppe and semi-desert taxa such as wolf, fox, hare, marmot, as well as of rodents and other small mammals indicative of mixed polar tundra – steppe biotopes (Stefanovskiy, 2006). Permafrost limits at the end of the Pleistocene (the Allerød warming) were still located at 55°N in the Trans-Urals / West Siberia (Velichko et al., 2002). Precipitation values during the late Last Glacial and the early Holocene (15-10 ka BP) amounted to ca. 60-65% on the East European Plains and ca. 80% in West Siberia comparing to the present-day values. The present geographic distribution of particular biotopes in the broader Ural territory reflects the Holocene climate development.

3.1.3 Pleistocene environments and early human peopling

The importance of the Trans-Urals lies in its specific geographic location at the westernmost limits of Siberia adjacent to the Ural Mountains in terms of study of prehistoric migration processes, adaptation strategies and cultural exchanges between Eastern Europe and Siberia. Interaction of the Pleistocene climatic variations with the regional relief modeling by the northern Arctic glaciations, interglacial sea-transgressions, and stages of river erosion and sediment weathering attest to the complexity of the Quaternary history of the broader Ural area. During most of the Pleistocene, the North and Central Trans-Urals, sloping east from the central mountain range, were ice-free. The bordering foothills and plains of West Siberia constituted parts of the periglacial super zone (Arkhipov, 1999) and the place of mass accumulation of aeolian deposits transported from the ablation zone of the NE European-Siberian ice-sheets. The geographic and topographic configuration of the Urals predisposed a long history of human occupation in this area, forming a natural gateway for the early human infiltration further east into the vast parklands and forest lowlands of West Siberia.

The initial peopling of the territory, presumably from the East European Plain and the northern Caucasus region, was a complex process governed by changing palaeoecology conditions in response to the Pleistocene climate variations. Warm and humid Middle Pleistocene interglacial, with rich biotic resources of forest-parklands, undoubtedly enabled early human migrations further north and east along the Ural range particularly through the major river systems (Volga, Kama, Pechora) draining the East European Plains. Mixed coniferous and broad-leaved forests with *Pinus silvestris, Betula pubescens, Quercus robur, Tilia cordata, Ulmus laevis* formed a mosaic vegetation cover during the late Middle Pleistocene (MIS 9 and MIS 7) interglacials (Bolikhovskaya & Molodkov, 2006). The oldest recorded (Early Palaeolithic) sites are mostly exposed along active river banks of the Kama reservoir by erosion of loess and loessic sediments overlying relics of the Middle Pleistocene (35-60 m) river terraces. The cultural material is represented by archaic and simply flaked core-and-flake stone industries (cores, retouched flakes, scrapers and other tools, partly bifacially worked) produced on cobbles from old river alluvia. Numerous fragmented animal bones with the anthropogenic working and use attributes (flaking, splitting, retouching) (Fig. 2B) indicate productive natural occupation habitats and complex behavioral activities. The formal technological uniformity of these Palaeolithic finds and their geological contexts related to the Middle Pleistocene alluvia suggests an intensive expansion of the Old Stone Age people across the middle latitudes of Eurasia prior to the last interglacial. (>130 ka BP) (Matyushin, 1994; Velichko et al., 1997; Chlachula, 2010a). Periglacial sub-Arctic tundra-steppe and open forest-tundra correlated with the late Middle Pleistocene (MIS 8 and MIS 6) glacial advances in the north-eastern part of Europe (the Pechora Urals) covered most of the non-glaciated territories occupied by periglacial fauna *(Mammuthus primigenius, Coelodonta antiquitatis* Blum.*, Rangifer tarandus, Bison* sp.*, etc.)*. These prominent Middle Pleistocene glacial stages brought a major geographical reduction of the early human occupation areas.

The pronounced climatic warming at the beginning of the Last Interglacial (MIS 5), recorded in the loess-palaeosol sections from the East European Plain to southern Siberia (Little et al., 2002; Chlachula, 2003), may have created inundations and ground saturation of the formerly glaciated areas and thus temporarily impeded free movements of early people. Increased continentality and MAT led to expansion of parkland-steppe also facilitating new migrations of the Middle Palaeolithic (Neanderthal/early *Homo sapiens*?) communities. High summer temperature (increased by ca. 4-5°C in respect to present values) and mean annual precipitation (by ca. 100 mm) led to the northward distribution of southern taiga along both sides of the Urals by 500-700 km beyond the present limits (Velichko et al., 1992; Velichko, 1993). Pine-spruce-birch forests with oak, hornbeam, lime, hazel and elm characterized the prevailing mosaic vegetation of the warm last interglacial climate sub-stages (MIS 5e, 5c). Recession of human occupation is linked with the onset of the early Last Glacial stage (MIS 4, 74-59 ka BP) reducing the formerly inhabited areas of the periglacial steppe zone around the Southern Urals and repeatedly during the late Last Glacial stage (MIS 2, 24-12 ka BP). The most intensive late Middle / early Upper Palaeolithic geographic expansion occurred during the mid-Last Glacial interstadial interval (MIS 3, 59-24 ka BP) reaching the northernmost regions of the Urals and the adjacent territories of northern Europe and NW Siberia (Pavlov, 1986; Pavlov et al., 2001, 2004).The progressive biological and cultural adaptation enabled to retain the new lands prior and shortly after the LGM. The Central Urals forming the free geographic passage connecting the East European Plain with the vast West Siberian territories most likely played a key role in this historical evolutionary process. The last glacial peopling of north-central Urals is principally associated with the cave complex in the Chusovaya River valley cutting E-W through the Ural mountain range. The multi-layer sites (e.g., Bezymyannaya Cave, Kotel Cave, Kamen Dyrovatyy, Bolshoy Glukhoy) distributed in a karstic formation over a 200 km distance encompass a long record spanning from the Late Pleistocene (Middle Palaeolithic) till the Iron Age (Fig, 3C). Apart of the main valley, the upper Chusovaya River tributaries (Koiva, Sylvitsa, Serebryanka) likely encompassed rich biotic niches even in glacial stages. A unique cultural record from the Kamen Dyrovatyy Cave treasured over 18,000 arrowheads made of various materials (stone, bone and metal) dating since the Late Pleistocene to Holocene, with the Late Palaeolithic level, dated to 13,757±250 yr BP (Serikov 2001). Other central Ural cave sites interpreted as cultic in view to archaeological inventories (Kumyshanskaya Cave, Ignatievskaya Cave, Grot Zotinskyy, Kotel Cave) as well as those in the southern Urals (Kapova Cave) (Bader, 1965; Matyushin, 1994) suggest a high socio-cultural and behavioral complexity. In the latter mountain zone, spruce and pine taiga forests persisted until the onset of the Last Glacial, ca. 22-21 ka BP (Danukalova & Yakovlev, 2006). A certain regional cultural non-uniformity across the Urals manifested in the lithic industry composition may indicate differences in environmental adaptations. The taxonomic diversity of hunted fauna is particularly evident at other Final Palaeolithic / Mesolithic sites, including reindeer- and mammoth-dominated sites in the Trans-Ural foothills and the eastern lowlands, respectively (Serikov, 1999; 2000). A multifunctional character and a variety of particular activities are well evident in the broad repertoire of stone and bone tools, including "hacking" instruments, barbed harpoons and arrowheads. A micro-lithization and stone tool polishing in the Mesolithic complexes distributed across the Urals and West Siberia indicate an adaptive adjustment and diversification of exploitation of the final Pleistocene/early Holocene environments.

Opposite to taiga forest of the Ural mountain valleys, the eastern Trans-Urals and the adjacent West Siberian Lowlands hosted a diversity of environments and settings related to seasonally water-saturated taiga forests with braided river streams, lakes and bogs. This vast and biotically diverse territory underwent a palaeogeographical and palaeoecological evolution reflecting large-scale environmental shifts, partly affected by the Arctic as well as mountain Pleistocene glaciations. The stratigraphic position of the mapped Pleistocene sites is determined by past geomorphic processes and biotic shifts of both the lowland and foothill areas with absence of thick loess deposits present on the eastern slopes of the Urals.

Fig. 3. (Trans-)Urals. A: the Pleistocene glacial mountain topography of the Northern Urals; B: humanly flaked bones and ivory of the Middle Pleistocene fauna (the upper Kama basin); C: karstic formations of the Chusovaya River valley with multilayer prehistoric cave sites; D: open parkland taiga of the Sosva River valley, Trans-Urals of West Siberia. E: frost-wedge casts incorporating fossil wood and charcoal of the last glacial landscape diagnostic of low MAT and a periglacial setting. F: Gari Site 1, Late Palaeolithic occupation with concentration of mammoth bones indicating cultural exploitations of the last glacial Siberian ecosystem.

The Palaeolithic in the Trans-Ural regions, predating the mid-Last Glacial, is poorly known contrary to the western regions of the Urals (Pavlov et al., 1995). The archaic character of lithic implements from the local Pleistocene occupation sites must not imply antiquity and may reflects a time-transgressive character of expedient stone industries of the Palaeolithic and Mesolithic complexes pre-determined by local raw material sources (*e.g.* sandstone and siltstones at Gorbunovskiy Torfianik; quartzite tuffs at Golyy Kamen) (Serikov, 1999; 2000). Geoarchaeology studies from the easternmost limits of the Trans-Ural area provide multiple lines of evidence of Palaeolithic peopling of this geographically marginal and still poorly explored territory of West Siberia following the mid-last glacial warming (35-24 ka). A specific cultural entity represents the Late Palaeolithic complex of open-air sites in the Sosva River basin on the periphery of the West Siberian Lowland contextually associated with massive concentrations of large Pleistocene fauna remains, previously interpreted as natural "mammoth cemeteries" known since the 19th century (Fig. 3D, F). Other fossil fauna with woolly rhinoceros, bison, horse, reindeer and other animals characteristic of periglacial tundra-steppe together with small (bird and fish) species indicate a diversity of exploitation of the local Late Pleistocene natural resources and adaptation to the last glacial sub-polar environments (Chlachula & Serikov, 2011). The skeletal remains from the main occupation localities suggest both active hunting and anthropogenic scavenging practices during the Last Glacial (with radiocarbon dates of ca. 25-12 ka BP). Human behavioral activities at the key Gari I site are also indicated by circular concentrations of mammoth and rhinoceros bones (including complete sculls, large bones and tusk fragments), some of them artificially modified, reminding primitive dwelling structures known in the Late Palaeolithic from the East European Plains (Pidoplichko,1998; Iakovleva & Djindjian, 2005).The geological context of the well-preserved fossil records, sealed in cryogenically distorted alluvial deposits of the meandering Sosva River, indicates complex taphonomic histories of the occupation sites buried in cold periglacial swampy-riverine settings. The palaeontological records show an extraordinary biotic potential of the Pleistocene periglacial West Siberian steppe marginally interspersed by the eastern Ural foothill ecosystems. The dominant exploitation of mammoth (98.5%) corroborates the faunal compositions from other Late Palaeolithic sites on both sides of the Urals (Byzovaya Cave, Mamontova Kurya in the Pechora basin, Volchiya Griva in west Siberia (Pavlov, 1996; Petrin, 1986; Savel'eva, 1997, Svendsen & Pavlov, 2003).

Due to the extreme environmental conditions (the high climate-continentality, surface saturation and natural water/stream barriers), the adjacent West Siberian Lowland is assumed to have been colonized rather late only after the Last Glacial Maximum at about 17-16 ka BP (Petrin, 1986). The Trans-Urals and probably the north Siberian Arctic regions remained largely ice-free even during the LGM and except for water-saturated and frozen boggy lands there were no major natural obstacles for free movement of Palaeolithic groups further east into NW Siberia prior and after that time interval. Fossil fauna displays a major biotic potential in the central northern Trans-Urals. The latitudinal zonal differentiation of the last glacial ecosystems of the broader Urals from the Arctic coast in the north to the Black/Caspian Sea was likely more uniform, largely covered by periglacial tundra-steppe in contrast to more mosaic interglacial parkland landscapes (Markova, 2007), facilitating a free migration of fauna as well as the Late Palaeolithic peoples particularly within the foothill corridors. Regional palaeoenvironmental deviations, differing from the overall territorial bio-geographic and climatic pattern (Panova et al., 2003), may have played a significant role in the process of early prehistoric peopling of the northern regions of Eurasia.

3.2 South-west Siberia (the Altai region and the Kuznetsk Basin)

The territory of SW Siberia shows a complex Quaternary history spanning the last 2.5 million years governed by climate changes leading to the present pronounced continentality of inner Eurasia in association with a regional Cenozoic orogenic activity. Dynamics of these processes is evident in formation of the relief of the southern Siberian mountain system. It is also witnessed in preserved geological records that may have a global relevance in view to the geographic location (such as deeply stratified loess-palaeosol sections indicative to high-resolution Pleistocene and Holocene atmospheric variations in the Northern Hemisphere). Reconstruction of the past climate dynamics that shaped the configuration of SW Siberian topography and local ecosystems is essential for understanding timing and adaptation of the initial peopling of the territory of northern Asia. Geoarchaeological investigations of the Palaeolithic and Mesolithic occupation have a long tradition particularly in the Altai region. The spatial distribution of the transitional geographic areas along the margin of SW Siberia believed to be the main gateway for early human migrations from the southern regions of Central Asia into Siberia is of utmost significance of multidisciplinary Quaternary studies.

3.2.1 Geography and natural setting

The territory of SW Siberia, including the Altai and the Kuzbass regions, is characterized by a rather diverse physiography with high mountain massifs in the south and east, and open lowlands in the north and west (Fig. 1A). The Altai Mountains (maximum elevation 4506 m asl.) and the Kuznetskiy Alatau (2171 m asl.) form a natural barrier from the South and East, respectively, connected by the Gornaya Shoria Mountains (1560 m asl.). The Salair Range (590 m) separates the adjacent continental depression in the north from the upper Ob Basin and the Kuznetsk Basin (150-300 m), representing the marginal parts of the Western Siberian Lowland. Hydrologically, the area belongs to the Ob River drainage system with the Katun', Biya, Irtysh and Tom' River being the main tributaries. The lower relief zone (>1200 m asl.) includes more then 50% of the mountain area and constitutes relics of old denudation surfaces covered by more recent Pleistocene deposits derived during past glaciations. The present climate is strongly continental with major seasonal temperature deviations between the northern lowlands and the southern mountains (Fig. 1B). In winter, climatic conditions in the montane zone are generally less severe than in the open northern steppes, and micro-climate prevails throughout the year in some protected locations in the Altai and West Sayan central valleys. Annual temperatures as well as the precipitation rate vary greatly according to particular topographic settings. Most of the precipitation falls on the W/NW slopes in the northern and central Altai, whereas the southern regions are more arid and increasingly continental with average July temperature +25°C. The average January temperature -33°C may drop to -60°C. Most of the area is underlain by perennial permafrost with the active thaw layer only 30-70 m thick. Vegetation is characterized by open steppe-parkland dominated by birch and pine. Mixed southern taiga (larch, spruce, pine, Siberian pine, fir and birch) is replaced by the alpine vegetation (with pine, larch and dwarf birch) at higher elevations. Semi-desert communities with admixture of taxa characteristic of the Mongolian steppes are established in upland depressions and on plateaus of the southern Altai. The present soil cover corresponds with the zonal vegetation distribution, with steppe chernozems in the northern lowland basins and plains, luvisolic soils at humid valley locations, brunisols and podzolic soils in the lower taiga forest zone, tundra regosols in the (sub)-alpine zone; and calcareous soils and solonets in the semi-desertic mountain basins.

3.2.2 Palaeogeography and climate history of SW Siberia

3.2.2.1 Southern mountain areas

Geological history of the broader territory of SW Siberia is linked with the formation of the southern Siberian mountain system initiated by the Miocene uplift of the Transbaikal region and reaching the Sayan-Altai area during the late Pliocene (ca. 3 Ma). This major orogenetic period that continued until the early Middle Pleistocene constructed a system of mountain ranges separated by deep depressions filled by large lakes (Fig. 1B-2). Following warm climatic conditions with landscape stability encompassing most of the Pliocene (5.3-2.5 Ma), the beginning of the Pleistocene period brought a major modification of the former relief as a result of dramatic climatic changes with progressive intra-continental cooling and aridity. The Pleistocene glaciations caused a regional topographic restructuring with intensive erosion in the glaciated alpine zones and accumulation of (pro)glacial, alluvial, proluvial, lacustrine and aeolian deposits in the intramontane depressions. Little is known about the earliest glacial events of the Altai, with evidence largely obliterated by erosional processes of subsequent glaciations. The earliest Quaternary (Early Pleistocene) non-glacial records are documented by gravelly alluvia and lacustrine sands in the Teleckoye Lake formation with pollen indicating a mixed taiga forest with warm broadleaf arboreal communities. A high-mountain forest-steppe stretched over the major southern Altai intermountain basins (the Chuya and Kuray Depressions) witnessed by pollen records with *Ephedra* and a large fossil fauna represented by *Hipparion* sp., *Coelodonta* sp., Elephantidae, Bovidae (Deviatkin, 1965).

The Middle Pleistocene glacial periods are evidenced by two glacial moraines correlated with the Kuyuss Glaciation and buried in the middle Katun' River basin. The pollen data from the lower till bear witness of a cold steppe environment (Markin, 1996). Warm interglacial conditions during the Middle Pleistocene from the Anui River valley, NW Altai, are evidenced by mixed pollen of pine, birch, oak, lime, maple and other arboreal taxa (Derevianko et al., 1992c). An analogous climatic development in the northern lowlands is documented in the Kuzbass Basin with a series of basal red soils indicative of warm and relatively humid forest-steppe followed by brown soils of cold tundra steppe (Zudin et al., 1982). Tectonic movements during the late Middle Pleistocene (0.3-0.2 Ma) initiated uplift of the mountain ranges to about the 1500-2000 m elevation, triggering intensive denudation processes of former unconsolidated deposits. The renewed orogenic activity caused a restructuring of the regional drainage system, with a dominant N-S oriented river-flow direction and a further deepening of the mountain river valleys by up to 100 m in the marginal areas (Anui, Charysh) and > 200 m in the central Katun River valley.

The regional climatic variations are documented by thick alluvial-proluvial sandy gravels of the Tobol (MIS 9) Interglacial separated by glacial till deposits (MIS 8) from the following Shirta (MIS 7) Interglacial (Markin, 1996). The warm periods are characterized by expansion of mixed taiga forests, and brown forest and forest-steppe soils in the Kuznetsk Basin (Zudin et al., 1982). The final Middle Pleistocene (MIS 6) glacial stage (Eshtykholskoye Glaciation) is believed to be the most extensive Pleistocene glacial event in the Altai and West Sayan area (Deviatkin, 1981) corroborated by the evidence from the Eastern Sayan Mountains (Nemchinov et al., 1999). The alpine glaciers presumably formed extensive coalescent ice fields supporting large glacial lakes in the central basins (Fig. 4A). Massive proglacial, ice-marginal deposits accumulated in the Chuya Basin with up to 50 m thick alluvial fans along the eastern flanks at the Kuray Range (4000 m asl.). In the foothills and the adjacent continental depressions, this climatic cooling is linked with appearance of forest-tundra soils (Zykina, 1999). Cold and dry climatic conditions in the extra-glacial areas are indicated by accumulation of loess, incorporating the typical loessic mollusc fauna (*Succinea oblonga, Pupilla muscorum, Vallonia tenuilabris, Columella columella, Vertigo alpestris*) and rodent species characteristic of tundra-steppe. An extensive geographic expansion of southern taiga forests dominated by spruce and the Siberian pine into the former steppe and upland region characterizes the Last Interglacial (MIS 5) (Shunkov & Agadzhanyan, 2000).

During the Late Pleistocene, the Altai, as well as other southern Siberian mountain ranges experienced two glaciation events (MIS 4 and 2) represented by two moraines and tills separated by non-glacial interstadial deposits. The first (Ermakovo/Chibit) glaciation followed the same topographic ice-expansion pattern as the previous Middle Pleistocene glaciation (MIS 6). Large glaciofluvial ice-marginal lakes repeatedly filled the upper reaches of the Chuya and Kuray Basins, with icebergs released from the surging glaciers (Deviatkin, 1965). Accumulation of coarse and massive glacigenic sediments and formation of erosional surfaces in the lower reaches at relative elevations of 200 m, 80-100 m, 50 m, 30-40 m, 4-6 m and 1-1.5 m followed the glacial lake drainage cycles (Fig. 4B), indicating periodicity of these processes during the two glacial events (Rudoy et al., 2000). Pollen from the lower Katun River terraces dominated by dwarf birch *(Betula nana)* documents very cold conditions with tundra-steppe expanded within the northern lowlands and depressions of SW Siberia.

The mid-last glacial (MIS 3) warming (59-24 ka BP) is associated with accumulation of gravelly alluvial sediment facies in the mountains, resulting from the former ice ablation, and river sands and lacustrine clays in the upper Ob and the Kuznetsk Basins (the Bachatsk Formation) (Markin, 1996; Zudin et al., 1982). Geological records display a wide range of environments and climates. Pollen from the 30 m Katun River terrace shows expansion of taiga forests with appearance of warm broadleaf flora *(Quercus, Tilia, Ulmus, Juglans)* whose distribution range is well south beyond the present Altai limits. A gradual transformation to forest steppe and tundra steppe characterizes the later part of this stage (the Ust'-Karakol section), reflecting cooling of the Konoshelskoye Stadial (33-30/29 ka BP). Analogous climatic trends are indicated by pollen records from the pre-Altai Plain.

The second (MIS 2) Late Pleistocene (Sartan/Akkem) glaciation (24-17 ka BP) was less extensive than the preceding two (MIS 6 and 4) glaciations. In the Altai, it is associated with terminal moraines at the >2000 m elevations. Glacial basins formed in the intramontane depressions of the Biya, Chulyshman, Bashkaus, Katun, Chuya and Argut River basins with the synchronous glacier expansion followed by cataclysmic flooding (Butvilovskiy 1985, Okishev, 1982; Baker et al., 1993; Chlachula, 2001a, 2011b), with the latest around 13 ka BP. Large alluvial fans overlying or laterally merging with the highest alluvial terraces attest to dramatic geomorphic processes during the Last Glacial. Periglacial conditions with cold-adapted fauna prevailed in the northern extra-glacial zone. Pollen data illustrate distribution of cold periglacial steppe in the Altai foothills and on the adjacent plains with isolated tree (birch, pine, spruce, willow) communities in more humid settings (Zudin et al., 1982). The broken mountain relief differentiated the regional climatic pattern during the Last Glacial, with microclimate conditions in the protected locations along the northern Altai foothills and the central Katun basin allowing survival of warm Pleistocene flora and other biota until the Holocene. Large periglacial stony polygons on the Plateau Ukok, formed as a result of permafrost dynamics, attest to very severe post-glacial climatic variations. Dramatically wasting mountain glaciers in the southern Altai (the Tabon Bogdo Ula Range, 4120 m) are linked with a progressing global warming despite very low MAT (-10°C) (Rudoy et al. 2000).

3.2.2.2 Northern lowland areas

The Pleistocene climatic evolution in the lowland extra-glacial areas of SW Siberia and the continental depressions north of the Altai-West Sayans are characterized by accumulation of aeolian (silty and sandy) deposits during glacial periods and surface stabilization with soil formation during interglacial periods (Arkhipov et al., 2000). Particularly the loess-palaeosol formations widely distributed in southern Siberia provide the most detailed multi-proxy chronostratigraphic correlation of past climatic cycles (Chlachula, 2003; Bábek et al., 2011).

In the Kuznetsk Depression, being a major tectonic basin of SW Siberia bordered by the Salair Range and the Kuznetskiy Alatau Mountains, the Quaternary climate evolution is evidenced by the unique 30-60 m thick stratigraphic sequence of loess and lacustrine deposits intercalated by up to 16 Early, Middle and Late Pleistocene palaeosols above the Pliocene red soils (Zudin et al., 1982). The contextually associated fossil fauna, illustrates diversity of the local ecosystems (open steppe and parkland-forest environments) in terms of the regional climatic evolution and the geographic relief configuration (Foronova, 1999). Intensive aeolian dynamics over the south-west Siberian parkland-steppe, being a part of the Eurasian loess belt, is eloquently documented by up to 150 m high loess sections on the Priobie (Ob River) Loess Plateau, with a series of prominent interglacial pedocomplexes of chernozemic steppe soils and luvisolic / brown forest soils, spanning from the Early to Late Pleistocene. The stages of aeolian reactivation of silty dust derived from up to 100 m deep deflation surfaces in the lowland areas of SW Siberia (Zykina, 1999) correlate with the main glacial cycles in the mountain regions. Because of earlier stratigraphic hiatuses, best-documented are the Late Quaternary palaeoclimate loess archives spanning over the last ca. 300 000 years.

The prominent last interglacial (MIS 5) pedocomplex (Fig. 2D), with three parkland-steppe chernozems disturbed by cryogenesis under humid and cold conditions, shows complexity of the interglacial climate evolution (130-74 ka BP) (Chlachula & Little, 2011). The MAT during the interglacial climatic optimum (MIS 5e, 125 ka BP) in SW Siberia was by 1-3°C higher than at present, with about a 100 mm increase in annual precipitation. Environmental conditions were broadly similar to the present ones during the following interstadials. Increased precipitation and annual temperature contributed to a northward expansion of southern taiga forests by about 500-700 km beyond their present distribution limits due to raised summer temperature by 4-5°C relative to the present values (Velichko et al., 1992). Climate during the mid-last glacial (Karginsk/MIS 3) interstadial optimum was likely warmer than at the present as witnessed by chernozemic soils dated to 35-31 ka BP. The overlying podzolic forest soils (26-22 ka BP) show a gradual cooling towards the early last glacial stage (MIS 2) succeeded by a cold periglacial tundra-steppe with humic gleysols analogous to the modern soils of central Yakutia (with January temperatures below -27°C). Following the LGM, pronounced aridization at the western margin of southern Siberia correlates with loess accumulation (19-14 ka BP) and the time-equivalent formation of large, ice-marginal glaciolacustrine basins in the Irtysh and Ob River valleys (Arkhipov, 1980).

In sum, both glacial and sub-aerial formations provide evidence of cyclic nature of the Pleistocene climates of south-western Siberia with pronounced shifts in past ecosystems. The Early Pleistocene interglacials, characterized by a higher heat balance and atmospheric humidity, promoted distribution of meadow-forests and mixed parklands. The progressing climatic continentality during the Middle and particularly Late Pleistocene led to establishment of the present-type forest-steppe vegetation zones during interglacial periods and periglacial steppe during cold stages correlated with the Altai mountain glaciations.

3.2.3 Pleistocene environments and early human peopling

The Quaternary climatic changes and transformations in natural environments in south-western Siberia are well manifested by geological, biotic, but also the early cultural records. The past glacial dynamics, controlled by the regional atmospheric air-mass circulation flows in conjunction with the neoteotectonic activity and river erosion, shaped the relief to form biotically productive river valleys and depressions. The cyclic nature of the glacial and interglacial stages led to a restructuring of landscapes and generation of specific ecosystems adjusted to particular topographic settings and responding to ongoing climatic variations.

The spatial distribution of the Palaeolithic sites on the territory of SW Siberia shows a location of most sites within the transitional zone between the southern mountain ranges and the northern lowlands. These geomorphological zones, about 75-150 km wide, form a topographic relief belt of the 300-1000 m altitude, narrowing or expanding in respect to the particular physiographic conditions and the configuration of the relief (Baryshnikov, 1992). Particularly the Altai region shows the high density of Pleistocene occupation localities. This pattern of the (palaeo)geographic site location, reflecting specific environmental adaptation strategies to local settings, applies for both open-air localities buried in alluvial, colluvial or aeolian deposits, as well as cave sites concentrated in the NW Altai (the Anui valley) (Fig. 4E). Formation of the latter relates to a progressive down-cutting by fluvial erosion through the Devonian-Carboniferous limestone bedrock during the late Middle Pleistocene.

The periodic Palaeolithic migrations in the broader SW Siberia were principally dependent of the Pleistocene climatic development. Intensified orogenic uplifts, triggering large-scale erosions in the river valleys, re-shaped natural occupation habitats. Relatively stable environmental conditions seem to have persisted in the central and northern Altai due to increased regional precipitation and a tempering atmospheric effect of the mountain ranges. Limestone caves excavated by a fluvial bedrock erosion provided shelters for a more permanent human inhabitation of the Altai area, particularly in the northern foothills and the central intramontane depressions (Derevianko & Markin, 1992) characterized by warmer microclimate conditions. Cataclysmic drainage of glacier waters, periodically dammed by the mountain glaciers during glacial stages, had temporarily a dramatic impact to the local Pleistocene ecosystems. Enormous erosion processes associated with these major events significantly reduced the visibility potential of occupation sites in the flooded areas, with localities preserved only at high topographic elevations above the glacial basin waterlines.

The initial peopling of the Altai - West Sayan region likely occurred in some of the early Middle Pleistocene interglacials in the process of the northern expansion of warm biotic communities. Mixed coniferous and broadleaf forests became established in the tectonically active mountain areas with maximum elevations about 1500-2000 m. Parklands covered most of the adjacent plains of West Siberia with continental depressions filled by lakes and drained by meandering rivers. Rudimentary core and flake ("pebble tool") stone collections scattered on high river terraces and the former lakeshore margins of the present arid basins (Kuznetsk and Zaisan Basin) attest to several stages of early human inhabitation and a relative environmental stability (Chlachula 2001a, 2010b). There is limited evidence on persistence of the Early Palaeolithic occupation during glacial stages on the territory of SW Siberia, although some intermittent semi-continuity in the southernmost areas is assumed in view to finds of weathered lithic artifacts from the old periglacial alluvia in association with cold-adapted megafauna. Mastering the technique of fire making was clearly the main precondition for early human survival in cold tundra-steppe and tundra-forest habitats.

Fig. 4. Altai, SW Siberia. A: former glacial valleys of central Altai; B: glacio-fluvial terraces (up to 200 m high) in the Chuya River basin formed by cataclysmic releases of glacial lakes dammed by valley glaciers; C: thick loess deposits along the Biya River, North Altai Plains; D: colluviated redsoils in NW Altai attesting to warm Middle Pleistocene climate conditions; E: the Anui River valley, NW Altai, with the Middle-Late Palaeolithic occupation cave sites; F: Denisova Cave, Anui valley. Mousterian (Neanderthal) tools of the Levallois tradition.

The last interglacial warming (130 ka BP), associated with the re-colonization of southern Siberia by coniferous taiga forests, is linked with the appearance of the Mousterian tradition. Changes in the relief configuration influenced a regional climate regime and opened new ranges of habitats for the Middle Palaeolithic population concentrated in the transitional zones (500-1000 m asl) in the karstic area of the NW Altai foothills (e.g., Ust'-Kanskaya, Strashnaya, Denisova, Okladnikova, Kaminnaya Caves) as well as at open-air sites in the central valleys (e.g., Kara-Bom, Ust'-Karakol, Tyumechin I and II) (Derevianko & Markin, 1999). The Middle Palaeolithic horizon, encompassing a time span of up to 140 000 years (180-40 ka BP), represents a marked cultural phenomenon in the Altai. Isolated teeth (2) from the Denisova Cave dated to MIS 5 and identified as *Homo neaderthalensis* support the model of biological evolution of pre-modern humans in Siberia (Derevianko & Shunkov, 2009). Major cooling during the early Last Glacial (MIS 4) led to establishment of full glacial conditions in the central and southern Altai, and a zonal geographic replacement of boreal forest by periglacial tundra-forest in the northern Altai and by arid tundra-steppe in the adjacent lowlands (the Ob River basin and the Kuznetsk Basin). Accentuated moderate climate fluctuations between cold stadials are evidenced by embryonic regosolic soils in the loess formations on the North Altai Plains and sparse cultural records in protected locations in the Altai foothills. Human occupation of the central and southern Altai during the early Last Glacial was impeded by harsh, ice-marginal environments and expansion of glaciers in the upper reaches of the Katun - Chuya valleys subsequently filled by large proglacial lakes.Progressive warming during the early mid-last glacial interstadial interval (MIS 3, 59-35 ka BP) caused a dramatic wasting of the ice fields accompanied by cataclysmic releases of ice-dammed lakes and large-scale mass-flow and slope erosional processes. The former valley glaciers either completely disappeared or receded to the highest elevations where they persisted as corie glaciers. The periodic outbursts of the glacial basins had a dramatic impact on the regional ecosystems, but also obliterating most of the earlier cultural records. Mixed forests dominated by birch, pine, spruce and fir invaded the former periglacial and ice-marginal landscape. The presence of broadleaf arboreal taxa (oak, lime, chestnut, maple) indicates a climate warmer than at the present time. Increased humidity and cooling during the later stage of the mid-last glacial interstadial interval (35-24 ka BP) initiated mass gravity slope processes and cryogenic deformations related to the Konoshelskoye Stadial (33-30 ka BP) followed by warmer oscillations with formation of podzolic forest gleysols, Appearance of the transitional early Late Palaeolithic cultural facies reflects human adaptation to mosaic interstadial habitats, including sub-alpine taiga, dark coniferous forest, mixed parklands and steppe with mixed non-analogue biotic communities. The identical geographical distribution of the Middle - Late Palaeolithic sites and the time-transgressive lithic technologies suggest a regional cultural (and biological?) continuity in the Altai area during the Late Pleistocene (Derevianko, 2010). A phalanx fragment from Denisova Cave dated to 40 000 years and interpreted on basis of DNA as an extinct human species (Dalton, 2010) reinforces this scenario. Re-establishment of cold tundra-steppe habitats correlates with dispersal of the developed Late Palaeolithic with blade-flaking techniques in stone tool production and associated with a periglacial megafauna that possibly survived in protected and milder microclimate locations in the northern Altai throughout the LGM (20-18 ka BP). Emergence of the microlithic stone tool assemblages with wedge-shape cores is linked with a new cultural adjustment in the final stage of the Palaeolithic development responding to natural transformations of the former periglacial ecosystems towards the end of Pleistocene.

3.3 South-central Siberia (the Yenisei Basin)

The upper Yenisei basin in the southern part of the Krasnoyarsk Region, south-central Siberia, is rather unique in view to its location near the geographical centre of Asia (Fig. 1A). It has been together with the upper Angara basin in the Irkutsk Region the key area of Quaternary stratigraphic studies mapping the Pleistocene climate evolution. The extensive loess cover with a suite of buried palaeosols from the Northern Minusinsk Basin, being a continuation of the southern Siberian loess belt, have provided the most complete, high-resolution Late Quaternary palaeoclimate record in the north-central Eurasia (Chlachula et al., 1997, 1998). Coupled with the pollen evidence, the continuous loess-palaeosol sequences document periglacial steppe-tundra established during cold stadial intervals, replaced by boreal forest and parkland steppe during the warm interstadials and interglacial stages. The associated Pleistocene landscape transformations, governed by marked climatic variations, are reflected by large diversity of fossil faunal species, including non-analogue communities to modern biota. Geoarchaeological investigations in the stratigraphic sections in the upper reaches of the Yenisei River basin, initiated after progressive erosion of the unconsolidated aeolian deposits, revealed a rich series of Early, Middle and Late Palaeolithic stone tools (industries) associated with an abundant Pleistocene fossil fauna (Drozdov et al., 1990, 1999; Chlachula, 2001b) (Fig.5). The main focus of the current studies is reconstruction of the Pleistocene ecology and chronology of the early human occupation of south-central Siberia. The unique loess-palaeosol records have a fundamental bearing for better understanding the past global climate and environmental history well beyond the limits of north-central Asia.

3.3.1 Geography and natural setting

The south-central Siberia, bordered by the western Mongolia from the South and the Central Siberian Plateau along the eastern margin of the West Siberian Lowland in the North, is a geographically extensive, and topographically and biotically diverse territory encompassing lowlands, high ranges of the Sayan Mountains and the transitional foothill areas transected by river valleys largely draining water discharge from the southern mountain massive. The broader southern Siberian continental basin is built by a system of regional tectonic depressions structured by the Hercynian orogenesis related to the formation of the central Asian mountain system. The upper Yenisei area, including the Northern and Southern Minusinsk Depression and the adjacent slopes of the Eastern and Western Sayan Mountains, lies in the southern Siberian loess zone along the upper reaches of the Yenisei River (52-56°N and 89-94°E). The central part of the Minusinsk Basin is structurally defined by a zone of tectonic breaks running in the north-south direction across the Batenevskiy Range (900 m asl.) in the north. From the east, it is bordered by ridges of the Eastern Sayan Mountains (1778 m), from the west by the Kuznetskiy Alatau (2178 m) and from the south by foothills of the Western Sayan Mountains (2735 m). In the northwest, the Yenisei basin broadens and joins a system of palaeo-valleys merging with the Nazarovskaya Depression of the West Siberian Lowland. The present continental climate (with MAT -0.4°C), with cold and dry winters with little snow cover and warm to hot summers, corresponds to the particular geographical location of the territory. Open steppe stretches over most of southern Central Siberia, with isolated mosaic mixed parkland forest-steppe communities dominated by birch and pine. A chernozemic soil cover underlies parklands, whereas brown forest and podzolic soils are developed under mixed and coniferous taiga forests, respectively. Sandy semi-deserts with saline lakes such as in Khakasia illustrate the marked regional climate diversity.

3.3.2 Palaeogeography and climate history of south-central Siberia

Pre-Quaternary geology of the broader southern Siberian territory was controlled by a series of tectonic events. Both relief and the geological structure encouraged intensive geomorphic processes in the past. The Cambrian and Proterozoic volcanism disturbed and subsequently dislocated the south Siberian pre-Cambrian crust composed of igneous and metamorphic rocks. The Middle and Late Palaeozoic orogenesis formed a system of high mountain ranges separated by deep intra-continental depressions. Later Hercynian and Oligocene tectonics in southern Siberia modified the original configuration of the Angara, Kuznetsk, Minusinsk and other basins north of the Altai, Salair and Sayan Mountains, which were filled by Devonian, Carboniferous, Jurassic and Palaeogene volcanic, lacustrine and alluvial deposits. Accumulation of extensive proluvia near the mountain fronts and lacustrine / alluvial formations in the main sedimentary basins continued throughout the Miocene and Pliocene. During the Late Pliocene and Early Pleistocene (ca. 3-1 Ma), an early fluvial system of the palaeo-Yenisei was established with relics old river terraces about 200 m above the present river level. A neotectonic movement during the early Middle Pleistocene divided the Minusinsk Basin into the northern and the southern parts. In response to climatic change in conjunction with the orogenic activity, a system of terraces gradually developed with the earlier, Early and Middle Pleistocene (70-90 m, 110-130 m, 150-170 m) preserved mostly in relics and buried by 10-50 m of aeolian loess as well as colluviated loessic deposits. The Late Pleistocene terraces (8-12 m, 15-20 m, 30-45 m) formed above river plains (Chlachula, 1999).

During glacial periods, the northward drainage of the Yenisei and the Ob River was diverted by the ice barrier of the northern inland (Arctic) ice-sheet to the southwest into the Aral and Caspian Sea (Arkhipov, 1998). The regional (palaeo)relief of southern Siberia with intermountain depressions preconditioned formation of large river dams blocked by glaciers periodically causing high-magnitude floods. These dramatic events resulted in an intensive syndepositional accumulation of alluvial and alluvio-lacustrine sediments in the river valleys adjacent to the Altai and Sayan Mountains (Rudoy & Baker, 1993; Yamskikh, 1996). Loess and loessic deposits, in average 10-20 m thick and derived from local alluvial plains during glacial periods, are distributed over most of the landscape, but mainly on the western, lee-side slopes where they reach up to 40 m (Chlachula, 2001b) (Fig. 5A). During warm interglacial and interstadial stages, the formation of alluvia as well as the aeolian sediment accumulation was interrupted by palaeo-surface stabilisation with formation of chernozemic and brunisolic palaeosols (Chlachula et al., 2004a). The geographical location of the Yenisei area close to the geographical centre of Eurasia contributed to extreme continental climatic conditions with widespread permafrost that developed during cold stadials. Permafrost was not preserved during the Last Interglacial (130-74 ka BP) in the southern part of Siberia, whereas it remained widely distributed farther north and east.

Detailed stratigraphic records of the last two glacial-interglacial cycles have been documented in the Kurtak area in the Northern Minusinsk Basin (Drozdov et al., 1990). Erosion of the Quaternary sedimentary cover overlying old terraces and igneous bedrock exposed the most complete loess - palaeosol sequences with a chronostratigraphic archive encompassing ca. the last 300 000 years. The unique geological record indicates pronounced and cyclic climatic variations during the Late Quaternary, displaying consistent correlation regularities with the deep-sea oxygen isotope records (Marine Oxygen Isotope Stages / MIS 7-1) also reflected by the corresponding patterned changes in the natural environments. Magnetic susceptibility of the fine aeolian dust has proven to be a very sensitive and useful proxy indicator of past climate change in this part of Siberia (Chlachula et al. 1997, 1998). A prolonged period of a relative environmental stability during the late Middle Pleistocene is manifested by a basal prominent chernozemic palaeosol (Kurtak 29 type-section) correlated with MIS 7a. The following cold period (MIS 6) experienced significant warm-cold climate fluctuations, expressed by an intensively colluviated loess unit traced across southern Siberia between two main cold and arid intervals defined by aeolian dust sedimentation. The last interglacial /(MIS 5) *sensu lato* (130-74 ka BP) is evidenced by a new period of landscape stabilization with intensified soil formation processes disrupted by intervals of loess deposition and periglacial surface deformation (Fig. 5B). The loess-palaeosol sequence attests to several relatively short, warm as well as very cold intervals (MIS 5e-5a), with a gradual shift from a strongly continental climate during the first half of the interglacial (130-100 ka BP) to cooler and more humid conditions. Reactivated loess sedimentation reflects a progressive cooling during the early stage of the last glacial (MIS 4), culminating in several cold and hyperarid stadials. Climatic amelioration during the early mid-last glacial (Karginsk) interval (MIS 3) is evidenced by the appearance of (gleyed) brunisolic soils. The interpleniglacial climatic optimum associated with the well-developed duplicate chernozem dated to ca. 31 ka BP was likely as warm, but more humid than the present (Holocene) interglacial. The secondary cryoturbation of the fossil chernozem shows a major drop of temperature and formation of underground permafrost. Gradual cooling leading to formation of gleyed brunisolic and later to regosolic horizons found in the upper Yenisei loess sections indicates that the transition towards the last glacial stage (MIS 2) was less dramatic than during the previous glacial interval (Chlachula, 1999). A further drop of annual temperatures associated with an intensive loess deposition marks the time around the Last Glacial Maximum / LGM (20 000 – 18 000 years BP). A progressive warming with several climatic oscillations expressed by an initial pedogenesis characterizes the Final Pleistocene climate development, eventually leading to establishment of the present (MIS 1) interglacial conditions. Frost wedges, filled with humus-rich sediments in the present chernozem, reflect cold winters with little snow cover and seasonally frozen ground.Pleistocene faunal remains are abundant in the larger southern Siberia and provide a significant line of proxy evidence on past natural habitats and biodiversity composition related to climate change. The most prolific and taxonomically rich large fauna collections originate from the Kuznetsk and Yenisei Basins (Fig. 5D). Taxonomically, more than 32 large fossil fauna species have been documented, with the evolutionary (Middle-Late Pleistocene) *Mammuthus/Elephas* lineage being the best-represented. Most frequent megafauna species include woolly rhinoceros *(Coelodonta antiquitatis* Blum.), mammoth *(Mammuthus chosaricus),* horse (*Equus hemionus* Pall. and *Equus mosbachensis - germanicus*), bison (*Bison priscus* Boj.), roe deer *Capreolus capreolus* L.), giant deer (*Megaloceros giganteus* Blum.), reindeer (*Rangifer tarandus*), saiga (*Saiga tatarica)* and sheep *(Ovis* sp.) among other animals (Foronova, 1999). Malacofaunal (fossil mollusc) assemblages represent one of most abundant and indicative biotic records in the loess-palaeosols formations with both interglacial and glacial taxa.

In sum, the diversity of the Pleistocene faunal communities in the Minusinsk and Kuzbass Basins documents particular evolutionary stages reflecting regional environmental transformations and zonal biotic shifts triggered by climate change during Late Quaternary. The presence of cold-adapted species in the fossiliferous sedimentary beds indicates a very high biotic potential of periglacial steppe that expanded during the stadial intervals and was replaced by parklands and mixed forests during the warmer interglacial/interstadial stages.

3.3.3 Pleistocene environments and early human peopling

Until the 1980s, Siberia was generally believed to have been colonized by the Late Palaeolithic people (Tseitlin, 1979). Systematic geoarchaeological investigations particularly in the Yenisei valley revealed much earlier traces of human peopling. Progressing erosion of the Krasnoyarsk Lake have exposed the nearly complete stratigraphic Late Quaternary geological stratigraphic record, enclosing a series of various Palaeolithic stone industries accompanied by a rich Middle and Late Pleistocene fauna (Drozdov et al., 1999). Abundant archaeological finds represented by simply flaked, but diagnostic stone artifacts originate, together with the taxonomically diverse fossil fauna remains, from basal alluvial gravels of the 60-70 m terrace beneath 20-40 m thick loess deposits (Chlachula, 2001b) (Fig. 5C).

The archaeological and palaeontological records incorporated in the fluvial and sub-aerial formations provide new insights on timing, processes and conditions of the initial peopling of this territory, as well as on biodiversity of the particular Pleistocene habitats. Abundant "pebble tool" assemblages bear witness of human occupation of the Yenisei area prior to the last interglacial (>130 ka BP) in agreement with other regions of southern and eastern Siberia (Fig. 1A). These earliest cultural records, classified on the basis of stone flaking technological attributes as the Early Palaeolithic, are age-corroborated by the contextual geological position (the Kamennyy Log, Sukhoy Log, Verkhnyy Kamen and Razlog sites). These archaic stone implements were largely redeposited by erosional processes, but are also documented *in situ* in the original geological context of the Middle Pleistocene alluvia. The formal variability of the Early Palaeolithic industries, displaying a differential degree of patination and aeolian abrasion, indicates several stages of the initial peopling of this area. These cultural finds, in conjunction with the Middle Pleistocene cold-adapted fossil fauna, suggest an early human adjustment to local periglacial environments. The archaeological records from the Minusinsk Basin, supported by analogous finds from the Kuzbass and the upper Angara River basins (Medvedev et al., 1990), show that parts of southern Siberia were occupied repeatedly at several stages during the Middle Pleistocene.

Evidence on the Middle Palaeolithic peopling of central Siberia is principally from the foothill areas of the Altai and Sayan Mountains and the local tributary river valleys (Derevianko & Markin, 1999). In the Minusinsk Basins, the Middle Palaeolithic is found in the Late Pleistocene loess strata and the interbedded soils, particularly of the last interglacial pedocomplex, but may be exposed on the present surface in low-sedimentation-rate areas of Khakhasia, Tuva, Gorno Altai, East Kazakhstan (Astakhov, 1986; Chlachula, 2001b, 2010b). Favorable climatic conditions promoted expansion of the Mousterian (Levallois) tradition during the Last Interglacial (MIS 5) that persisted until the early last glacial stage (MIS 4).

An exceptional Middle Palaeolithic site producing a unique archaeological as well as palaeontological locality is Ust'-Izhul' exposed by erosion at the Krasnoyarsk lakeshore (Chlachula et al., 2003). Numerous concentrated fossil skeletal remains were incorporated *in situ* on top of the last interglacial chernozem IRSL dated of 125 ka±5 ka (Sib-1) and overlying the 65 m Middle Pleistocene Yenisei terrace (Fig. 5E) The fauna included an early form of mammoth (*Mammuthus primigenius* Blum.), woolly rhinoceros (*Coelodonta antiquitatis*), bison (*Bison priscus*), horse (*Equus mosbachensis*), elk (*Cervus elaphus*), as well as small mammals such as marmot (*Marmota* cf. *baibacina*), beaver (*Castor sp.*) and badger (*Meles meles*). The most abundant species -- mammoth -- was represented by several hundreds bone fragments and 42 molar teeth of at least 12 individuals, mostly juveniles, parts of which were recorded in anatomical position. The associated mollusc fauna shows a xerotheric grassland setting. A perfectly preserved stone tool assemblage (ca. 200 pcs) was found in association with the

Fig. 5. Central Siberia. A: eroded late Quaternary loess sections along the Krasnoyarsk Lake; B: the key section (Kurtak 29) with the interglacial (MIS 7, 244-170 ka; MIS 5, 130-74 ka BP) and interstadial (MIS 3 59-24 ka BP) pedocomplexes; C: exposed gravel alluvia of the Middle Pleistocene (65 m) Yenisei River terrace incorporating fossil fauna and archaic pebble-tools; D: Kuznetsk Basin, mammoth teeth (left: *M. trogonterii, M.* sp., *M. primigenius*); E: the Middle Palaeolithic site Ust'-Izhul' with remains of early *Mammuthus primigenius* and stone artifacts (F) *c.* 125 000 year old, representing a most unique and complete early occupation in Siberia.

fossil fauna and three fireplaces with a fir-wood charcoal radiocarbon dated to >40 ka BP. The stone artifacts mostly represented by simple flakes (Fig. 5F) were used for processing the slaughtered animals. Human activity is also manifested by flaked and cut bones of mammoth, rhinoceros, bison and elk, and mammoth tusk fragments. The palaeogeographic site configuration with the concentrations of the skeletal remains suggest that the game was hunted by people in the nearby area across the 100-110 m terrace over a cliff onto the present 65-70 m terrace (which formed the floodplain) and transported in dissected pieces to the habitation place. The composition of the inventories reinforces the interpretation that it was a short-term occupation / processing campsite. In respect to the high age and the contextual completeness, the Ust'-Izhul' locality is at present without parallel in Siberia. Together with other Pleistocene megafaunal assemblages from the Minusinsk Basin, the associated fauna record bears witness to the rich biological potential of the area for early human occupation. There is no consensus if the Middle Palaeolithic tradition in southern Siberia can be associated with the European and Near Eastern cultural milieu, although some "classical" Mousterian influences in the Altai cave sites are evident (Derevianko & Markin 1992) (Fig. 4F). The Middle Palaeolithic stone flaking technology, especially the Levallois technique, is still reminiscent in the Late Palaeolithic traditions, suggesting a certain continuity of the cultural and possibly biological human evolution in Siberia during the Late Pleistocene.

The Late Palaeolithic occupation in the southern Central Siberia is documented at both open-air and cave sites. Warming during the mid-last glacial interstadial stage (MIS 3) accelerated formation of the Late Palaeolithic cultures characterized by the developed blade flaking techniques. The upper Yenisei basin is one of the major loci of the Late Palaeolithic sites in Siberia (the Krasnoyarsk – Kanskaya forest-steppe, the Northern and Southern Minusinsk Basin, and the Western Sayan foothills) (Astaknov, 1986; Drozdov et al., 1990; Larichev et al., 1990; Vasiliev, 1992; Yamskikh, 1990). The earliest occupation sites have been found buried in the Karginsk (MIS 3) Pedocomplex (31-20 ka BP), in the early Sartan loess and the intercalated, weakly developed, humic loamy interstadial forest-tundra soils (dated to 25-22 ka BP). Intervals of significant climatic deterioration and onset of full glacial (MIS 2) conditions are manifested by the absence of any archaeological record except for biotic refugia in the northern foothills of the Western Sayan Mountains. During the LGM (22 000 – 19 000 yr BP), the Yenisei area, as most of southern Central Siberia, seems to have been vacated due to very cold and arid environmental conditions with the biologically productive mid-last glacial parkland-steppe gradually replaced by a harsh and inhospitable steppe-tundra.

More recent Final Palaeolithic finds from the steppe and foothill areas (Abramova, 1979a, 1979b; Drozdov et al., 1990; Vasiliev et al., 1999) provide evidence of re-colonization of the upper Yenisei region during warm climatic oscillations at the end of the Pleistocene. A pre-dominant smaller fauna (reindeer, red deer, argali sheep) hunting, composite bone tools (spears/harpoons) and a micro-blade stone flaking illustrate major shifts in adaptation strategies in response to changing environments during the warm interstadial (16-13 ka BP), preceding the last cold climatic interval of the Final Pleistocene (Younger Dryas, 12.9 ka BP). In sum, the present archaeological records from the larger southern Central Siberia provide eloquent evidence that this territory was occupied repeatedly by early people at several stages during the Pleistocene, including glacial stages with harsh periglacial natural conditions. Human occupation may have persisted throughout the Last Glacial in protected southernmost locations despite severe hyperarid climates over most of the northern plains.

3.4 East Siberia (the Angara, Lake Baikal and the Lena Basin)

The palaeogeographical and palaeoenvironmental evolution over East Siberia during the Quaternary Period is characterized by a dynamic interaction of past global climatic changes and a regional neotectonic relief modeling. Sub-rifting orogenic regimes of the Mongolian-Siberian mountain zone gave rise to a diversity of geomorphic settings throughout the Pleistocene. The Cenozoic tectonics in conjunction with mountain glaciations continuously shaped the topography of Eastern Siberia structured by a system of mountain ranges (North Baikalskyy, Verkhoyanskyy, Cherskego, Kolymskyy) separated by the major (Angara, Lena, Viluy, Aldan, Indigirka, Yana, Kolyma) river valleys and their tributaries. The geographical isolation with a partly reduced atmospheric circulation regime contributed to a pronounced climatic continentality of this vast territory. Pleistocene glaciations over the East Siberian mountain ranges, though largely localized to isolated ice-caps, played a significant role in the landscape development and related environmental shifts in the extraglacial depressions. Increased accumulation rates of sub-aerial (sandy and loessic) sediments and periglacial palaeosol surface deformations point to a gradually progressing climate cooling during the Late Quaternary. Fossil pollen, fauna and early cultural records from stratified geological contexts provide evidence of pronounced palaeoecology changes in natural habitats. Study of the palaeogeographical processes is essential for reconstruction of the Quaternary climate history, geomorphic and biotic transformations of ecosystems as well as human occupation. The key regions for the Quaternary studies are located in the southern part of East Siberia.

3.4.1 Geography and natural setting

The south-eastern Siberia (Fig. 1A),,, encompassing the Pribaikal area (the upper Angara and Lena basins) and the SE Transbaikal area (the Selenga River basin), is a territory of contact of the major tectonic structures of the Siberian Platform and the adjacent mountain massifs. The southern part of the territory is formed by the Irkutsk Depression (800-1200 m asl.), geographically confined by the Eastern Sayan Mnts. and the Baikal Range (2500-3000 m asl.). Lake Baikal (-1620 m max. depth) occupies a tectonic basin of 636 x 26-79 km or 31,500 km². A system of palaeolakes existed in this area since the late Oligocene (25 Ma ago), although formation of the present lake (Fig. 6A) relates to a major orogenic activity initiated during the Pliocene/Pleistocene (the Baikal Tectonic Phase spanning for the last 3.5 Ma). The Baikal rift zone, 150 km wide in average, extends for about 2,500 km from Lake Khubsugul in Mongolia to the upper Aldan River in south-east Yakutia, and includes a series of tectonic depressions (Khubsugul, Tunkin, Baikal, Barzugin, and the upper Angara depression).

Most of the relief is shaped by smooth mountain ranges (800-1300 m asl.) separated by river valleys and shallow basins. Taiga forests occupy the northern lowland areas, while open steppe grasslands (Pribaikalye) and semi-deserts (Transbaikalye) broadly extend in the southern regions. Present climate is strongly continental, reflecting the geomorphological configuration of East Siberia, forming an orographic atmospheric barrier strengthening the influence of the Siberian Anticyclone. Average temperatures -20°C (January) and +23°C (July), and low annual precipitation values (250 mm) characterize most of the territory. Maximum precipitation falls on the north-exposed slopes (700-800 mm), covered by coniferous (mostly pine) forests. More humid regional climate prevails along the margins of the Irkutsk and upper Lena basins due to a warming effect of the Lake Baikal (350-450 mm). Mixed southern taiga is largely represented by spruce, larch and birch. Most arid interior valleys (250-300 mm *per* year) include xenotheric steppe and mosaic parkland communities.

3.4.2 Palaeogeography and climate history of East Siberia

Both orogenesis and climate have shaped the present relief of East Siberia. The pre-Cenozoic geological formations are represented by relics of the Palaeogene palaeo-relief preserved on plateaus tectonically raised to 300-500 m asl above the basal igneous and metamorphic rocks and sedimentary, partly lithified deposits. The Oligocene and Miocene uplift of the marginal regions of the Irkutsk Basin (including the upper Lena area) initiated formation of the united morphostructural province of the broader Baikal region. Tectonic processes, significantly reactivated during the Cenozoic Period, generated massive colluvial deposits that filled the former riverine basins and shallow depressions originating by the Late Miocene pre-rifting (Rezanov, 1988).The orogenic movement around the Pliocene/Pleistocene boundary (2.5 Ma ago), opened the Baikal rift zone and caused a major uplift of the Pribaikal and West Transbaikal regions. The raised topographic gradient lead to intensive river erosion followed by the establishment of the present drainage of the major East Siberian rivers (Angara, Lena, Selenga). Formation of the surrounding mountain ranges precipitated the accumulation of fluvial deposits and a gradual filling of the Pliocene rift graben (Bazarov, 1986). In the Lake Baikal surrounding, the geomorphic dynamics is evidenced by 6-10 raised lake terraces related the former lake water-stands and elevated up to 200 m above the present lake level. During the Early-Middle Pleistocene, the Lake Baikal discharge followed a passage through the Irkut River valley into the former Yenisei basin. The present regime was established during the Late Pleistocene as a result of a new tectonic movement opening an outlet into the present Angara River drainage system (Fig. 6B).

The neotectonic movements during the early Middle Pleistocene (after 750 ka BP) contributed to an accumulation of thick deltaic, fluvio-lacustrine and slope deposits filling the major river valleys and intermountain depressions, particularly in the Selenga River basin (Rezanov, 1988, 1995). A reactivation of the orogenic activity during the initial stage of the late Middle Pleistocene resulted in the Lake Baikal transgression. A system of 40-50 m-high river and lake terraces and adjacent alluvial fans formed during the Tobol (MIS 9) interglacial and continued in the following Samarovo (MIS 8) glacial period. During this maximum Middle Pleistocene glaciation, initiated by global climate cooling coupled by a regional uplift, mountain glaciers extended to 1000-1300 m elevation with terminal moraines along the foothills of the Baikal, Barguzin and Angarsk Ridges (Rezanov & Kalmikov, 1999). A renewed regional tectonic activity of the Baikal rift at the beginning of the Late Pleistocene (130 ka BP) triggered an accelerated erosion of the Selenga River, and the upper Angara and Lena River (Bazarov, 1986). Marked climatic shifts characterize the last glacial-interglacial cycle, with ice-cap formations in the East Siberian mountains during the cold stages (MIS 4 and 2). The Ermakovo (MIS 4) glacial stage, with a major glaciation in the Baikal Range and the East Sayans, is evidenced by moraine ridges and altiplanation terraces at >1000 m elevations (Nemchinov et al., 1999). Increased aridity and decrease of annual temperatures during the cold stadials together with strong NW winds stimulated formation of deflation and drifting sandy surfaces on top of the Early and Middle Pleistocene terraces, and aeolian dust deposition in main river valleys and depressions (Medvedev et al., 1990). Climate fluctuations of the early Last Glacial (MIS 4) caused intensive solifluction processes. The mid-last glacial (MIS 3) optimum (dated to ca. 31 ka BP) was thermally approaching the last interglacial (MIS 5e) climate conditions. A reduction of the glaciated areas is observed for the following Sartan (MIS 2) glacial stage. The present geomorphological configuration and a regional environmental zonation became gradually established during the Holocene.

3.4.2.1 Quaternary climate change and environments of the glacial mountain areas

Pleistocene glaciations in the East Siberian mountains, bordering the Angara and Lena basin from the south and east played a major role in the landscape development and related shifts in ecosystems of the extra-glacial depressions (Fig. 6E). The main glacial events documented in the Eastern Sayan Mountains occurred during the late Middle Pleistocene and Late Pleistocene, with the penultimate (Ermakovo, MIS 4) glaciation (74-50 ka) been the most extensive (Nemchinov et al., 1999). This evidence corroborates the glacial records from the northern part of the Baikal Depression and the Baikal Range as witnessed by relics of terminal moraines assigned to the Samarovo (MIS 8) glacial stage (Bazarov, 1986; Rezanov & Kalmikov, 1999). Warmer interglacial / interstadial periods correlate with a mountain ice retreat and expansion of thermophilous biotic communities over large geographical areas.

The Late Pleistocene climate variations caused a major landscape restructuring with a deep bedrock frost weathering, sediment deposition, solifluction and cryogenesis during cold stages, and pedogenic development during warm interstadials, particularly in the loess area of the Irkutsk Basin. Renewed tectonic activity contributed to intensive river incision, leading to formation of the present river valleys, but also of passages for the last glacial ice expansion. The initial Late Pleistocene interglacial (MIS 5) promoted spread of pine-birch forests including some warm arboreal broad-leaved taxa *(Tsuga, Tilia, Corylus)*. A progressing climatic cooling at the beginning of the early Last Glacial (Zyriansk Glaciation, MIS 4) gave rise to major montane glaciers in the Baikal Range and the most intensive Pleistocene glaciation in the Sayan and Altai Mountains. Mountain valley glaciers, expanding from the north-eastern slopes of the Bolshoy Sayan Ridge, formed mighty piedmont glaciers penetrating up to100 km north into the extra-glacial zone. Accumulation of massive colluvia (up to 20 m thick) was apparently triggered by the reactivated regional neotectonic movements towards the end of the early last glacial stage. A climatic warming during the following (Karginsk) mid-last glacial interval (MIS 3) is evidenced by pollen records form alluvial and lacustrine sedimentary formations. Apart of coniferous species *(Pinus silvestris, Pinus sibirica, Abies, Picea* and *Tsuga)*, deciduous trees *(Quercus, Ulmus, Corylus* and *Alnus)* and shrubs *(Salix, Alnaster)* were present. The East Siberian mountain glaciers largely wasted or retreated into high topographic locations to form isolated ice-caps. The late Last Glacial (Sartan Glaciations, MIS 2) in the Eastern Sayan Mountains was largely confined to active corrie glaciers above the 2700 m elevation (Nemchinov et al., 1999). On the contrary, the Pribaikal Highlands likely experienced the most extensive Pleistocene glaciation during the LGM (22-19 ka BP), with piedmont glaciers advancing far down into the foothill valleys (Rezanov & Kalmikov, 1999). Only isolated ice-caps formed on the NE Siberian mountain ranges (Verkhoyanskyy, Cherskogo, Kolymskyy) despite very low temperature, but due to a very high aridity and a lack of winter precipitations. Cold Late Pleistocene climatic intervals are also documented by cryogenic surface deformations, with the most prominent (up to 5 m deep) at the contact of the Pleistocene/Holocene formations assigned to Younger Dryas (12.9-11.5 ka BP). Cold periglacial tundra with isolated shrub and dwarf trees occupied the adjacent northern foothills until the warm climatic oscillations sparked off a complete zonal vegetation rearrangement with the approaching end of the last glacial stage. Cold and strongly continental climatic conditions prevail in the East Siberian mountain region at the present time, with MAT of -4°C. A considerable part of the territory is covered by mountain tundra with larch *(Larix sp.),* Siberian pine *(Pinus sibirica)* and dwarf birch *(Betula nana).* Most of the area is underlain by active perennial permafrost.

3.4.2.2 Quaternary climate change and environments of the extra-glacial areas

Quaternary geology of the extra-glacial areas of the Baikal territory (the Irkutsk Depression, the Transbaikal and upper Lena River areas) shows a complex structure resulting from the interaction of past climatic changes and a topographic modeling by the neotectonic activity. This encouraged mass deposition of alluvial, aeolian, colluvial and other clastic materials. The iron-stained gravels and reddish strongly weathered lacustrine clays attest to warm climatic conditions during the late Pliocene in corroboration with the incorporated pollen of thermophylous taxa *(Quercus, Carya, Tsuga)* of mixed coniferous-broadleaved forests.

The Pliocene/Pleistocene transition around 2.5 Ma BP defines the first major cooling evidenced by prominent cryogenic features. The Early Pleistocene climate trend of progressing continentality caused a complete zonal palaeogeographic rearrangement (Medvedev et al., 1990). The Early and Middle Pleistocene climatic variations are witnessed by a wide range of palaeosols (kashtanozems, chernozems, luvisols, brunisols, gleysols), indicating a mosaic vegetation and relief zonation. Stages of climatic deterioration correlate with weakly developed initial soils (regosols), solifluction horizons, cryogenic deformations and relic periglacial landforms. A seasonal strengthening of temperate and humid climate promoted a latitudinal zonal vegetation distribution. The Middle Pleistocene interglacial periods were warm and humid, allowing a northern expansion of a mixed pine-spruce-fir taiga with broad-leaved species *(Quercus, Fagus, Ulmus, Corylus, Pterocarya, Juglans* and *Tilia amurensis)* with up to a 50% increase in precipitation in the warmest Tobol Interglacial (MIS 9) (Rezanov & Kalmikov, 1999). The following climate deterioration during the late Middle Pleistocene is manifested by progressive fossil fauna forms of open periglacial parklands.

Diversity of soils developed during the Last Interglacial on alluvial and aeolian deposits (tundra gleysols, gleyed podzolic soils of northern taiga, podzols of central taiga, brown soils of southern taiga and chernozems of parkland-steppe). Polygons with up to 3-4 m deep frost wedges on top of the last interglacial (MIS 5e) chernozem indicate a dramatic cooling during the following stage (MIS 5d) with temperature minimum around ca. 115 ka BP, suggesting a major glaciation in eastern Siberia corroborated by the record from Lake Baikal and the loess-palaeosol sequences from other parts of southern Siberia (Karabanov et al., 1998; 2000; Prokopenko et al., 1999, Chlachula 2003, Chlachula et al., 2004b) (Fig. 2B).

Onset of the early Last Glacial stage (Ermakovo Glaciation) is associated with an intensive erosion of the exposed land surfaces leading to a massive transfer of sediments, particularly fine sands from the major denudation areas /deflation surfaces along the margin of the Irkutsk Depression. Solifluction of loessic deposits underwent during the following time intervals as a result of a moderate warming. The mid-last glacial Karginsk interstadial stage (MIS 3, 59-24 ka BP) included an earlier (cold and arid) phase of loessic sediment deposition succeeded by a later (warm and humid) phase of progressive soil formation (the Osinsk Pedocomplex in the Angara – upper Lena region). The first phase of the late Last Glacial (MIS 2), radiocarbon dated to 24-17 ka BP, is linked with marked climatic oscillations evidenced by solifluction and cryoturbation of the preceding (MIS 3) soils, succeeded by loess sedimentation with the approaching Last Glacial Maximum (22 ka BP). Cold climates corroborate appearance of cold-adapted fauna characteristic of periglacial steppe *(Saiga tatarica, Alces alces, Equus hemionus)*. A warming interval after the LGM (17-16 ka BP) is manifested by an intensified gleying and degradation of permafrost in the southern part of the territory due to a global climatic amelioration also reflected by the expansion of coniferous taiga forests, and by the Palaeolithic re-colonization of south-eastern Siberia.

3.4.3 Pleistocene environments and early human peopling

Present cultural evidence attest to several stages of human occupation of Eastern Siberia during the Pleistocene, with the oldest represented by the Early (?) and Middle Pleistocene records, implying very early hominid migrations into the mid- and high-latitudes of Asia. Systematic investigations at the occupation sites in the upper Angara, Lena, Aldan, Viluy and Selenga River basins, along the Baikal Range and Eastern Sayan foothills, contextually associated with diverse palaeo-geomorphic zones and geological (alluvial, colluvial, aeolian, karstic) settings, deliver detailed multi-proxy information on the Pleistocene climate change and associated palaeoenvironmental processes in the particular areas (e.g., Medvedev et al., 1990; Konstantinov, 1994; Mochanov, 1992; Mochanov & Fedoseeva, 2002; Chlachula, 2001c). Principal Quaternary research in the Pribaikal and Trensbaikal regions (the upper Angara, Lena and Selenga basins and their tributary valleys) reflects a limited transport accessibility of the north-eastern territories. Archaeological sites may be partly obliterated or poorly preserved around the Lake Baikal due to neotectonic activity triggering erosional processes. The cultural finds, chronologically defined by the stratigraphic geological position and diagnostic technological attributes of stone flaking, include: 1) the Early Palaeolithic from Early(?) / Middle Pleistocene alluvial deposits (>130 ka BP) (Fig. 6B); 2) the Middle Palaeolithic buried in the last interglacial (MIS 5) pedocomplex and the early last glacial (MIS 4) gleysol horizons; 3) the early Late Palaeolithic (42-30 ka BP) from the mid-last glacial (MIS 3) soils; 4) the "classical" Late Palaeolithic from the late mid- and early last glacial (MIS 2) gleyed soil horizons (30-17 ka BP); 5) the final Palaeolithic (17-12 ka); and f) Mesolithic (12-8 ka BP) from diverse geo-contexts (Medvedev et al., 1990; Chlachula et al., 2004b).

The earliest Pleistocene sites with definite cultural occurrences are associated with old alluvial formations in the Lena and Angara River basins. Presently one of the oldest sites in East Siberia was found at the Diring Uriah within coarse sandy deposits of a 200 m-high terrace of the Lena River near Yakutsk. Despite its unclear chronological assignment ranging from 2.5 Ma to 400 ka and the postulated model of the extra-tropical origin of humans in NE Siberia (Mochanov, 1992) this site eloquently demonstrates a very early Pleistocene peopling of northern parts of Asia in the principal river valleys much earlier than previously believed probably during some of the Early or Middle Pleistocene interglacials (Waters et al., 1995). Temperate and drier conditions of the late Middle Pleistocene interglacial interval (MIS 7) promoted expansion of mixed taiga (*Pinus, Picea, Alnus, Salix, Betula*) and open parklands. A Middle Palaeolithic occupation is documented at the Mungkharyma site (64°N) located on the 70 m alluvial terrace of the middle Viluy River with a diagnostic stone industry, including elaborated bifaces, bifacial knives and side-scrapers, found in association with Pleistocene fauna (mammoth, woolly rhinoceros). Luminescence dates 150±38 ka BP (RTL-958) in the overlying sandy-silt layer and 600±150 ka BP (RTL-957) in the basal part suggest a (late) Middle Pleistocene age of this unique locality (Mochanov & Fedoseeva, 2001) (Fig. 6C-D). These as well as other pre-last glacial (>74 ka BP) sites from the central and southern Lena basin and its major tributaries (Viluy, Aldan) also have the principal implications in respect to the initial prehistoric colonization of the American continent (Chlachula, 2003b).

The archaeological records from the upper Lena and Angara River areas dated to the Middle Pleistocene, represented by analogous stone artifacts made on quartz / quartzite cobbles with the typical archaic tool forms (choppers, bifaces, scrapers on flakes, polyhedral cores), display strongly abraded surfaces as a result of aeolian weathering. The age of these earliest cultural assemblages from the Angara region exposed along the Bratsk Lake (Igetei locality), referred to as of the Acheulian-Mousterian tradition, is estimated to ca. 200 ka BP, corresponding to the Shirta Interglacial /MIS 7 (244-170 ka BP) (Medvedev et al., 1990). Climatic cooling, leading to expansion of mosaic steppe and pine-larch parklands, is linked with the onset of the Samarovo (270-244 ka BP) glacial period. A drop of annual temperature and increase in aridity during the Tazov Glacial (170-130 ka BP) hastened a degradation of interglacial forests and extension of open periglacial tundra landscapes in the broader Baikal region (Nemchinov et al., 1999). Human survival in south-east Siberia during the Samarovo (MIS 8) and Tazov (MIS 6) glacial stages presumes knowledge of fire-making. The marked final Middle Pleistocene environmental deterioration is indicated by records of cold-adapted fossil fauna from river terrace formations, including more progressive taxa typical of open periglacial landscapes, such as mammoth, wooly rhinoceros, bison, horse, kulan, giant deer, argali, and some species specific to the Trans-Baikal area - Kiakhta antelope *(Spirocerus kiakthensis)*, the Baikal yak *(Poephagus baikalensis)*, dzeren/gazelle *(Procapra gutturosa)*, camel *(Camelus* sp.) (Kalmikov, 1990), in congruence with the preserved glacial landscape relics in the Northern Baikal Range and the Eastern Sayan Mountains (Rezanov & Nemchinov, 1999). During the Last Interglacial (MIS 5), mixed taiga forests with Siberian pine *(Pinus sibirica)*, spruce, larch, birch, willow and hemlock were widely distributed in the mountain areas, indicating a temperate continental climate similar to the present one (Nemchinov et al., 1999). Brown fossil chernozems at the Mal'ta and Igetei Middle Palaeolithic sites (dated to MIS 5e and 5c, respectively) illustrate open parkland-steppe settings of the Angara basin. Marked cooling, increased aridity accompanied by a strong aeolian activity contributed to accumulations of extensive sandy deposits derived from drying up river beds in the Irkutsk Depression and Lena Basin during the early last glacial (MIS 4), also witnessed by an intense abrasion of the contextually associated stone industries. In the western Baikal region, this stimulated drifting sandy surfaces and accumulation of fine, aeolian dust on lee slopes in the form of a loess-like series at the eastern margin of the Eurasian loess belt. Cold intervals with increased atmospheric humidity correlate with cryoturbation processes, solifluction horizons and pollen records of invading periglacial grasslands and open pine-birch tundra. A local Middle Palaeolithic (Mousterian?) occupation may have survived under such harsh environmental conditions in the upper Angara at the Igetei Site III (Medvedev et al., 1990).

The mid-last glacial interstadial interval (MIS 3), with an early cold and arid loess sedimentation phase (59-40 ka BP) followed by a warm and humid pedogenic phase (40-24 ka BP) brought a major change in the distribution of vegetation zones in East Siberia, with a northern expansion of mixed taiga forests and pine-birch dominated parklands. Broad-leaved arboreal taxy (oak, beech, elm, hazel) distributed in more humid river valleys of the Baikal region testify to mosaic habitats with climate conditions possibly warmer than today (Rezanov & Kalmikov, 1999). The pedogenic record in the Transbaikal region (the lower hor. Kamenka site) indicates very high MAT of +8/10°C, thus considerably exceeding the present day temperature values (Dergachova et al., 1995). This warming correlates with a transgression of Lake Baikal, triggering deposition of thick deltaic and estuarine sediments. Cultural finds from the principal occupation sites (Ust'-Kova in the Angara basin, Mezin in the Kana valley, Kamenka in the Selenga basin, the Aldan River complex) dated to 30-24 ka BP display a poor preservation due to cryogenesis and solifluction processes persisting until the onset of late Last Glacial (Sartan) stage (MIS 2). Remains of exploited Pleistocene fauna (horse, antelope, woolly rhinoceros, mammoth, bison, sheep-argali and camel) similarly as rodent taxa are indicative of an open steppe and parkland habitat. At the Kamenka site in

Fig. 6. East-Central Siberia. A: the central west coast of Lake Baikal with marked orogenic structures; B: Upper Angara Basin (Igetei Site), eroded fossiliferous Late Quaternary loess-palaeosol sections overlying a Middle Pleistocene alluvium incorporating cultural remains; C: the Mungkharyma Site (64°N) on the 70 m Viluy River terrace, east-central Yakutia; D: Middle Palaeolithic quartzite tools (Mungkharyma II Site; courtesy Yu. A. Mochanov); E: coarse gravelly alluvial fans in the formerly glaciated Verkhoyanskyy Range, NE Yakutia; F: the meandering Yana River valley was one of corridors for peopling of the Arctic Siberia.

southern Buryatia, the most represented animal species, Mongolian gazelle *(Procapra gutturosa)*, lives today in dry grassland steppes and semi-deserts in eastern Altai, Mongolia and Inner Mongolia in gently rolling landscapes with a reduced snowfall. Several specific site functional complexes related to game processing, woodworking, stone and bone tool production and manufacturing of mineral paints, and other behavioral cultural (ritual) activities were documented (Lbova 1996). The taxonomic fossil fauna variety from different (mountain) ecotones shows a wide (>100km) mobility range of the local Palaeolithic hunters. Expansion of the Late Palaeolithic occupation ambit into the extreme parts of the East Siberian Arctic is documented at the Yana RHS site located 100 km south of the Laptev Sea coast (70°43′N, 123°25′E) (Pitulko et al., 2004). The cultural evidence (stone and bone industry and fossil fauna), sealed in frozen and cryogenically distorted silt blocs on a 18 m Yana River terrace radiocarbon dated to ca. 28-26 ka BP, illustrates warm interstadial (MIS 3) climates and a relative environmental stability of floodplain meadows of the Yana River delta (Fig. 6F). This exceptional site provides eloquent evidence of humans migrating along the ice-free northern coast of Siberia (the exposed continental Arctic shelf) across the exposed Bering land-bridge during the mid-last gladial interstadial (50-24 ka BP), opposing the model of "initial" peopling of Alaska by the East Siberian Diyuktai Culture at 12 ka BP.

The late Last Glacial stage (MIS 2) - the Sartan Glaciation (24-12 ka BP) - is the best-studied Pleistocene time interval in eastern Siberia in view to contextual cultural records. Tundra-steppe covered most of the territory occupied by the Late / Final Palaeolithic people with the famous sites Mal'ta and Buret' in the upper Angara basin (Tseitlin, 1979). Climate deterioration with sparse Arctic vegetation and progressive loess accumulation around the LGM (18 ka BP) caused a major decline in the population density in East Siberia, although some adaptation to extreme periglacial environments was suggested at the Krasnyy Yar Site, with animal bones and fossil coal used as fuel (Medvedev et al., 1990). A periglacial fauna (horse, wooly rhinoceros, mammoth, bison, giant deer, elk, saiga) implies cold periglacial tundra steppe across vast areas across the Angara, upper Lena and the Trans-Baikal basins. Climate amelioration after the LGM (18 ka BP) is best-evidenced by micro-mammal (rodent) faunas from the Baikal-Angara-Lena Palaeolithic sites (Buret', Krasnyy Yar, Igetei, Mal'ta, Bolshoi Jakor) that indicate a gradual transition from tundra-steppe and meadow-steppe to forest-steppe landscapes corresponding to shifts from a cool and dry climate to milder and humid conditions (Medvedev et al., 1990, Khenzykhenova, 1999). This warming trend fostered dispersal of the Final Palaeolithic complexes with micro-blade stone and bone technologies during the late Last Glacial (18-12 ka BP). Re-colonization of broad areas of East Siberia regions reached the marginal sub-polar regions of the NE Russian Arctic (e.g, Berelekh site in northern Yakutia at 70°N, dated to 14-13 ka BP) (Vereshagin, 1977; Vasil'ev, 2001). The broad geographical distribution of new technologies to the most distant parts of NE Siberia and the Russian Far East Islands (Vasilevskyy, 2008) documents a successful prehistoric adaptation to the Final Pleistocene environments. This process culminated in a spatial spreading and presumably a major population increase by the end of the Pleistocene, represented by the early Mesolithic (11,000-10,000 yr BP) hunters and gatherers.

In sum, palaeoenvironmental (geological and biotic) proxy records document well the evolutionary development of Quaternary climate change in East Siberia as reflected in the gradual cultural adaptation of people and the geographic expansion of occupied areas. A great diversity of the early Quaternary biotic communities became reduced as a result of climatic cooling, leading to a relative taxonomic faunal uniformity in the Late Pleistocene.

4. Conclusion

Geological and palaeoecological evidence across western, south-central and eastern Siberia, including the Ob, Irtysh, Kuznetsk, Yenisei, Angara and upper Lena basins and the adjacent mountain regions of the Trans-Urals, Altai, Western and Eastern Sayans Mountains, Baikal Range and the NE Siberian mountains, shows patterned cyclic climatic changes during the Quaternary leading to establishment of the present natural conditions. A progressing global trend towards a strongly continental climate, with increased aridity and high seasonal temperature fluctuations, can be traced since the late Pliocene. At that time, northern and north-eastern Siberia experienced the first major regional glaciations (Laukhin et al., 1999) in response to changes in solar radiation and atmospheric circulation over high latitudes of the northern Hemisphere. The subsequent development during the Quaternary Period (last 2.5 Ma) is evidenced by zonal geographic shifts in the vegetation distribution, with expansion of boreal (taiga) forests northward during the interglacial periods and warmer interstadials, succeeded by sub-arctic periglacial forest-steppe, and tundra-steppe during cold glacial intervals, when the tree cover became much reduced and spatially limited to biotic refugia in the southernmost portions of central Siberia (the Altai and Sayan Mountain foothills).

Complex palaeoenvironmental evolution archives are stored in the high-resolution loess-palaeosol sequences, pollen records and fossil fauna remains from deeply stratified alluvial and loessic formations. The loess-palaeosol sections on the Northern Altai Plains and in the upper Yenisei basin have provided most complete information on the past climatic variations, the landscape development and the associated changes in the Pleistocene biotic communities on the territory of southern Siberia. The high-resolution stratigraphic records coupled with pollen and palaeontology data indicate marked Pleistocene ecosystem transformations, with arctic tundra and forest-tundra during cold stadial intervals, being replaced by boreal forest and later by parkland - steppe during warm interstadial intervals.

The Early and Middle Pleistocene climates brought a major modification of natural habitats, facilitating the northward dispersal of Palaeolithic people from the southern areas of Central Asia and Mongolia, and their environmental adaptation to the Siberian habitats and regional settings. The earliest unequivocally documented Middle Pleistocene (Early Palaeolithic) occupation centered in the southern continental basins and river valleys north of the Altai-Sayan Mountains. The human dispersal into East Siberia is assumed to have principally occurred during warm interglacials in the processes of the northern expansion of mixed parkland forests and associated fauna communities, whereas only local movements of early human groups are envisaged during cold stages. The Tobol (MIS 9) Interglacial (390-270 ka BP), when the MAT was by ca. 3-4°C higher than at present, is likely to have been (one of) the most favorable time period for the main initial migration to northern Asia through the major Siberian river valleys, reaching as far north as 60°N latitude. The Early and Middle Palaeolithic finds bear witness of repeated inhabitation of the Irtysh, Ob, upper Yenisei, Angara, Viluy, Aldan and the upper Lena River basins prior to the Last Interglacial. During the Middle Pleistocene glacial periods, glaciers in the Western and Eastern Sayan ranges expanded into the foothills to about 300-400 m altitude preceded by a down-slope retreat of dark coniferous taiga forests. In the Northern and Southern Minusinsk Depression, in the Kuznetsk Basin and some extraglacial locations in the northern Altai valleys protected by mountains from the west and east, and separated by high ranges from the frigid arctic tundra in the north, propitious (although periglacial) conditions with great biomass concentrations of steppe-parklands are likely to have persisted during the glacial stages. This is evidenced by the abundant fossil remains from the old (60-80 m) Yenisei and Angara alluvia, as well as the colluvial reworking of the older (late Middle Pleistocene) loess cover, indicative of a relatively humid, moderately cold and fluctuating climate regime.

Within the Last Interglacial (MIS 5; 130-74 ka BP), most of Siberia was covered by coniferous or mixed forests, with forest-steppe distributed at lower elevations and in river valleys. At that time, the Middle Palaeolithic (Neanderthal or early *Homo sapiens*) people entered the territory from Central Asia and/or the East European Plains. Expansion of the occupation habitat into the mountain areas, following the last interglacial climatic optimum, likely occurred in the later (MIS 5c and 5a) interstadials. During the early Last Glacial (Zyriansk) stage (MIS 4; 74-59 ka BP), cold periglacial tundra or tundra-steppe expanded across Siberia. The approaching glacial maximum disrupted human occupation, although this may have persisted in some protected southern locations. Following the interval of intensive loess deposition at the end of the glacial, renewed warm climate pulses during the mid-glacial (Karginsk) interval (MIS 3; 59-24 ka) preconditioned formation of zonal soils contextually associated with the transitional Middle/early Late Palaeolithic stone industries, suggesting a certain regional cultural (and possibly biological) evolutionary continuity in the Late Pleistocene Siberia. Moderately cold and stable environments during the second half of the interstadial interval (30-24 ka BP) promoted a major enlargement of habitats marking a climax of the Palaeolithic peopling in Siberia associated with the emergence of the "classical" Late Palaeolithic cultures. Productive interstadial ecosystems with mixed parkland-forest vegetation were gradually transformed into periglacial tundra with the approaching last (Sartan) glacial stage (MIS 2; 24-12 ka BP). A reduced population density is assumed around the LGM (22-19 ka BP) hindered by extremely harsh climate conditions. Some occupation continuity persisting until the end of the Pleistocene may have applied just for biotic refugia in the protected southernmost locations along the Altai-Sayan foothills.

Overall, the spatial and temporal distribution of the cultural records documents climatic instability over large parts of Siberia during the Quaternary Period (the last 2.5 Ma). Specific geographical and contextual locations of early sites indicate that environmental conditions during the earlier periods were generally more stable and favorable for peopling than during the later periods. On the other hand, increased continentality and gradual shifts towards cold and arid conditions accelerated adaptation of Palaeolithic populations to harsh periglacial climates promoting progressive development of sophisticated survival strategies. Timing and evolutionary processes related to the initial colonization of northern Asia are still insufficiently mapped, although ongoing archaeological investigations supply new and often surprising evidence about particularities and general trajectories of this evolutionary process. The traditional views, assuming only a Late Pleistocene inhabitation of Siberia and Beringia, have been definitely challenged. The archaeological discoveries disprove the long-held assumption of a late penetration (by Late Palaeolithic people) into the middle and high latitudes of northern Asia. Instead, glacial-interglacial and stadial-interstadial climate cycles regulated a geographic movement of early people northwards, predetermining the inhabitability of particular geographical areas. During glacial maxima, most of Siberia seems to have been vacated, especially during the earlier periods, because of the expansion of continental glaciers in the north, and harsh and inhospitable environments in the southern extra-glacial regions. Gradual adaptation to cold natural habitats accelerated during the Late Pleistocene in connection with the advanced cultural and biological adjustment, enabling people to establish permanently in the vast and geographically diverse Siberian territory.

5. References

Abramova, Z.A. (1979a). *The Palaeolithic of the Yenisei. The Kororevo Culture.* Nauka, Novosibirsk, 199 p. (in Russian).

Abramova, Z.A. (1979b). *The Palaeolithic of the Yenisei. The Afontovskaya Culture.* Nauka, Novosibirsk, 156 p. (in Russian).

Arkhipov, S.A., Ed. (1980). *Paleogeography of West-Siberian Plain at the Late Zyryanka Glaciation Maximum.* INQUA Project: Quaternary Glaciations of the Northern Hemisphere. Nauka, Novosibirsk, 108 p. (in Russian).

Arkhipov, S.A. (1998). Stratigraphy and paleogeography of the Sartan glaciation in West Siberia. *Quaternary International,* Vol.45/46, pp. 29-42.

Arkhipov, S.A. (1999). Natural habitat of early man in Siberia. In: Chlachula, J., Kemp, R.A., Tyráček, J. (Eds), *Quaternary of Siberia, Anthropozoikum,* Vol. 23, pp. 133-140.

Arkhipov, S.A.; Bespaly, V.G.; Faustova, M.A.; Glushkova, O.Yu.; Isaeva, L.L., Velichko, A. (1986). Ice-sheet reconstructions. Quaternary Science Reviews, Vol. 5, pp. 475-482.

Arkhipov, S.A.; Gnibidenko, Z.N. & Shelkoplyas, V.N. (2000). Correlation and paleomagnetism of glacial and loess-paleosol sequences on the West Siberian Plain.*Quaternary International,* Vol. 68-71, pp. 13-27.

Astakhov, S. N. (1986). *The Palaeolithic of Tuva. Nauka,* Novosibirsk, 174 p. (in Russian).

Astakhov, V. (1997). Late glacial events in the central Russian Arctic, *Quaternary International,* Vol. 41-42, pp.17-25.

Astakhov, V. (2001). The stratigraphic framework for the Upper Pleistocene of the glaciated Russian Arctic: changing paradigms, *Global and Planetary Change,* V.31, pp. 283- 295.

Bader, O.N. (1965). *The Kapova Cave – Palaeolithic Paintings.* Nauka, Moskva.

Baker, V.C.; Benito, G. & Rudoy, A.N. (1993). Paleohydrology of Late Pleistocene super-flooding, Altai Mountains, Siberia, *Nature* ,Vol. 259, pp. 348-350.

Baryshnikov, G.Ya., (1992). *Relief Evolution in Transitional Zones of Mountain Areas during the Cenozoic Era,* University of Tomsk Press, Tomsk, 182 p. (in Russian).

Bazarov, D.B.(1986). *Pribaikal and West Transbaikal Cenozoic.* Nauka, Novosibirsk (in Russian).

Bolikhovskaya, N.S. & Molodkov, A.N. (2006). East European loess-palaeosol sequence: Palynology, stratigraphy and correlation, *Quaternary International,* Vol. 149, pp. 24-36.

Butvilovskiy, V.V. (1985). Catastrophic releases of waters of glacial lakes of the south-eastern Altai and their traces in relief, *Geomorphology,* Vol. 1985, No. 1, pp. 65-74.

Chlachula, J. (1999). Loess-palaesol stratigraphy in the Yenisei Basin, southern Siberia. In: Chlachula, J.; Kemp, R.A. & Tyráček, J. (Eds.), *Quaternary of Siberia. Anthropozoic* , Vol. 23, pp. 55-70.

Chlachula J. (2001a). Pleistocene climates, natural environments and palaeolithic occupation of the Altai area, west Central Siberia, *Lake Baikal and Surrounding Regions* (S. Prokopenko, N. Catto and J. Chlachula, Eds.), *Quaternary International,* Vol. 80-81, pp. 131-167.

Chlachula J. (2001b). Pleistocene climates, natural environments and palaeolithic occupation of the upper Yenisei area, south Central Siberia. *Lake Baikal and Surrounding Regions,* (S. Prokopenko, N. Catto and J. Chlachula, Eds.), *Quaternary International,* Vol. 80-81, pp. 101-130.

Chlachula J. (2001c). Pleistocene climates, natural environments and palaeolithic occupation of the Angara-Baikal area, east Central Siberia.*Lake Baikal and Surrounding Regions*

(S. Prokopenko, N. Catto and J. Chlachula, Eds.), *Quaternary International*, Vol. 80-81, pp. 69-92.

Chlachula J. (2003a). The Siberian loess record and its significance for reconstruction of the Pleistocene climate change in north-central Asia, *Dust Indicators and Records of Terrestrial and Marine Palaeoenvironments (DIRTMAP)* (Edward Derbyshire, Editor). *Quaternary Science Reviews*, Vol. 22. No. 18-19, pp. 1879-1906.

Chlachula J. (2003b). Controverse sur l'age des premiers Américains, *La Recherche* (Paris), Vol. 370, pp. 42-46.

Chlachula, J.; Evans, M.E. & Rutter, N.W. (1998). A magnetic investigation of a late Quaternary loess/palaeosol record in Siberia., *Geophysical Journal International*, Vol. 132, pp. 399-404.

Chlachula, J. (2010a). Environmental context and human adaptation of Palaeolithic peopling of the Central Urals, *Eurasian Perspectives of Environmental Archaeology* (J.Chlachula & N.Catto, Editors). *Quaternary International.*, Vol. 220, pp. 47-63. Chlachula. J. (2010b). Pleistocene climate change, natural environments and Palaeolithic peopling of East Kazakhstan, *Eurasian Perspectives of Environmental Archaeology* (J.Chlachula and N.Catto, Editors). *Quaternary International*, Vol. 220, pp. 64-87.

Chlachula, J.; Rutter, N.W. & Evans, M.E. (1997). A late Quaternary loess-palaeosol record at Kurtak, southern Siberia. *Canadian Journal of Earth Sciences* , Vol. 34, pp. 679-686. Chlachula J.; Drozdov N.I. & Ovodov N.D. (2003). Last interglacial peopling of Siberia: the Middle Palaeolithic Site Ust'-Izhul', the Upper Yenisei area. *Boreas*, Vol 32 (2003), pp. 506-520.

Chlachula J.; Kemp R.A.; Jessen .; Palmer A.P. &Toms P.S. (2004) (Chlachula et al. 2004a). Landscape development in response to climate change during the Oxygen Isotope Stage 5 in the southern Siberian loess region, *Boreas*, Vol. 33, pp. 164-180.

Chlachula J.; Medvedev G.I. & Vorobyova , G.I., 2004 (Chlachula et al. 2004b). Palaeolithic occupation in the context of the Pleistocene climate change in the Angara region, Central Siberia. *Anthropozoikum*, Vol. 25, pp. 31-49.

Chlachula, J. & Serikov, Yu. (2011). Last glacial ecology and geoarchaeology of the Trans-Ural area: the Sosva River Upper Palaeolithic complex, West Siberia, *Boreas*, Vol 40. No. 1, pp. 146-160.

Babek, O.; Chlachula, J. & Grygar, J. (2011). Non-magnetic indicators of pedogenesis related to loess magnetic enhancement and depletion from two contrasting loess-paleosol sections of Czech Republic and Central Siberia during the last glacial-interglacial cycle, *Quaternary Science Reviews*, Vol. 30, pp. 967-979.

Baryshnikov, G.Ya. (1992). *Relief Evolution in Transitional Zones of Mountain Areas during the Cenozoic Era*, University of Tomsk Press, Tomsk, 182 p. (in Russian).

Butvilovskiy, V.V. (1985). Catastrophic releases of waters of glacial lakes of the SE Altai and their traces in relief, Geomorphology, Vol.1985, No. 1, pp. 65-74 (in Russian).

Chlachula, J. & Little, E. (2011). A Late Quaternary high-resolution loess-palaeosol record from Iskitim, south-western Siberia, *The Second Loessfest (2009)*, (Markowic, S.B., Catto, N., Smalley, I.J., Zoeller, L), *Quaternary International*, Vol. 240, pp. 139-149. Dalton, R. (2010). Fossil finger points to new human species. DNA analysis reveals lost relative from 40.000 years ago, *Nature*, Vol. 464, pp. 472-473.

Derevianko, A.P. (1990). *Palaeolithic of North Asia and the Problem of Ancient Migrations*, SB AS SSSR, Novosibirsk, 122 p.

Derevianko A.P. (2010). Three Scenarios of the Middle to Upper Paleolithic Transition: Scenario 1: The Middle to Upper Paleolithic Transition in Northern Asia. *Archaeology, Ethnology and Anthropology of Eurasia,* Vol. 38, No. 3, pp. 2-32.

Derevianko, A.P. & Markin, S.V. (1992). The Mousterian of the Gorno Altai, Nauka, Novosibirsk, 224 p. (in Russian).

Derevianko, A.P. & Markin, S.V. (1999). The Middle and Upper Palaeolithic of the Altai, *Quaternary of Siberia* (Chlachula, J., Kemp, R.A. & Tyráček, J., Eds), *Anthropozoikum,* Vol. 23, pp. 157-166.

Derevianko, A.P. & Shunkov, M.V. (2009). Development of Early Human Culture in Northern Asia, *Paleontological Journal,* Vol. 43, No. 8, pp. 881-889.

Dergacheva, M.I., Fedeneva, I.K. & Lbova, L.V. (1995). A pedogenic evidence of the Upper Palaeolithic occupation of the Varvarina Mountain, *Ecology and Culture of the peoples of Siberia and America, horizons of complex research,* Chita's Pedological Institute, Chita, pp. 88-89.

Deviatkin, E.V. (1965). Cenozoic Deposits and Neotectonics of the South-Eastern Altai, *Proceedings GIN AN SSSR,* Vol. 126, 244 p. (in Russian).

Deviatkin, E.V. (1981). *The Cenozoic of the Inner Asia,* Nauka, Moskva, 196 p. (in Russian).

Drozdov; N.I., Laukhin, S.A.; Chekha, V.P.; Kol'tsova, B.G.; Bokarev, A.A. & Vikulov, A.A. 1990 (Drozdov et al., 1990). The Kurtak Archaeological Region, *Chronostratigraphy of Palaeolithic of North, Central and Eastern Asia, and America,* SB RAS, Krasnoyarsk.

Drozdov, N.I.; Chlachula, J. & Chekha, V.P. (1999). Pleistocene environments and Palaeolithic occupation of the Northern Minusinsk Basin, Krasnoyarsk Region *Quaternary of Siberia* (Chlachula, J., Kemp, R.A. & Tyráček, J., Eds), *Anthropozoikum,* Vol. 23, pp. 141-155.

Evans, M.E.; Rutter, N.W.; Catto, N.; Chlachula, J. & Nyvlt, D. (2003). Magnetoclimatology: teleconnection between the Siberian loess record and North Atlantic Heinrich events. *Geology,* Vol. 31, No. 6, pp. 537-540.

Foronova, I.V., 1999. Quaternary mammals and stratigraphy of the Kuznetsk Basin (southwestern Siberia). *Quaternary of Siberia* (Chlachula, J., Kemp, R.A. & Tyráček, J., Eds), *Anthropozoikum,* Vol. 23, pp. 71-97.

Grosswald, M.G. & Hughes, T.J. (2002). The Russian component of an Arctic ice-sheet during the Last Glacial Maximum, *Quaternary Science Reviews,* Vol. 21, pp. 121-146.

Grygar; T., Kadlec, J.; Pruner, P.; Swann, G.; Bezdička, P.; Hradil, D.; Lang, K.; Novotná, K. & Oberhansli, H. (2006). Paleoenvironmental record in Lake Baikal sediments: Environmental changes in the last 160 ky, *Palaeogeography Palaeoclimatology Palaeoecology,* Vol. 237, pp. 240-254.

Kalmikov, N.P. (1990). *Large mammalian fauna of the Pribaikal and West Transbaikal Pleistocene.* Proceedings of the Buryat Scientific Centre, Ulan-Ude, 116 p (in Russian).

Khenzykhenova, F.I.. (1999). Pleistocene disharmonius faunas of the Baikal regiona (Russia, Siberia) and their implication for palaeogeography, *Quaternary of Siberia* (Chlachula, J., Kemp, R.A. & Tyráček, J., Eds), *Anthropozoikum,* Vol. 23, pp. 119-124.

Konstantinov, M.V. (1994). *L'age de Pierre de la Région est de l'Asie de Baikal.* Ulan-Ude, 180 p.

Larichev, V.; Khol'ushkin, U. & Laricheva I., (1990). The Upper Palaeolithic of Northern Asia: Achievements, Problems, and Perspectives. II. Central and Eastern Siberia. *Journal of World Prehistory,* Vol. 4, No. 3, pp. 347-385.

Laukhin, S.A.; Klimanov, V.A. & Belaya, B.V. (1999). Late Pliocene and Pleistocene palaeoclimates in northeastern Chukotka, *Quaternary of Siberia* (Chlachula, J., Kemp, R.A. & Tyráček, J., Eds), *Anthropozoikum*, Vol. 23, pp. 17-24.

Lbova, L. (1996). The Palaeolithic site Kamenka, *New Palaeolithic Sites of the Zabaikalye*(Konstantinov, M., Ed.), Nauka, Chita, pp. 24-47 (in Russian).

Little, E.C.; Lian, O.B.; Velichko, A.A.; Morozova T.D.; Nechaev, V.P.; Dlussky, K.G. & Rutter, N.W. (2002). Quaternary stratigraphy and optical dating of loess from the East European Plain (Russia), *Quaternary Science Review*, Vol. 21, pp. 1745-1762.

Mangerud, J.; Svendsen, J.I.. & Astakhov, V.I. (1999). Age and extent of the Barents and Kara Sea ice-sheets in northern Russia, *Boreas, Vol.* 28, No. 1, pp. 46-80.

Mangerud, J.; Astakhov, V.I. & Svedsen, J.I. (2002). The extent of the Barents-Kara ice sheet during the Last Glacial Maximum. *Quaternary Science Reviews*, Vol. 21, pp. 111-119.

Mangerud, J., Gosse, J., Matiouchkov, A. & Dolvik, T. (2008). Glaciers in the Polar Urals, Russia, were not much larger during the Last Global Glacial Maximum than today, *Quaternary Science Reviews*, vol. 27, pp. 1047-1057.

Markin, S.V. (1996). *The Palaeolithic of the North-West of the Altai-Sayan Mountain Region*, SB RAS, Novosibirsk, 58 p. (in Russian).

Markova, A.K. (2007). Pleistocene mammal faunas of Eastern Europe, *Quaternary International*, Vol. 160, pp. 100-111.

Matyushin, G.N. (1994). *Stone Age of the Southern Urals*. Nauka, Moskva (in Russian).

Medvedev, G.I.; Savel'ev, N.A. & Svinin, V.V. (1990). *Stratigraphy, Palaeogeography and Archaeology of Southern Central Siberia*, AN SSSR, Siberian Branch. Irkutsk, 164 p.

Mochanov, Y.A. (1992). *The Early Palaeolithic at Diring and the Question of Extra-Tropical Origin of Man*, SB RAS, Novosibirsk.

Mochanov, Y.A. & Fedoseeva, S.A. (1996). Berelekh, Allakhovsk Region, *American Beginnings: The Prehistory and Palaeoecology of Beringia* (F. H. West, Ed.), University of Chicago Press, Chicago., pp. 218-222.

Mochanov, Yu.A. & Fedoseeva, S.A. (2001). *Neosphere and Archaeology*, Nauka i Tekhnika v Yakutii, Vol. 2001, No. 1, pp. 28-33 (in Russian).

Mochanov, Yu.A. & Fedoseeva, S.A. (2002). Archeology, Palaeolithic of North-East Asia, Ex-Tropical Homeland of Humankind and Early Human Occupation Stages of America, Polyrnyj Krug, Yakutsk, 59 p.

Nemchinov, V.G.; Budaev, R.T. & Rezanov, I.N. (1999). Pleistocene glaciations of the eastern Sayan Mountains), *Quaternary of Siberia* (Chlachula, J., Kemp, R.A. & Tyráček, J., Eds), *Anthropozoikum*, Vol. 23, pp. 11-15.

Okishev, P.A. (1982). *Dynamics of glaciations in the Late Pleistocene and Holocene*, Tomsk State University Press, Tomsk. 209 p. (in Russian).

Panova, N.K.; Jankovska, V.; Korona, O.M. & Zinov'ev, E.V. (2003). The Holocene dynamics of vegetation and ecological conditions in the Polar Urals. *Russian Journal of Ecology*, Vol. 3, No. 4, pp. 219-230.

Pavlov, P.Yu. (1996). *Palaeolithic Monuments of the North-Eastern European Part of Russia*, Ural Department RAS, Syktyvkar, 194 pp. (in Russian).

Pavlov, P.; Svendsen, J.I. & Indrelid, S. (2001). Human presence in the European Arctic nearly 40,000 years ago, *Nature*, Vol. 413, pp. 64-67.

Pavlov, P.; Roebroeks, W. & Svendsen, J.I. (2004). The Pleistocene colonization of north--eastern Europe, *Journal of Human Evolution*, Vol. 47, No. 1-2, pp. 3-17.

Petrin, V.T. (1986). *Palaeolithic sites of the West Siberian Lowland,* Siberian Branch RAS, Novosibirsk, 142 p. (in Russian).

Pitulko, V.V.; Nikolsky, P.A.; Girya, E. Yu.; Basilyan, A.E.; Tumskoy, V.E.; Koulakov, S.A.; Astakhov, S.N.; Pavlova, E. Yu. & Anisimov, M.A. (2004). The Yana RHS Site: Humans in the Arctic before the Last Glacial Maximum, *Science, Vol.* 303, pp. 52-56.

Prokopenko, A.A.; Hinnov, L.A.; Williams, D.F. & Kuzmin, M.I. (2006). Orbital forcing of continental climate during the Pleistocene: a complete astronomically tuned climatic record from Lake Baikal, SE Siberia, *Quaternary Science Reviews,* Vol. 25, pp. 3431-3457.

Rezanov, I.N. (1988). *Cenozoic Deposits and Morphostructure of East Pribaikal.* Nauka, Novosibirsk, 128 p. (in Russian). Rezanov, I.N. (1995). The role of the neotectonic support in the formation of sandy series of the West Transbaikal valley systems, *Geomorphology,* Vol. 1, pp. 80-87 (in Russian).

Rezanov, I.N. & Kalmikov, N.P. (1999). Palaeogeography of the Pribaikal and Transbaikal Quaternary), *Quaternary of Siberia* (Chlachula, J., Kemp, R.A. & Tyráček, J., Eds), *Anthropozoikum,* Vol. 23, pp. 43-48.

Rudoy, A.N. & Baker, V.R. (1993), Sedimentary effects of cataclysmic Late Pleistocene glacial outburst flooding, Altay Mountains, Siberia, *Sedimentary Geology,* Vol. 85, pp. 53-62.

Rudoy, A.N., Lysenkova, Z.V., Rudskiy, V.B. & Shishin, M.Yu. (2000). Ukok. The Past, Present and Future. Altai State University Press, Barnaul. 174 p. (in Russian).

Rutter N.W.; Rokosh C.E.; Evans M.E.; Little C.E.; Chlachula J. & Velichko A.A. (2003). Correlation and interpretation of paleosols and loess across European Russia and Asia over the last interglacial cycle, *Quaternary Research,* Vol. 60, No. 1, pp. 101-109.

Savel'eva, E.A., Ed. (1997). *Archaeology of the Komi Republic.* DiK, Moskva, 758 p. (in Russian).

Serikov Yu.B. (1999). *Palaeolithic of the Central Urals,* Nizhniy Tagil University Press, Nizhniy Tagil, 103 p. (in Russian).

Serikov Yu.B. (2000). *Palaeolithic and Mesolithic of the Central Trans-Urals,* University Press, Nizhniy Tagil', 271 p. (in Russian).

Serikov Yu.B. (2001). The Cave Palaeolithic of the Chusovaya River valley and problems of the initial peopling of the Central Urals, *Problems of Prehistoric Culture,* Gilem Press, Ufa, pp. 117-135 (in Russian).

Serikov, Yu.B. (2007). *The Palaeolithic Occupation Site Gari and Some Problems of the Ural Palaeolithic,* Nizhniy Tagil' University Press, Nizhniy Tagil', 137p. (in Russian).

Stefanovskiy, V.V. (2006). *Pliocene and Quaternary of the Eastern Slope of the Urals and Trans-Urals,* Ural Branch RAS, Ekaterinburg, 223 p. (in Russian).

Shun'kov, M. V. & Agadzhanyan, A.K. (2000). Palaeogeography of Palaeolithic of Denisova Cave, *Archaeology, Ethography and Anthropology of Eurasia,* Vol. 2000, No. 2, pp. 2-20.

Svendsen, I.G. & Pavlov, P.Yu. (2003). Mammontova Kurya: an enigmatic, nearly 40 000 year-old Paleolithic site in the Russian Arctic, *Chronology of the Aurignacian and the Transitional Technologies, Dating Stratigraphies, Cultural Implications* (Zilhao, J. & d'Errico, F., Eds), *Trabalhos de Arquelogia,* Vol. 33, pp. 109-120.

Tseitlin, S.M. (1979). *Geology of Palaeolithic of Northern Asia,* Nauka, Moskva, 286 p

Vasiliev, S.A. (1992). The Late Paleolithic of the Yenisey: a new outline, *Journal of World Prehistory,* Vol. 6, pp. 337-383.

Vasiliev, S.A.; Yamskikh, A.F.; Yamskikh, G.Y.; Svezhentsev, Y.S. & Kasparov, A.K. (1999). Stratigraphy and palaeoecology of the Upper Palaeolithic sites near the Maima village (Upper Yenisei valley, Siberia), *Quaternary of Siberia* (Chlachula, J., Kemp,

R.A. & Tyráček, J., Eds), *Anthropozoikum*, Vol. 23, pp. 29-35. Vasil'ev, S.A. (2001), Man and mammoth in the Pleistocene Siberia, The World of Elephants. Proceedings of the First International Congress. Consiglio Nazionale delle Ricerche, Rome. (Cavarretta, G., Gioia, p., Mussi, M., Palombo, M.R., Eds.), pp. 363-366.

Vasilevskyy, A.A. (2008). Stone Age of the Sakhalin Island, Yuzno Sakhalinsk Press, 412 p.

Velichko, A.A., Ed. (1993). Evolution of Landscapes and Climates of Northern Eurasia. Late Pleistocene - Holocene. Elements of Prognosis. 1. Regional Palaeogeography. RAN, Institute of Geography RAN, Nauka, Moskva, 102p. (in Russian).

Velichko, A.A., Ed. (1993). Evolution of Landscapes and Climates of Northern Eurasia. Late Pleistocene - Holocene. Elements of Prognosis, 1. Regional Palaeogeography. RAN, Institute of Geography. Nauka, Moscow.

Velichko, A.A.; Catto, N.; Drenova, A.N.; Klimaov, V.A.; Kremenetski, K.V. & Nechaev, V.P. (2002). Climate changes in East Europe and Siberia at the Late glacial-Holocene transition, *Quaternary International*, Vol. 90, pp. 75-99.

Velichko, A.A.; Grichuk, V.P.; Gurtovaya, E.E.; Zelikson, E.M.; Borisova, O.K. & Barash, M.S. (1992). Climates during the last interglacial, *Atlas of Paleoclimates and Paleoenvironments of the Northern Hemisphere, Late Pleistocene – Holocene* (Frenzel, B., Péczi, M. & Velichko, A.A., Eds.), Geographical Institute, Hungarian Academy of Sciences and Gustav Fisher Verlag, Budapest - Stuttgart, pp. 86-89.

Velichko, A.A.; Kononov, Yu.M. & Faustova, M. (1997). The last glaciation of the Earth: Size and volume of ice-sheets, *Quaternary International*, Vol. 41/42, pp. 43-51.

Vereschagin, N.K. (1977). The Belerekh mammoth „cemetery". *Trudy Zoologicheskogo Instituta*, Vol. 72, pp. 5-50 (in Russian).

Vereschagina, V.S. (2001). Alluvial deposits of the high antropogenic terraces of the Urals, *Antropogen Urala*, Perm State University Press, Perm, pp. 74-95 (in Russian).

Yakhimovich, V.L.; Nemkova, V.K. & Sidnev, A.V. (1987). *Pleistocene of the Fore-Urals*, Nauka, Moskva, 113 p. (in Russian).

Waters, M.R., Forman, S. &. Pierson, J. (1995). Diring Yuriakh: A Lower Palaeolithic Site in Central Siberia and its implications to the Pleistocene peopling of the Americas. – *Abstracts*, Annual Meeting of the Society for American Archeology, Minneapolis, May 3-7, SAA, p. 194.

Yamskikh, A.F. (1990). Late Paleolithic ecology of the Upper and Middle Yenisei valley, *Early Man News*, Vol. 15, pp. 39-41.

Yamskikh, A.F. (1996). Late Quaternary intra-continental river palaeohydrology and polycyclic terrace formation: the example of south Siberian river valleys, *Global Continental Changes: the Context of Palaeohydrology* (Branson, J., Brown, A.G. & Gregory, K.J., Eds.), Geological Society, London, Vol. 115, pp. 181-190.

Zenin, V.N., (2002). Principal stages of peopling of the West-Siberian Plain by Palaeolithic man, *Archaeology, Ethnography and Anthropology of Eurasia*, Vol. 12, pp. 22-44.

Zudin, A.N.; Nikolaev, S.V.; Galkina L.I.; Butkeeva, O.Yu.; Efimova, L.I.; Panychev, V.A. & Ponomareva, V.A. (1982). Stratigraphic scheme of the Neogene and Quaternary deposits of the Kuznetsk Basin, *Problems of Stratigraphy and Palaeogeography of the Pleistocene Siberia*, Nauka, Novosibirsk, pp. 133-149 (in Russian).

Zykina, V. S. (1999). Pleistocene pedogenesis and climate history of western Siberia. *Quaternary of Siberia* ((Chlachula, J., Kemp, R.A. & Tyráček, J., Eds), *Anthropozoikum*, Vol. 23, pp. 49-54.

9

Accumulation of Bio Debris and Its Relation with the Underwater Environment in the Estuary of Itanhaém River, São Paulo State

Fresia Ricardi Branco[1], Sueli Yoshinaga Pereira[1],
Fabio Cardinale Branco[2] and Paulo R. Brum Pereira[3]
[1]D. Geologia e Recursos Naturais, Instituto de Geociências,
Universidade Estadual de Campinas -UNICAMP, Campinas, SP
[2]Environmentality, São Paulo, SP
[3]Instituto Florestal, SP
Brazil

1. Introduction

The following chapter presents some considerations regard to the interrelation between the bioclastic accumulations (plants and animals) in a coastal estuarine environment, and the influence of sub surface water in its preservation in the geological record. To it, the plants and zoo debris accumulations and the aquifers in the Itanhaém River basin, south coast of São Paulo State, Brazil were studied and characterized.

The estuary Itanhaém River is located on the southern coast of São Paulo State as part of Santos - Peruibe Quaternary Coastal Plain. It has characteristic of tropical environments, mainly by the exuberance of the Atlantic Rain Forest (Brazilian biome characterized by its high biodiversity which grows associated with the Brazilian coastal portion), largely in primary conditions of conservation, by the Forest of Restinga or dense ombrophyle forest of low altitudes (forest characterized by a high biodiversity that develops in sandy soils of the coastal regions of Brazil) also mostly in primary conditions for conservation and by the Mangrove, which is considered one of the most well protected area of the state.

The Coast of São Paulo has two climatic determiners that influence the dynamic of atmospheric water: (1) the line of the Tropic of Capricorn crosses the region, and (2) the presence of the Serra do Mar with altitudes (maximum of about 1000 meters) very close to the coast of the Atlantic Ocean, and covered by dense tropical vegetation, the Atlantic Forest. Thus, polar air masses and tropical, have interactivity with elements of the relief. The altitudes near the coast, the moisture of ocean evaporation, the evapotranspiration, the slope orientation are the factors responsible to make this region as one of the most humid in Brazil.

The characterization of climate dynamics is based on information from weather stations used in the regional analysis presented in Table 1.

The weather stations of São Paulo city was taken as a reference to the availability of climate information for long period to high altitudes. The station of the Research Center of the Itanhaém river (CPeRio), despite the short period of observation, is a complete station, situated in the Itanhaém river basin and was considered as a local reference.

Stations (name)	Period (years)	Latitude (degrees and Min.)	Longitude (degrees and Min.)	Altitude (m)
Santos	1961-1990	23°56'	45°20'	14
São Paulo	1961-1990	23°30'	46°37'	792
Ubatuba	1961-1990	24°26'	45° 06'	08
Paranaguá	1961-1990	25°31'	48°31'	05
Iguape	1977-1986	24°42'	47°33'	03
CPeRIO*	2002-2004	24°10'	46°47'	02

Table 1. Weather stations used for the study (the south coastal region of the state). * Climate Station of the Research Center of Itanhaém River.

The Central and Southern coast region of São Paulo has average annual temperatures between 19.6 and 21.8 ° C. The highest values are found during January, February and March, while the lowest values are during June, July and August. The absolute maximum temperatures on record occurred in the summer and their values ranged between 39.0 and 40.0 ° C. The absolute minimum temperatures were recorded in the winter months and ranged from 2.4 to 6.4 ° C.

The rates of average annual total rainfall range between 1,932.2 and 2,080.8 mm. The rainiest quarter is January, February and March with rates accrued between 741.6 and 816.2 mm, making a percentage in relation to the total annual rates between 35.9 and 42.2%. The less rainy quarter is in June, July and August with combined rates between 271.9 and 283.0 mm, present in relation to the total annual rates ranging between 13.6 and 14.1%. The maximum daily rainfall recorded ranged between 206.7 and 289.9 mm.

The average values of atmospheric pressure range between 867.7 and 968.6 hPa. The highest values are observed in July (range between 917.2 and 1,018.8 hPa), while the lowest values are found in November (776.0 hPa) and February (1,009.9 hPa).

In the area, the total annual evaporation ranges from 736.9mm 968.6 mm. The maximum values occur in the summer with rates ranging between 83.4 and 93.3mm (January), while the lowest values are observed in winter, where rates range from 45.6 (June) and 70.4mm (July).

Annual insolation values range between 1,227.6 and 1,494.1 hours. Each month, the highest values are found in summer and range between 142.7 and 149.6 hours (January), while the lowest values are found in the spring and ranges between 73.0 and 88.7 hours (September).

The monthly average rates of cloudiness range between 6.3 and 6.4 tenths. The highest rates are found in late winter and spring, ranging between 6.1 and 7.3 tenths, while the lowest values are found in autumn / winter, ranging between 5.2 and 6.0 tenths.

The average annual rates of relative humidity range between 80.0 and 84.0%. The highest rates are observed in late summer with rates between 83.0 and autumn winter further south in the area, with rates of 86%, however, the lower values in the central portion, in winter, with a rate of 75%.

The hydrographic basin of the Itanhaém River is the second largest on the coast of the state of São Paulo, with an area of 930 km², mostly located in the municipality of Itanhaém, in the southern coast of the state (Camargo et al., 2002; Ricardi-Branco et al., 2009, 2011). The Quaternary coastal plain of the Itanhaém is limited by the city of Peruíbe to the southwest and the city of Mongaguá to the northeast, extending for some 50 km, it stretches 15-16 km to the west of the base of the Serra do Mar mountain range (Suguio and Martin, 1978;

Accumulation of Bio Debris and Its Relation
with the Underwater Environment in the Estuary of Itanhaém River, São Paulo State

193

Camargo et al., 2002; Amaral et al., 2006). The upper portion of the basin is contained within the Serra do Mar State Park with its sources well preserved. In the middle and lower portions of the basin, the anthropogenic influence is more intense. The main variations in the basin over time are thus the consequence of the patterns of variation of the tides and the intensity of the rain (Camargo et al., 2002). The tidal range is relatively low (less than 1 m, Giannini et al., 2009), but these microtides influence all of the river flowing across the narrow plain.

The main rivers draining the basin are the Itanhaém, formed by the junction of Branco and Preto rivers, and the Mambu, Aguapeú and Guaraú rivers, all tributaries of the Branco River. Studies have been conducted in Itanhaém, Preto and Branco rivers as described next. The Itanhaém river is formed by the confluence of Branco and Preto rivers, presenting estuarine features and drainage area of approximately 26 km² (Camargo et al., 2002). On its banks are still present the Restinga Forest and Mangrove Forest, as well as ornamental species introduced in recent years (e.g. *Terminalia*). Its course was modified in the first half of the twentieth century (the 40s) by opening a channel (artificial cut levees, joining two different points of the river after a meander), then about 1 m, now more than 100 meters wide. The opening of this artificial channel caused the abandonment of a portion of the meander, less active today and known as Acima River.

The Branco River originates in the Serra do Mar and runs much of its extension in the Precambrian terrains (Camargo et al., 2002) until it reaches the coastal plain. It is the largest river of the basin with 68 km in length. At its mouth there is a mixture of salt and fresh water under the influence of saline. In its margins are present the Atlantic Forest and the Restinga Forest, both considerably well preserved, as well as plantations (banana, cassava, etc.). Finally the Preto River has about 40 km long, originating in the Serra do Mar, although its greatest extent is located in the Quaternary plain (Camargo et al., 2002). It has little influence of salt water, mostly in its estuary with the Itanhaém River and its banks covered by the Atlantic and the Restinga Forests.

2. Bio debris taphonomy

2.1 Plant debris accumulations

Besides being a classical sedimentary environment of fossilization, the macro plant debris accumulations were studied in the middle and coastal portion for being the ones with best conditions for the biodebris accumulations. These accumulations were studied in detail by Ricardi-Branco et al. (2009, 2011) employing the original analytical methodologies for taphonomic studies. In addition, a systematic survey of the vegetation along the river basin was developed to draw comparisons with the leaves found in the accumulations. Despite the accumulation of plant debris are to be composed of leaves, twigs and fronds, preference was given to the study of angiosperm leaves, since they are the most abundant elements of the accumulations. The branches were small, often in the larger petioles. Finally, in the accumulations studied, none skeletal remains of invertebrates or vertebrates were found.

The survey was conducted during the years of 2002-2009, and were preferentially studied and / or accompanied with nine sites with leaves accumulations (Fig. 1), selected because they are in facies with the best potential for preservation and therefore fossilization. These accumulations are spatially distributed on the Quaternary plain as follows:

- in the levees in the interior of the plain on the banks of the Branco River (P1);
- in the middle portion of the plain in point bars - Inclined heterolithic structure desposits (IHS deposits) of the Preto River (P2 - P5);

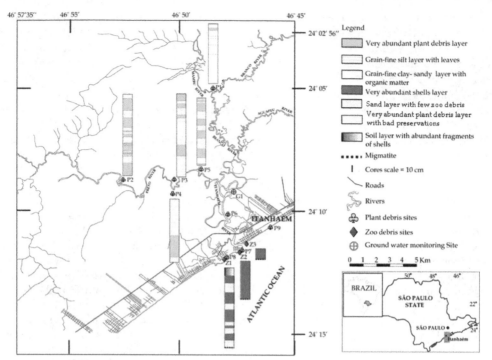

Fig. 1. Location map of the sites studied to plant debris (P1-P8) and to zoo debris (Z1-Z3). It was also plotted the cores developed to P1 to P5, in order to illustrate the constant presence of leaves packages in the margins of those rivers that drain the Itanhaém plain. The cores developed in Sites Z1 – Z3 show the studies to the shell accumulations collected in the coast of the basin.

-   in the lower portion of the plain within the Mangrove of the Itanhaém River (P6) and
-   in the coastal portion (Sibratel beach - Poço de Anchieta - P8, Mouth of the Itanhaém River - P7 and Itanhaém - Suarão beach -P9). These were followed up and did not have a detailed study of their components, because they are very scattered and still, quickly fragmented by wave action.

With the purpose of carrying out the identification of angiosperm leaves, the methodology described by Ricardi-Branco et al. (2009) was applied. First, the leaves collect were included in morphotypes (Figs. 3, 4 and 5) and later compared to the leaves of plants collected on the banks of the study sites. The results of plant debris analysis clearly showed their differential distribution in the quaternary plain of the Itanhaém River. So within the plain (Branco River, Site P1), in the middle portion (Preto River, Sites P2-P5), and the low portion (Itanhaém River Site P6) the accumulation of leaves were characterized as parautochthonous, primarily reflecting the surrounding vegetation of the study sites. In the Preto and Branco rivers the species corresponded to the Restinga Forest, and in the Mangrove of the Itanhaém River the leaves corresponded to that forest, the latter being the least biodiverse accumulation of those studied. In the coastal portion (Itanhaém River Mouth, Site P7; Sibratel Beach, Site P8 and Suarão Beach/Itanhaém, Site P9), plant debris were characterized as allochthonous composed by various elements from the mainland and from the ocean as well, and these were the only ones with plant and animal origins Ricardi-Branco et al. (2009).

Fig. 2. Sites of plant debris studies in the Itanhaém River Basin. 1, panoramic view of the Itanhaém River near its mouth; the arrow indicates the location of the Mangrove; 2 and 3, Site P1 overview and hand collection of leaves for study, 4 and 5, Preto River, meanders where layers of plant debris are deposited; 4 low tide and 5 high tide. 6 and 7, outcropping level of plant debris and 7 level of plant debris in core developed at P2.

Morphotipe	C1 - B1	C2 - B2	C5 - B3	C6 - B4	Ac1 - B5	Ac2 - B6	P1 - B7
Venation	Campt	Campt	Campt	Campt	Acro	Acro	Parallelo
Leaf shape	Obovate	Elliptic	Elliptic	Oblong	Obovate	Elliptic / Ovate	Ovate / Oblong
Base Shape	Cuneate	Cuneate	Cuneate	Convex	Cuneate	Cuneate/Convex	Convex
Apex shape	Acuminate	Acuminate	Convex	Rounded	Convex	Straight	Straight
Margin shape	Entire	Entire	Entire	Entire	Entire	Entire	Entire
Quantity	160	121	7	140	25	72	5
Related family	Fabaceae *(Inga)*	Lauraceae	Myrtaceae	Fabaceae *(Senna)*	Melastomataceae	Melastomataceae	Poaceae
Image	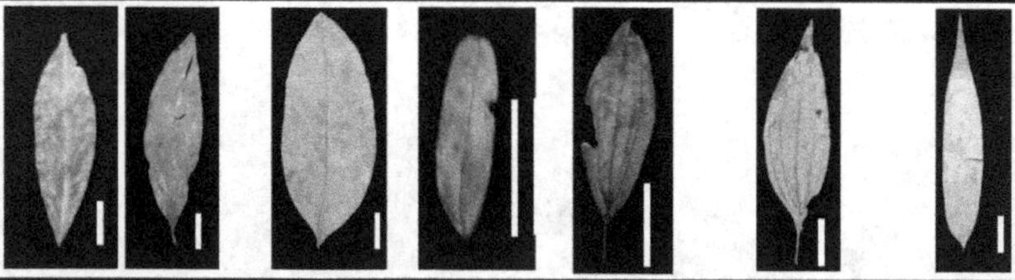						

Fig. 3. Morphotypes of the Branco River and their botanical affinity. Site P1. Campt = camptodromous; Acro= Acrodromous and Parallelo = Parallelodromous.

Morphotipe	C1-P1	C2-P2	C7-P3	C8-P4	Cl1-P5	Cr3-P6	P1-
Venation	Campt	Campt	Campt	Campt	Campt	Craspedo	Parallelo
Leaf shape	Elliptic	Elliptic	Oblong	Ovate	Elliptic	Elliptic	Elliptic
Base Shape	Complex	Cuneate	Cuneate	Complex	Convex	Convex	Cuneate
Apex shape	Straight	Convex	Convex	Acuminate	Convex	Acuminate	Straight
Margin shape	Entire	Entire	Entire	Entire	Entire	Crenate	Entire
Quantity	4	48	25	81	19	73	3
Related family	Lauraceae	Fabaceae *(Inga)*	Rubiaceae	Fabaceae *(Inga)*	Clusiaceae	Sapindaceae	Poaceae
Image	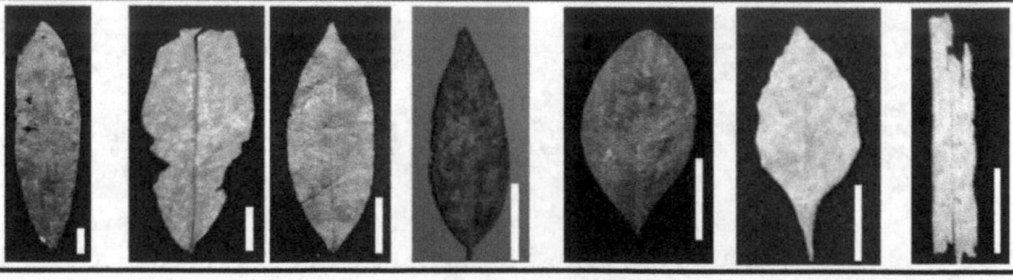						

Fig. 4. Morphotype of the Preto River and its botanical affinity. Site P2. Campt = camptodromous; and Parallelo = Parallelodromous.

Accumulation of Bio Debris and Its Relation
with the Underwater Environment in the Estuary of Itanhaém River, São Paulo State

197

At the same time, shallow cores were conducted with aluminum tubes of 7cm, manually buried during low tide up to 2m depth (Ricardi-Branco et al., 2009 , 2011), in three of the sites monitored for assembly of plant debris to follow the changes and evolution of the accumulations selected after the burying. The localities selected were: one on the margin of the Branco River inside the quaternary plain (P1), four on the margins of the Preto River in IHS deposits, in the middle portion of the basin (P2-P5) and the third in the mangroves in the lower portion of the plain. The best results regarding the preservation of the leaves were found in the cores of the Preto River (Fig. 1). The accumulations of plant debris along the meandering course of the Preto River to be characterized in greater detail with the aid of geophysical methods, were very frequent and associated with virtually all the meanders of the river (Fig. 4), featuring IHS deposits with an estuarine system with micro seas, where the monthly variations of the tide deposit thin laminated layers or tidal coupplets the cause of variations in river level. So the more thick layers (around 20 cm) of plant debris observed in the study correspond to climatic events that defoliated the trees of the Restinga Forest (Ricardi-Branco et al., 2011).

Morphotipe	M1	M2
Venation	Campt	Campt
Leaf shape	Obovate	Elliptic
Base Shape	Cuneate	Convex
Apex shape	Convex	Convex
Margin shape	Entire	Entire
Quantity	8	30
Related family	Rizophoraceae (*Rizophora*)	Combretacaea (*Laguncularia*)
Image	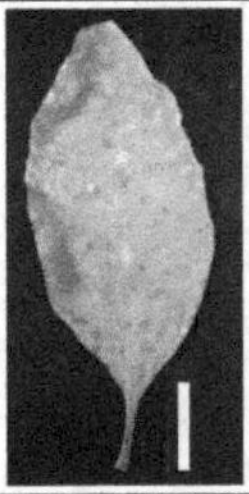	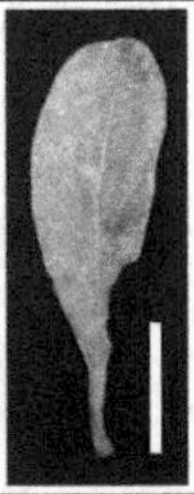

Fig. 5. Morphotype of the Itanhaém river and its botanical affinity. Site P6. Campt = camptodromous.

2.2 Zoo debris accumulations

From the second half of 2005 until the first half of 2006, natural deposits of zoo debris were characterized and described in the coastline portion of the Itanhaém River Basin (Zincone et al., 2007), showing the change in relative sea level during the late Holocene. These are the deposits where the presence of shell accumulations were detected (Fig. 1):

- Cibratel beach – beachrock of Poço de Anchieta (Site - Z1);
- Conchas beach in Costão de Paranambuco (Site - Z2) and
- the beach of Cabras Island located at a distance of approximately 140 m from the mainland (Site - Z3).

Pits were made in every site of the studied area up to the contact with the accumulations with the basement (Migmatite). The profiles were described and collected approximately 5 kg of each level has been taken to the laboratory for study and dating by C^{14}.

Site - Z1. Cibratel beach - beachrock of Poço de Anchieta (24 ° 12 '05.1 "S / 46 ° 48' 41.3" W): consisting of composed zoo debris cemented accumulation with approximately 1.30 m depth (Table 2). This kind of accumulation can be considered a beachrock, chenier or conglomerates of beach, usually consisting of quartz grains bound by carbonate cement, located in the intertidal zones and containing biogenic intraclasts (Suguio, 1992, 2004).

Site - Z2. Conchas beach, (24 ° 11 '46.7 "S / 46 ° 48' 06.01"): the accumulation is associated with a small shore stuck on the left margin of the coast, situated between the walls of migmatites, that features 13 meters wide in low tide and 22.5 meters wide at the posterior border with the coast. During the sampling collect, excavations were conducted at three different points of this beach. The deposit of bioclasts (mollusc shells) has 60 cm depth up to the base where it borders the migmatite.

Site Z3- Cabras Island (24 ° 11 '38.1 "S / 46 ° 47' 33.1" W): as the site is located on the Cabras Island at approximately 140m far from the Pescadores beach, the access occurs during low tide when the tide allows crossing the canal. In this place, three trenches were made in order to know the thickness of the shells deposit (Fig. 1 and tables 2 and 3).

The systematic composition of the deposits studied can be seen in Table 2. As the composition of sites Z1, Z2 and Z3 is very similar (Fig. 6).

Sites Bioclasts	Site - Z1 Beachrock (Anchieta beach) (24°12'05.1" S. / 46°48'41.3" W)	Site -Z2 Conchas beach (24°11'46.7" S / 46°48'06.1" W)	Site Z3 Cabras island (24° 11'38.1" S / 46°47'33.1"W)
Molusca - Bivalvia	Anachis sp., Perna perna, Tellina sp., Tivela sp. and Trachycardium sp.	Anadara sp., Macoma sp., Tivela sp. and Trachycardium sp.	Anadara sp., Brachidonte sp., Mactra sp., Mytella sp., Ostrea sp., Perna perna, Pisania sp., Tivela sp. and Trachycardium sp.
Molusca - Gastropoda	Colissella sp., Diodora sp., Dorsanum sp., Hastula sp., Littorina sp., Olivancillaria sp., Thais sp. and microgastropds	Colissella sp., Crepidula sp., Hastula sp., Janthina sp., Littorina sp., Pisania sp. e Thais sp. and microgastropods	Colissella sp., Diodora sp., Hastula sp., Olivancillaria sp., Pisania sp. and Thais sp.
Polychaeta -	Tubes	Tubes	Tubes
Artropoda	Chelae of crabs and barnacles	Chelae of crabs and barnacles	Chelae of crabs and barnacles
Vertebrata			Fish vertebrae
Fragments	Shells and sea urchin spines	Shells and sea urchin spines	Shells and sea urchin spines

Table 2. Systematic composition of the studied deposits with zooclasts.

Accumulation of Bio Debris and Its Relation
with the Underwater Environment in the Estuary of Itanhaém River, São Paulo State

199

Fig. 6. Zoo debris found in points Z1, Z2 and Z3. 1. *Trachycardium* sp.; 2, *Tivela* sp.; 3, *Perna perna*; 4, *Thais* sp.; 5, fragments of gastropods, 6, chelicerae crab, 7, tubes of polychaetes and 8, fragments of spines of sea urchins. Scale bar= 1, 2, 3, 4, 5 and 6 = 1 cm. 7 = 5mm e 8 = 0.5 mm

Regard to the taphonomy in points Z1, Z2 and Z3 the bioclasts (shells) are disconnected, densely packed and poorly chosen, since they vary in size and polymodal distribution, which does not observe a preferential orientation of bioclasts. The polymodal distribution indicates activity a turbulent flow during the formation of fossiliferous assemblage as in the case studied here where the waves remove the accumulations daily, the paleoecological fashion is polytypic (Holz & Simões, 2002). Regarding the presence of taphonomic signatures, the most obvious ones found were drilling marks by predation, mainly gastropods, and fragmented valves. These signatures were observed in the valves of mollusks collected at all study sites.

The age obtained for the beachrock (Table 3) indicates that although it is a deposit partially eroded and emerged, it reflects the environmental conditions during the deposition in intertidal zone (Angulo et al., 2002). On the other hand it is interesting to mention that as stated above, the composition in terms of bioclasts among sites Z1, Z2 and Z3, despite having different age composition is very similar, showing that at 1,185 cal yr BP the malacofauna of the Itanhaém coastal region was very similar to that currently inhabits the coast.

Age Sample/Depth (cm)	Conventional age (^{14}C years AP)	Calibrated Age (2 σ) (cal yr BP)	Historical Age (yr AD/BC)	Samples Numbers ^{14} Cena - USP
Z2 –Conchas beach 30 - 50	570+/- 70	510 - 665	766 – 1,024 AD	975 – CENA 547
Z1 – Anchieta Beach 92 - 98	1,130 +/- 70	930 – 1,185	1,286 – 1,442 AD	976 – CENA 549

Table 3. Ages obtained from deposits of shells Z1 and Z2.

3. River – Ground waters

3.1 Regional aspects

The aquatic environment presents a great sensitivity in regions where the contacts between seawater and fresh water outflow are related. This estuarine environment that sustains unique living beings is a region that has developed under a dynamic that involves relief, tides, river flows and climate dynamics of masses of tropical and polar fronts.

The river belongs to the Southeast Atlantic Hydrographic Macro-region of Santos Coastline and has an area of 102.57 km². The formation rivers, Branco River and Preto River, have areas of 411.66 km² and 426.46 km², respectively. Yet, this report estimates of average flows for long period of 4.10 m³/s, 12.9 m³/s, 18.9 m³/s, and minimum flows ($Q_{7,10}$) of 1.0 m³/s, 3.15 m³/s and 4.63 m³/s for the Itanhaém , Preto and Branco rivers, respectively (Comitê de Bacias Hidrográficas/ Baixada Santista [CBH-BS], 2000).

The region is influenced by the tide and consequent saline intrusion. According to Camargo et al. (2002) seawater at high tide penetrates through the river channel of the Itanhaém and Branco rivers, occurring predominantly in the lower river basin. The average values reported are of 2,252.6 ppm salinity in the Itanhaém River. The Branco River has a small influence of saline water with maximum records of 200.0 ppm salinity and no influence on the Preto River (average of 4.3 ppm and maximum value of 20.0 ppm). However, the influence of pressure of the tide is noticed in the entire length of the rivers mentioned, with the lifting and lowering of the water level. It is interesting to note the differences in the aquatic environments between rivers Itanhaém and Branco, and the Preto River. In the latter one, its water is dark attributed by the presence of organic matter. The clear water of the other rivers may be attributed to poor soils, the large slope of the terrain, turbulence and the small number of primary producers, consumers and decomposers (Camargo et al., 2002).

The contrast of the steep topography of the Serra do Mar and the coastal plain provide different aquifer systems with dynamics of complex flow and interactivity of groundwater. The Crystalline Aquifer System is mainly composed of granites, gneisses, migmatites, schists and phyllites, Açungui Group (granites and schists varied), the Embu Complex and Coastal Complex (heterogeneous migmatites), and eluvial - colluvium continental sediments, bordering such rocks and being their alteration products. It is characterized by its large regional extension, fractured, free to semi-confined, heterogeneous, discontinuous and anisotropic. In the study area are characterized by the Serra do Mar, the hills that exist in the plains, and in the subsurface with basement of the sediments deposited which makes up the Cenozoic Coastal System.

The Cenozoic Coastal System is composed by sands interbedding with clay and siltstone and characterized by having a limited extension, granular, free and semi-confined, discontinuous, heterogeneous and anisotropic, usually with shallow water level. It is

Accumulation of Bio Debris and Its Relation
with the Underwater Environment in the Estuary of Itanhaém River, São Paulo State

201

composed by Holocene sediments of Mangrove and swamp (sand and clay) by Santos Formation of fluvial sediments - lagoon (sand and clay), coastal marine sediments (sands), and reworked coastal marine sediments and coastal marine sediments (sands) of Cananéia Formation.

3.2 Groundwater monitoring

An area of six thousand square meters (6,000 m^2) was selected for installation of eight monitoring wells for groundwater, located in the meander on the left bank of the Itanhaém River (Fig. 7), in order to understand the influence of tides on the river and in groundwater. Information from local people revealed the existence of at least three aquifers in this area: the first, superior, unconfined, with 7 meters deep; the second, confined between two clay layers, between 8 and 20 meters deep; and the third, below the clay layer from 20 meters deep.

Thus, monitoring wells were installed in the first two aquifers, four in the shallower aquifer and three in the second aquifer, whose main characteristics are shown in Table 4. Conductivity Hydraulic essays (slug tests) were made and the results and hydraulic conductivity (Hvorslev, 1951, as cited in Fetter, 2001) and transmissivity are presented in Table 4 and Fig. 7 shows the study area and location of the monitoring wells.

Monitoring wells	Coordinates Latitude/ Longitude	Topographic elevation (m)	Well depth (m)	K (m/s)	T (m^2/s)	Aquifer
P-1	24° 8' 44" 46° 47' 10"	2.020	6	1.6×10^{-4}	8.8×10^{-4}	unconfined
P-2	24° 8' 43" 46° 47' 9"	1.854	6	1.6×10^{-4}	8.8×10^{-4}	unconfined
P-3	24° 8' 44" 46° 47' 9"	2.109	6	1.2×10^{-4}	6.6×10^{-4}	unconfined
P-4	24° 8' 44" 46° 47' 9"	2.005	6	1.3×10^{-4}	7.2×10^{-4}	unconfined
PII-1	24° 8' 44" 46° 47' 8"	2.261	21	1.8×10^{-6}	4.6×10^{-6}	confined
PII-2	24° 8' 45" 46° 47' 7"	1.926	21	5.1×10^{-6}	1.2×10^{-5}	confined
PII-3	24° 8' 45" 46° 47' 10"	1.452	21	1.8×10^{-4}	4.4×10^{-4}	(semi) confined

Table 4. Localization of monitoring wells. K – hydraulic conductivity (m/s); T – transmissivity (m^2/s)

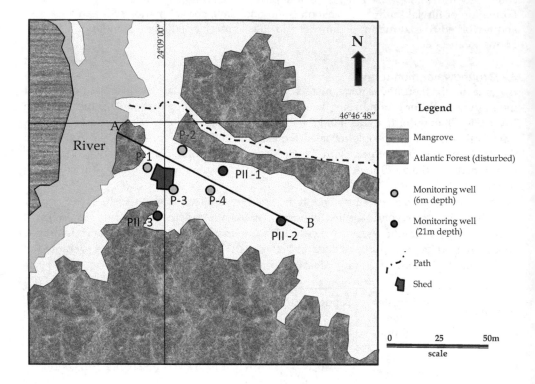

Fig. 7. Localization of monitoring wells and cross section (A – B) in the study area

The study area shows organic sandy soil in the upper portion, with fine-grained, and black color due to the presence of organic matter (up to half a meter deep). Then the soil becomes sandy with fine texture and a yellowish color (up to 1 meter deep), becoming dark gray after 2.5 meters deep. In the depth of 3.5 meters it presents light gray sand with the presence of plant debris to a depth of 7.0 meters. A 1- meter- thickness layer of silty clay is found in this interval, and between depths of 8 to 18.5 meters is found fine sand, with light gray color, but without the presence of plant debris. In the interval of 18.5 to 20 meters deep is found light gray fine sand with remains of calcareous shells. Finally, below 20 meters was found a silty clay layer, compact, with dark gray color (Fig. 8).

Two aquifers were identified: (a) the shallowest, whose bottom is impervious is at 7 meters deep (silty clay) and (b) the confined aquifer located between two silty clay layers, and with a 12-meter- thickness. So the silty clay layers correspond to the aquifuges of the site.

The shallow aquifer, unconfined and homogeneous, presents hydraulic conductivity between 1.6×10^{-4} to 1.2×10^{-4} m/s (transmissivity between 8.8×10^{-4} and 6.6×10^{-4} m2/s). The deeper aquifer presents lower hydraulic conductivity, between 4.6×10^{-6} and 1.2×10^{-5} m/s (transmissivity between 4.6×10^{-6} and 1.2×10^{-5} m2/s). The PII-3 well presented values of K and T intermediate (1.8×10^{-4} m/s to 4.4×10^{-4} m2/s) indicating the possibility of semi-confinement.

Accumulation of Bio Debris and Its Relation
with the Underwater Environment in the Estuary of Itanhaém River, São Paulo State

203

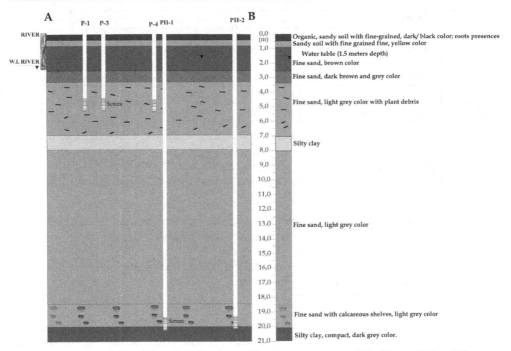

Fig. 8. Cross section (A – B) and geologic profile description (modified from Batista Filho, 2006)

3.3 Groundwater dynamics under tidal fluctuations

Two field campaigns were conducted to monitoring the dynamics of groundwater during the daily fluctuations of the tide. The first campaign was conducted between March-April 2005 (03/28 to 03/04/2005 - rainy season), and the second in August 2005 (08/20 to 08/26/2005 - less rainy season). The monitoring consisted of continuous measurements (every 30 minutes) of water levels from monitoring wells, pH and electrical conductivity (E.C.) during the fluctuation of the tide.

Table 5 shows the average results of monitoring of the river water (water level, pH, Eh and E.C.). The waters sampled during the rainy season are quite distinct from the waters of the less rainy period, as in the differences in tidal height variation as the values of pH, Eh and E.C.

The fluctuation measured of the level of river water is related to the location point of the tide gauge, located four kilometers from the Itanhaém river, downstream of the study area. The other physical-chemical parameters were measured in the study area, where the monitoring was conducted. The tide measured in CPeRio takes 1 hour and 30 minutes to reach the study area, through the river channel.

The differences of the variation in height of water level indicates a higher or lower influence of fresh water from the river (higher flow), combined with the fluctuation of the tide; in the rainy season there is a decrease in the difference between the maximum and minimum values of height. During the dry season, there is a reduction of river flow and consequently greater influence of the tides and greater height variation of the water level monitored.

RIVER	Water level (m) *		pH		Eh (mV)		E.C. (mS/m)	
	03/28 to 03/04/05 Rainy season	08/20 to 08/26/05 Less rainy season	03/28 to 04/04/05 Rainy season	08/20 to 08/26/05 Less rainy season	03/28 to 04/04/05 Rainy season	08/20 to 08/26/05 Less rainy season	03/28 to 04/04/05 Rainy season	08/20 to 08/26/05 Less rainy season
Average	0.51	0.74	7.5	6.4	224	260	27	827
Minimum	-0.19	-0.20	6.5	3.4	-169	182	10	186
Maximum	1.08	1.22	8.2	6.9	112	357	278	1100

Table 5. Average results (maximum and minimum) found in the monitoring of water level parameters, pH, Eh and E.C. of the river, from 03/28 to 03/04/2005 and 08/20 to 08/26/2005. * Change the water level in CPeRio (located four kilometers from the study area).

During the rainy season (January to March), the waters are less acidic, reducing environment and lower concentrations of salts and less influence of seawater in the study area.

During the less rainy season (June to August), the waters become more acidic due to the increasing concentration of humic acids. There is the increase of the concentration of dissolved salts, the lower river flow and higher facility of inflow of sea water in the channel at great distances.

In both monitoring periods the accumulated rainfall was 18.8 mm in 03/28 to 03/04/2005 and 3.0 mm in 08/20 to 08/26/2005. The rains in the first monitoring occurred on 03/28/2005 (7.4 mm) and on 03/04/2005 (11.4 mm); in the second monitoring, the rainfall occurred on 08/21/2005 (0.2 mm), 08/23/2005 (0.2 mm), 08/24/2005 (0.8 mm), 08/25/2005 (1.6 mm) and 08/26/2005 (0.2 mm).

Potentiometric surface maps for the two aquifers, shallow and confined, were prepared at various times of the rise and fall of the tide. Monitoring the values of pH, Eh and E.C. were made in conjunction with measurements of water level.

Figures 9 to 12 presents three moments each monitored day representing the dynamics of groundwater in accordance with the fluctuations of the tide. The days selected, representative of each period (wet and dry) were 03/30/2005 and 08/20/2005, at the same times of water sampling for physical and chemical analysis.

The potentiometric surface maps of the day 03/30/2005 of the shallow aquifer at high tide presents S-N direction, SE- NW and W-E, with hydraulic head higher in PM 3, followed by a hydraulic head of PM 1 (Fig. 9). This flow pattern seems to indicate the pressure of high tide in the aquifer near the river in the opposite movement of freshwater into the river (effluent). During the falling tide and low tide the flow of groundwater toward the river has lower hydraulic gradients, but realize also that the flow directions of S-N and SW-NE are predominant. In the deep aquifer (Fig. 10) the direction of groundwater flow is the opposite (NE-SW and NE-SE), however, the distribution of hydraulic head at various times of tide fluctuations indicates vertical oscillations of the aquifer potentiometry. At high tide, the hydraulic gradient is higher and decreases with low tide.

On 08/20/2005 (Figs. 11 and 12) the movement of groundwater in the shallow aquifer has lower hydraulic gradient and the predominant SW-NE direction. The influence of the tides (high and low) in the groundwater is noted with the increase in hydraulic head at high tide. In the deep aquifer groundwater, the direction of water flow is NE-SW (toward the river).

Accumulation of Bio Debris and Its Relation
with the Underwater Environment in the Estuary of Itanhaém River, São Paulo State

205

The higher hydraulic head moves at high tide and at the beginning of the fall (Fig.12). In the deep aquifer, the groundwater flows toward the river (NE-SW), the hydraulic head increases with the rising tide and decrease during the fall of the tide.

The dynamics of groundwater flow in the shallow aquifer has a direct influence of fluctuations of the tide. In March 2005, the potentiometric maps presented a strong northeast flow direction that is intensified at high tide, with greater hydraulic gradients than at low tide. The same behavior is observed in August 2005. Yet the prevailing flow direction to the northeast, but with lower gradients. The hydraulic heads, however, present higher values in August than in March.

In the deep aquifer, the groundwater flow has a less variable behavior, flowing to the southwest toward the river. However, there is a shift in the area of higher hydraulic head from north to northeast.

The major direction northeast flow of groundwater in the shallow aquifer is consistent with the situation of the river channel of the Itanhaém river, which greatly influences the movement of groundwater. In the case of deep aquifer the behavior is different and has SW direction toward the river. There is influence of the tide, but to a lesser extent than the shallow aquifer.

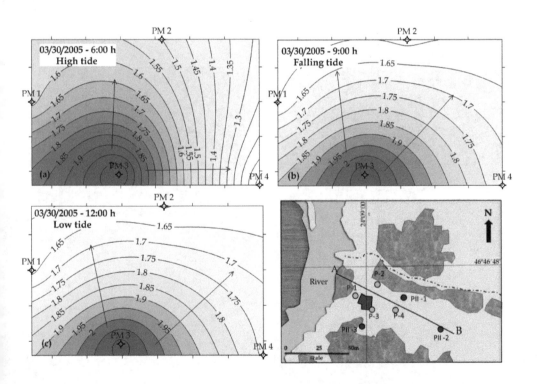

Fig. 9. Potentiometric surface maps representing groundwater flows in the following fluctuations: (a) high tide; (b) falling tide and (c) low tide - Shallow Aquifer (03/30/2005).

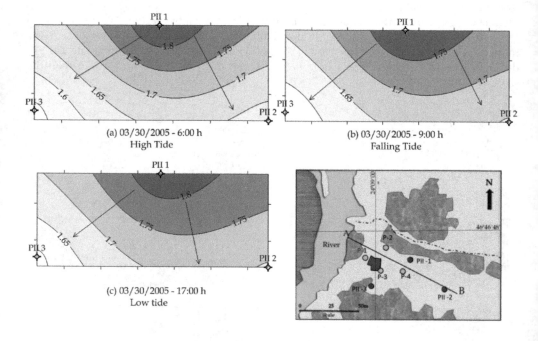

Fig. 10. Potentiometric surface maps representing groundwater flows in the following fluctuations: (a) high tide; (b) falling tide and (c) low tide - Deep Aquifer (03/30/2005).

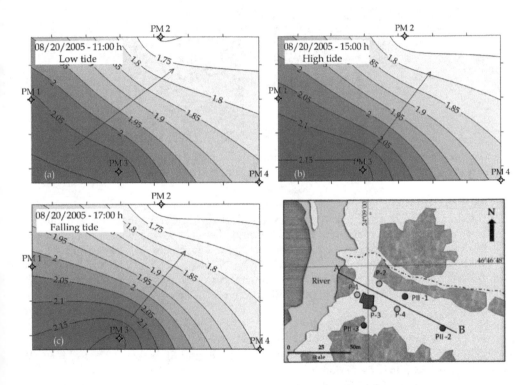

Fig. 11. Potentiometric surface maps representing groundwater flows in the following
fluctuations: (a) low tide; (b) high tide and (c) falling tide - Shallow Aquifer (08/20/2005).

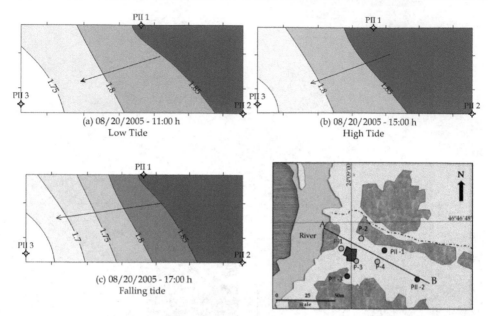

Fig. 12. Potentiometric surface maps representing groundwater flows in the following fluctuations: (a) high tide; (b) falling tide and (c) low tide - Deep Aquifer (03/30/2005).

3.4 Hydrochemistry

Samples of river water and groundwater were collected at different times of oscillation of the diary tide to study the variations of physical and chemical composition. The parameters pH, Eh and E.C. were determined by portable field equipment *in loco*, and HCO_3^- by titrations. The chemical analysis F^-, Cl^-, Br^-, $N-NO_3^-$, $N-NO_2^-$, SO_4^{2-} e PO_4^{2-} and PO_4^{2-} were performed by ion chromatography and the elements Na, K, Si, Mg, Ca, Sr, Co, Mn, Cu, Zn, Pb, Al, Ba, Cd, Ni, Fe and Cr by ICP-OES (Inductively Coupled Plasma Optical Emission Spectrometry).

The test results are presented in tables 6 and 7. The elements Cd, Co, Cr, Cu, Ni and Pb showed no concentrations above the detection limit of the equipment.

In 03/30/2005 the river presented alkaline pH (7.5 to 7.8) and Eh between 227 and 259 mV. The values of E.C. in this period ranged between 12 and 13 mS/m. In general, low concentrations are presented for all elements analyzed. The element zinc is not present in these waters, and the concentrations of phosphate and barium are very low. The waters have low concentrations of calcium (average of 4.4 mg/L), magnesium (2.2 mg/L), sodium (13.4 mg/L), potassium (1.8 mg/L), fluoride (0.05 mg/L) chloride (23.5 mg/L), sulfate (3.78 mg/L), nitrate (0.26 mg/L) and bicarbonate (16.9 mg/L). The concentrations of elements between the different moments of the tide almost did not change with the fluctuation of the tide in the river in the study area.

The waters of the shallow aquifer had acid pH (between 4.8 and 7.3, average 5.4), Eh between 222 and 271 mV and low E.C. values (between 7 and 8 mS/m). The concentrations of all elements of water from each well hardly varied over the period monitored, however the well PM-2 showed higher concentrations of calcium ions and bicarbonate and lower concentrations

Accumulation of Bio Debris and Its Relation
with the Underwater Environment in the Estuary of Itanhaém River, São Paulo State

209

of potassium and sulfate in relation to other wells (PM-1, PM-3 and PM-4). In general, calcium ions, magnesium, sodium and potassium ranged from 1.1 to 4.0 mg/L, 0.4 to 0.8 mg/L, 5.5 to 8.3 mg/L and 0.8 to 2.8 mg/L, respectively. Bicarbonate concentrations ranged from 6.7 to 10.1 mg/L, sulfate between 4.8 and 8.3 mg/L, and chloride between 11 and 13 mg/L. There is presence of fluoride (up to 0.06 mg/L), bromide (from 0.46 to 0.88 mg/L), aluminum (0.082-0.400 mg/L), barium (0.017-0.042 mg/L), iron (0.20 to 0.93 mg/L), manganese (0.009 to 0.017 mg/L), strontium (0.016-0.044 mg/L) and zinc (up to 0.009 mg/L).

The water of the deep aquifer is more mineralized with E.C. between 57 and 270 mS/m, alkaline pH (between 6.6 to 7.5), and lower Eh (from 129 to 187 mV). However, there is significant variation in concentrations among the water of the wells (PMII-1, and PMII-2 and PMII--3). The waters of the well-PMII-1 had E.C. of 270 mS/m, followed by the waters of PMII-2, E.C. between 79 and 95 mS/m, and PMII-3 with E.C. of 58 mS/m. It is notorious the concentration of sodium and bicarbonate in these waters (between 58 and 501 mg/L, and 322 and 1,152 mg/L, respectively), as well as chloride (40 to 260 mg/L). Among the elements of lower concentrations are the fluoride (between 0.17 and 0.45 mg/L), bromide (between 0.10 and 2.23 mg/L). The concentrations of calcium, magnesium and potassium ranged from 25.0 to 39.9 mg/L, 10.3 to 41.5 mg/L, and 8.4 to 41.2 mg/L, respectively. Sulfate, nitrate and phosphate presented low concentrations of up to 2.8 mg/L, from 3.30 to 21.90 mg/L and up to 2.61 mg/L respectively. Other elements like aluminum and barium showed negligible concentrations; iron, manganese, strontium and zinc presented variations between 0.02 to 0.43 mg/L, 0.006 to 0.091 mg/L, and absent to 0.056 mg/L, respectively. The three wells in the deep aquifer presented distinct compositional aspects, while the lowest mineralized water was in PMII-3 and the one with highest concentration water was in PMII-1.

The water of the shallowest aquifer had lower concentrations of chemical elements and higher compositional homogeneity in relation to the deepest aquifer. The river water had low concentrations of elements and less reduced environment. The monitoring in this period showed a predominance of fresh water in the estuary at that location, since the river flow had a strong contribution of rainwater from the entire drain area of the basin.

According to the monitoring of August (08/20/2005), the groundwater analyzed presented low compositional variation in relation to the monitoring results of March (03/30/2005). However, the river water presented high mineralization, with E.C. between 860 to 1,100 mS/m, acidic pH (6.5 and 6.8) and Eh from 240 to 289 mV. The surface water had low calcium concentration (between 5.6 and 7.4 mg/L), magnesium (23.26 and 16.90 mg/L) and potassium (between 6.93 and 9.76 mg/L), and high concentrations of sodium (between 151.5 and 192.4 mg/L). Regarding the anions, the concentrations vary as follows: bicarbonate, between 35.58 and 39.04 mg/L, sulfate, 362.4 and 470.6 mg/L, chloride, between 2,654.4 and 3,446.3 mg/L. There is an absence of phosphate, nitrite, fluoride, barium and zinc. Nitrate occurs at concentrations between 0.986 and 1.243 mg/L.

The waters of the shallow aquifer presented more acidic pH than in the first monitoring (4.2 to 4.8), with Eh values between 254 to 341 mV and E.C. between 6 and 10 mS/m. Elements of calcium, magnesium, sodium and potassium concentrations presented between 1.5 and 3.8 mg/L, 0.71 to 1.19 mg/L, 6.4 to 9.2 mg/L and 0.84 and 4.14 mg/L. Bicarbonate (between 4.9 and 17.1 mg/L), sulfate (4.5 to 13.2 mg/L), chloride (9.1 to 15.0 mg/L) presented at low concentrations, and absence of nitrate and nitrite. Fluoride (0.02 mg/L) occurs only in the first and second sampling of PM-1. The other elements such as bromide, aluminum, barium, iron, manganese, zinc and strontium in their respective concentration ranges: 0.35 to 0.52 mg/L, 0.248 to 0.513 mg/L, 0.021 to 0.062 mg/L, 0.421 to 0.934 mg/L, 0.011 to 0.025 mg/L, 0.023 to 0.047 mg/L and 0.009 to 0.025 mg/L.

The deep aquifer presented slightly acidic to neutral water (6.7 and 7.2), Eh between 109 and 196 mV, and EC 49 and 230 mV. As in the shallow aquifer, the waters of the deep aquifer did not show significant compositional differences between one period and another. The concentrations of calcium, magnesium, sodium and potassium presented respective values of 23.6 and 49.0 mg/L, 12.28 and 44.49 mg/L, 54.3 to 442.4 mg/L, 302.56 and 1,073.6 mg/L. High concentrations of bicarbonate (302.6 and 1,073.6 mg/L) and chloride (38.0 and 262.0 mg/L) are also present, as well as phosphate (2.86 and 30.90 mg/L). There is no presence of barium, and low concentrations of nitrate and zinc (occurring on average concentration of 0.045 mg/L in PMII-1). There is nitrite concentration up to 60.0 mg/L. Other elements such as fluoride, bromide, aluminum, iron, manganese, strontium and zinc occurred in the respective concentrations: 0.17 and 0.41 mg/L, 0.85 and 4.16 mg/L, 0.008 and 0.017 mg/L, 0.267 and 0.519 mg/L , from 0.037 to 0.145 mg/L, 0.278 and 0.287 mg/L, and absent at 0.049 mg/L.

The Piper Plot (Fig. 13) shows the compositions of the water analyzed and monitored in the river in the shallow aquifer, and in the deep aquifer. The chemical groundwater type does not change with the fluctuations of the tide.

The river water presents Na-Cl water type, typical of marine waters, although in the sample of 03/30/2005 waters have mixed behavior in relation to the sampling results of 08/20/2005, which are predominantly Na-Cl water type.

The waters of the shallow aquifer had mixed composition, Na-Cl-SO_4 water types.

The waters of the deep aquifer had Na-HCO_3 water type, and high concentrations. However, there are samples of Ca-HCO_3 water type. The presence of bicarbonate may result from the calcite dissolution, present in the organic matter but also on the shells found in this aquifer. The increased presence of sodium in these coastal environments over calcium may indicate exchange of bases where calcium can be adsorbed and sodium released, resulting Na-HCO_3 water type. It happens when fresh water flushes salts that are in the aquifer (Appelo & Postma, 2007).

The fraction of seawater (*f*sea) was calculated using the chloride concentration of the water sample. The chloride molar relation from the sample with the chloride from the sea water results in the contribution of salinity in fresh water (Appelo & Postma, 2007).

It is assumed that chloride is a conservative parameter and its only contribution is the sea water. Thus, Equation 1 was used.

$$f\text{sea} = m_{CL-,\ sample}/\ 566\ (\text{mmol}/L) \tag{1}$$

*f*sea - fration of seawater

$m_{CL-,\ sample}$ - concentration of Cl^- (mmol/L)

566 mmol/L - concentration of 35‰ (grams of salt per kilogram) seawater

The results of *f*sea of the waters from the shallow aquifer ranged between 0.05 and 0.07%. *f*sea of the waters from the deep aquifer ranged from 0.20 to 1.29%, and the waters of PMII-1 are those with the highest rate.

There was no variation in these rates in groundwater in the daily tidal fluctuations.

The river water presented variations in the fraction. The waters on 03/30/2005 showed low values, 0.11 and 0.12. On 08/20/2005, the waters had rates of 13.2, 17.2 and 16.1%, indicating influence of seawater into the river in the study area.

Accumulation of Bio Debris and Its Relation
with the Underwater Environment in the Estuary of Itanhaém River, São Paulo State

211

03/30/2005	pH	E.C. (mS/m)	Eh (mV)	Ca (mg/L)	Mg (mg/L)	Na (mg/L)	K (mg/L)	HCO3 (mg/L)	SO4^2 (mg/L)	(mg/L)	PO4^3- (mg/L)	(mg/L)	NO3 (mg/L)	(mg/L)	Al (mg/L)	Ba (mg/L)	Fe (mg/L)	Mn (mg/L)	Sr (mg/L)	Zn (mg/L)
PM-1 (6:00 h)	5.0	8	252	1.6	0.8	6.8	2.7	6.7	7.6	13	< 0.05	0.02	<0.013	0.62	0.230	0.041	0.60	0.017	0.021	0.005
PM-1 (9:00 h)	4.9	8	271	1.6	0.8	6.6	2.8	6.7	7.8	15	< 0.05	0.02	0.043	0.88	0.210	0.042	0.53	0.015	0.020	< 0.005
PM-1 (12:00 h)	5.0	8	269	1.5	0.8	6.0	2.7	6.7	8.2	13	< 0.05	0.06	<0.013	0.52	0.180	0.034	0.42	0.013	0.018	0.005
PM-2 (6:00 h)	5.3	7	233	4.0	0.7	6.0	0.8	10.1	5.0	11	< 0.05	0.01	<0.013	0.47	0.190	0.017	0.47	0.009	0.043	< 0.005
PM-2 (9:00 h)	5.1	8	245	3.9	0.7	5.5	0.9	10.1	5.1	11	< 0.05	0.02	<0.013	0.46	0.082	0.014	0.20	0.008	0.044	< 0.005
PM-2 (12:00 h)	5.3	8	256	4.1	0.8	6.8	0.8	10.1	4.8	11	< 0.05	0.03	<0.013	0.49	0.220	0.017	0.41	0.008	0.044	0.005
PM-3 (6:00 h)	4.9	8	248	1.5	0.7	6.6	2.0	6.7	8.3	12	< 0.05	<0.001	<0.013	0.57	0.270	0.039	0.65	0.015	0.021	0.007
PM-3 (9:00 h)	4.8	8	256	1.6	0.7	6.8	1.9	6.7	7.8	12	< 0.05	0.02	<0.013	0.52	0.270	0.040	0.59	0.013	0.021	< 0.005
PM-3 (12:00 h)	5.2	7	280	1.6	0.6	6.1	1.7	6.7	6.2	11	< 0.05	<0.001	<0.013	0.59	0.270	0.028	0.54	0.012	0.020	< 0.005
PM-4 (6:00 h)	7.3	8	222	1.2	0.4	7.3	1.3	6.7	6.4	13	< 0.05	0.02	<0.013	0.69	0.360	0.038	0.92	0.011	0.016	0.005
PM-4 (9:00 h)	5.7	8	227	1.1	0.4	8.3	1.3	6.7	6.5	13	< 0.05	0.03	<0.013	0.76	0.400	0.041	0.93	0.011	0.016	0.009
PM-4 (12:00 h)	6.5	8	228	1.2	0.5	7.7	1.3	6.7	6.4	13	< 0.05	<0.001	<0.013	0.74	0.390	0.039	0.92	0.010	0.016	0.005
PMII-1 (6:00 h)	7.2	270	170	25.1	39.0	440.0	32.2	1102.9	<0.01	245	18.60	0.37	<0.013	0.65	< 0.010	<0.0005	0.43	0.033	0.240	0.056
PMII-1 (9:00 h)	7.1	270	175	25.0	41.1	501.0	41.2	1151.7	0.9	259	19.50	0.45	<0.013	1.01	< 0.010	<0.0005	0.41	0.036	0.240	0.049
PMII-1 (12:00 h)	7.2	270	187	24.4	41.5	473.0	35.8	1132.2	2.8	260	21.90	0.33	2.61	2.23	< 0.010	<0.0005	0.38	0.032	0.240	0.044
PMII-2 (6:00 h)	7.5	81	147	22.6	14.7	121.0	14.0	409.9	1.3	64	3.60	0.29	0.09	1.01	< 0.010	<0.0005	0.02	0.006	0.170	< 0.005
PMII-2 (9:00 h)	7.5	79	163	25.9	18.4	118.0	14.3	429.4	<0.01	67	5.40	0.30	<0.013	0.18	< 0.010	<0.0005	0.04	0.012	0.180	< 0.005
PMII-2 (12:00 h)	7.5	95	161	28.1	21.1	165.0	15.6	507.5	1.8	85	8.10	0.31	0.043	2.31	< 0.010	<0.0005	0.05	0.018	0.200	< 0.005
PMII-3 (6:00 h)	6.7	58	129	39.9	10.3	59.4	8.5	327.0	0.4	40	3.30	0.17	<0.013	<0.005	< 0.010	<0.0005	0.31	0.077	0.230	< 0.005
PMII-3 (9:00 h)	6.7	58	138	39.5	10.6	58.9	8.4	322.1	1.6	40	3.60	0.17	<0.013	0.10	< 0.010	<0.0005	0.32	0.091	0.220	< 0.005
PMII-3 (12:00 h)	6.6	57	147	39.0	10.4	58.0	8.5	331.8	1.6	40	3.30	0.17	<0.013	0.10	< 0.010	<0.0005	0.30	0.074	0.220	< 0.005
River (6:00 h)	7.6	12	259	4.7	2.2	12.6	1.8	16.8	3.8	21	0.17	0.05	0.26	0.24	0.260	0.006	1.09	0.058	0.037	< 0.005
River (9:00 h)	7.8	13	233	3.8	2.2	13.9	1.8	16.8	3.5	25	< 0.05	0.05	0.26	0.23	0.230	0.008	0.77	0.032	0.032	< 0.005
River (12:00 h)	7.5	12	227	4.7	2.3	13.7	1.9	16.8	4.0	24	< 0.05	0.05	0.26	0.35	0.150	0.004	0.89	0.046	0.037	< 0.005

Table 6. River and ground water compositions under tidal fluctuations in the study area (03/30/2005)

08/20/2005	pH	E.C. (mS/m)	Eh (mV)	Ca (mg/L)	Mg (mg/L)	Na (mg/L)	K (mg/L)	HCO3 (mg/L)	SO4^2 (mg/L)	Cl (mg/L)	PO4^3 (mg/L)	NO2^- (mg/L)	NO3^- (mg/L)	F^- (mg/L)	Br^- (mg/L)	Al (mg/L)	Ba (mg/L)	Fe (mg/L)	Mn (mg/L)	Sr (mg/L)	Zn (mg/L)
PM1 (11:00 h)	4.8	10	287	1.8	1.19	7.8	2.96	8.5	6.9	13.8	< 0.05	<0.013	<0.017	0.02	0.36	0.285	0.062	0.934	0.025	0.026	0.022
PM1 (15:00 h)	4.5	7	308	1.8	1.16	7.7	4.14	7.3	8.5	13.0	< 0.05	<0.013	<0.017	0.02	0.47	0.304	0.055	0.748	0.021	0.025	0.020
PM1 (17:00 h)	4.4	8	315	1.6	1.00	7.4	2.88	7.3	8.8	13.1	< 0.05	<0.013	<0.017	<0.001	0.35	0.308	0.053	0.737	0.020	0.024	0.015
PM2 (11:00 h)	4.7	7	342	3.5	0.88	6.4	0.95	14.6	4.6	9.1	0.22	<0.013	<0.017	<0.001	0.37	0.248	0.023	0.588	0.013	0.047	0.013
PM2 (15:00 h)	4.8	6	341	3.6	0.85	6.4	0.84	17.1	4.5	9.5	< 0.05	<0.013	<0.017	<0.001	0.37	0.251	0.021	0.421	0.011	0.046	0.009
PM2 (17:00 h)	4.8	6	299	3.8	0.85	6.5	0.87	17.1	4.6	9.6	< 0.05	<0.013	<0.017	<0.001	0.39	0.251	0.021	0.430	0.011	0.047	0.011
PM4 (11:00 h)	4.2	9	266	1.5	0.71	9.2	1.50	4.9	13.2	14.4	< 0.05	<0.013	<0.017	<0.001	0.50	0.533	0.058	1.367	0.018	0.023	0.025
PM4 (15:00 h)	4.3	8	265	1.5	0.72	9.1	1.45	5.5	12.5	14.8	< 0.05	<0.013	<0.017	<0.001	0.52	0.513	0.054	1.357	0.018	0.023	0.019
PM4 (17:00 h)	4.4	9	254	1.5	0.74	9.2	1.45	5.5	12.4	15.0	< 0.05	<0.013	<0.017	<0.001	0.51	0.496	0.054	1.360	0.017	0.023	0.017
PMII-1 (11:00 h)	7.1	230	135	23.9	44.49	439.0	29.93	1073.6	<0.01	258.6	30.21	<0.013	58.62	0.41	3.50	0.012	<0.006	0.519	0.039	0.278	0.049
PMII-1 (15:00 h)	7.1	230	134	23.6	44.30	435.4	29.99	1073.6	<0.01	260.1	29.94	<0.013	<0.017	0.41	4.16	0.017	<0.006	0.509	0.037	0.280	0.042
PMII-1 (17:00 h)	7.2	230	109	24.1	43.08	442.4	30.20	1073.6	<0.01	262.0	30.90	<0.013	60.00	0.34	3.99	0.015	<0.006	0.500	0.038	0.285	0.044
PMII-3 (11:00 h)	6.7	49	196	49.0	12.28	55.5	9.36	302.6	0.1	38.0	2.86	<0.013	15.17	0.17	0.93	0.008	<0.006	0.267	0.143	0.281	<0.01
PMII-3 (15:00 h)	6.6	49	190	46.8	12.35	54.3	9.36	302.6	0.1	38.7	2.89	0.013	15.86	0.17	1.82	0.009	<0.006	0.361	0.142	0.278	<0.01
PMII-3 (17:00 h)	6.9	49	150	45.3	12.40	55.0	9.37	312.3	0.1	38.7	2.90	<0.013	15.52	0.17	0.85	0.005	<0.006	0.282	0.145	0.287	<0.01
River (11:40 h)	6.5	860	240	5.6	16.90	151.5	6.93	35.4	362.4	2654.4	< 0.05	0.986	<0.017	<0.001	9.80	0.007	<0.006	0.046	0.007	0.114	<0.01
River (16:10 h)	6.5	1100	289	7.4	23.26	192.4	9.76	39.0	470.6	3446.3	< 0.05	1.243	<0.017	<0.001	12.89	0.008	<0.006	0.016	0.009	0.151	<0.01
River (18:15 h)	6.8	1000	256	7.0	22.28	187.2	9.52	39.0	451.0	3333.3	< 0.05	1.113	<0.017	<0.001	11.16	0.006	<0.006	0.014	0.008	0.146	<0.01

Table 7. River and ground water compositions under tidal fluctuations in the study area (08/20/2005)

Accumulation of Bio Debris and Its Relation
with the Underwater Environment in the Estuary of Itanhaém River, São Paulo State

213

LEGEND

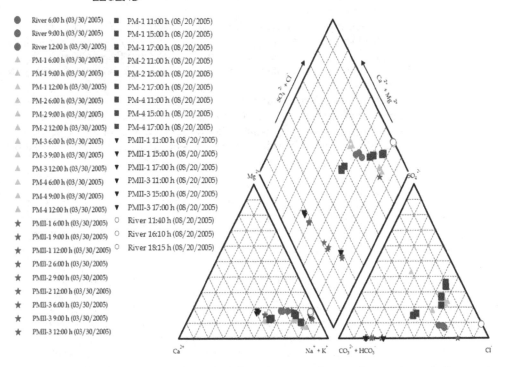

● River 6:00 h (03/30/2005)	■ PM-1 11:00 h (08/20/2005)
● River 9:00 h (03/30/2005)	■ PM-1 15:00 h (08/20/2005)
● River 12:00 h (03/30/2005)	■ PM-1 17:00 h (08/20/2005)
▲ PM-1 6:00 h (03/30/2005)	■ PM-2 11:00 h (08/20/2005)
▲ PM-1 9:00 h (03/30/2005)	■ PM-2 15:00 h (08/20/2005)
▲ PM-1 12:00 h (03/30/2005)	■ PM-2 17:00 h (08/20/2005)
▲ PM-2 6:00 h (03/30/2005)	■ PM-4 11:00 h (08/20/2005)
▲ PM-2 9:00 h (03/30/2005)	■ PM-4 15:00 h (08/20/2005)
▲ PM-2 12:00 h (03/30/2005)	■ PM-4 17:00 h (08/20/2005)
▲ PM-3 6:00 h (03/30/2005)	▼ PMII-1 11:00 h (08/20/2005)
▲ PM-3 9:00 h (03/30/2005)	▼ PMII-1 15:00 h (08/20/2005)
▲ PM-3 12:00 h (03/30/2005)	▼ PMII-1 17:00 h (08/20/2005)
▲ PM-4 6:00 h (03/30/2005)	▼ PMII-3 11:00 h (08/20/2005)
▲ PM-4 9:00 h (03/30/2005)	▼ PMII-3 15:00 h (08/20/2005)
▲ PM-4 12:00 h (03/30/2005)	▼ PMII-3 17:00 h (08/20/2005)
★ PMII-1 6:00 h (03/30/2005)	○ River 11:40 h (08/20/2005)
★ PMII-1 9:00 h (03/30/2005)	○ River 16:10 h (08/20/2005)
★ PMII-1 12:00 h (03/30/2005)	○ River 18:15 h (08/20/2005)
★ PMII-2 6:00 h (03/30/2005)	
★ PMII-2 9:00 h (03/30/2005)	
★ PMII-2 12:00 h (03/30/2005)	
★ PMII-3 6:00 h (03/30/2005)	
★ PMII-3 9:00 h (03/30/2005)	
★ PMII-3 12:00 h (03/30/2005)	

Fig. 13. Piper plot showing compositions of river water and groundwater from shallow and deep aquifers

3.5 Groundwater dynamic and the tide influence

The waters of the shallow and deep aquifers do not suffer direct interference of tides and saline intrusion that enters by the river channel. However, the potentiometric surface of the shallow aquifer has groundwater flow direction from SW to NE, or the river to the inland area of study, both in the rainy season (represented by day 03/30/2005) as during dry season (08/20/2005). During the rainy season, the river flows are high and the daily tide has less influence on the dynamics of groundwater, due to a lower pressure of the tide (the flow direction of the river to the mouth is predominant). However, in the less rainy period, one notices the presence of more saline waters in the river, and an influence of saline intrusion in relation to the fresh waters of the river as the river flows are lower. Thus, the potentiometry of the water in the shallow aquifer can be related to this dynamic. The daily fluctuation of the tide makes the potentiometry of the water in the shallow aquifer, both vertically and laterally.

The waters of the deep aquifer have the flow towards NE-SW, toward the river with lateral variations between one period and another, and vertical (especially during the rainy season).

The waters of the shallow and deep aquifer almost do not suffer the influence by tidal changes and seasonal climate. The waters of the shallow aquifer have low mineralization, are acidic and less reduced, have mixed composition, Na-Cl-SO4 water type, and are similar to the composition of the river water.

The waters of the deep aquifer already show compositional differences, and are characterized as Na-HCO3 water type and Ca-Na-HCO3 water type. These waters have calcium and bicarbonate resulting from the dissolution of calcite from the shells and organic matter in the geological formation. In contact with saline waters, there is ion exchange with sodium and change to Na-HCO3 water type.

The waters of the aquifer do not have significant contribution from the sea water. However, the river waters suffer significant compositional changes between the rainy and less rainy period. During the rainy season the waters are slightly mineralized, although in the dry season, there is an increase of dissolved salts in the waters of the river and its composition becomes typical of sea water (Na-Cl water type).

4. Final considerations

The preservation of plant and zoo debris accumulations is directly related to the sediment and aquatic environments. These characteristics have in turn influenced by the dynamic seasonal climate variations and the tides throughout the Quaternary plain of the Itanhaém River.

Along the estuary there are two clear taphonomy trends differentiated by the joint action of the depositional and aquatic environments. The Preto River has a low oxygen content and a high humic acid content, low salinity and higher stability in the chemical composition of water, due to little influence of the saline intrusion. These factors joined to low energy of the river present an ideal location for the preservation of plant biomass that is deposited in the IHS deposits. However, the preservation of zoo debris is minimal, since the skeletons of carbonate composition are easily dissolved in these conditions.

On the other hand, the Itanhaém and Branco rivers show a greater variability in hydrodynamics and in chemical composition. This variation may be a result of: (1) the Branco river rises in the Serra do Mar in steep areas and its journey through the Quaternary plain is shorter than the Preto River, a fact that favors greater water oxygenation; (2) Branco River presents greater flow than the Preto River, softening the chemical characteristics of the Preto River in the Itanhaém River, and finally (3) the Itanhaém River also receives the direct influence of the saline intrusion.

So the rivers Preto, Branco and Itanhaém have high variability in their dynamics, chemical and sedimentological characteristics that are reflected in the nature of bio debris accumulations.

Unlike the waters of rivers, sub aquatic environment have no significant compositional changes, however, there is influence of tidal oscillations in the groundwater flow dynamics. The potentiometry of the shallow aquifer has a strong influence on the daily fluctuation of the tide; however, it was not observed compositional changes, not with seasonality.

In the shallow aquifer, the waters have the same water type of the river waters in the rainy season. However, in the less rainy period, the composition is maintained and the river becomes more saline. The more acidic and oxidizing environment of groundwater are favorable to the preservation of logs rich in lignin. Yet in the waters of the deep aquifer, carbonate skeletons are preserved, since the water is more basic and saline.

These unique features of the sea water are the conditions that preserve the shells in the coastal assemblies of zoo debris.

The integrated and multidisciplinary studies are important tools in studying environment and paleoenvironmental, also contributing to better understand the processes that can lead or not to the preservation of fossil records. The aquatic environment where the records are being fossilized or suffering the epigenesis is crucial to the preservation of fossils or not.

5. Acknowledgements

The authors would like to acknowledge the financial support of the Foundation for the Support of Research of the State of São Paulo (FAPESP, Process 01/09881-2 and FAPESP, Process 2007/07190-0) and the Brazilian National Council for Scientific and Technological Development (CNPq; post-doctoral fellowship), as well as the collaboration of the Research Center of the Estuary of the Itanhaém river (CPeRio) for making their installations available and assisting with the research in the rivers of the basin, including the provision of a boat and the invaluable assistance of the excellent pilot and guide, José Machado.

6. References

Amaral, P.G.C., Ledru, M.P., Ricardi-Branco, F. & Giannini, P.C.F. (2006). Late Holocene development of a mangrove ecosystem in southeastern Brazil (Itanhaém, state of São Paulo): *Paleogeography, Palaeoclimatology and Palaeoecology*, Vol. 241, No. 3-4, (November 2006) pp. 608-620, ISSN 0031-0182

Angulo, R.J., Souza, M.C. & Pessenda, L.C.R. (2002). O significado das datações ^{14}C na reconstrução de paleoníveis marinhos e na evolução das barreiras quaternárias do litoral paranaense. *Revista Brasileira de Geologia*, Vol. 32, No. 1, pp. 95-106, ISSN 0375-7636

Appelo, C.A.J. & Postma, D. (2007). *Geochemistry, groundwater and pollution*. 2nd Edition. A.A. Balkema Publishers. ISBN 0415364213, Amsterdam, the Netherlands.

Batista Filho, J.J. (2006). A dinâmica das águas subterrâneas no estuário do rio Itanhaém, litoral sul do estado de São Paulo, 06/28/2011, Available from: http://www.bibliotecadigital.unicamp.br/document/?code=vtls000393402&opt=1

Camargo, A.F.M., Pereira, L.A. & Pereira, A.M.M. (2002). Ecologia da bacia hidrográfica do rio Itanháem. In: *Conceitos de bacias hidrográficas: teorias e aplicações*, A. Schiaretti & A.F.M. Camargo (Eds.), 239-256, Editora da UESC. ISBN 857455099X, Ilhéus, Brazil.

Comitê de Bacias Hidrográficas/Baixada Santista (CBH-BS). (2000). Relatório Zero: Minuta Preliminar do Relatório de Situação dos Recursos Hídricos da UGRHI 7. São Paulo-SP, In: *http://www.sigrh.sp.gov.br/sigrh/ARQS/RELATORIO/CRH/CBH-BS/218/relbsseg.pdf*, 05/10/2011, Available from: < http://www.sigrh.sp. gov.br>.

Fetter, C.W. *Applied Hydrogeology*. 4th Edition. Prentice Hall, ISBN 013088239-9, New Jersey, United States of America.

Giannini, P.C.F., Guedes, C.F., Nascimento, D.R., Tanaka, A.P.B., Angulo, R.J., Assine, M. & Souza, M.C. (2009). sedimentology and mophological evolution of the ilha comprida barrier system, southern São Paulo coast. In: *Geology and Geomorphology of Holocene Coastal barriers of Brazil*, S. Dillenburgand & P., Hesp (Eds.), 177-224, Springer-Verlag. ISBN 978-3-540-25008-1 Berlin, Germany.

Holz, M & Simões, M.G. (2002) *Elementos fundamentais de Tafonomia*. Editora da Universidade UFRGS, ISBN 8531406730, Porto Alegre, Brazil

Lopes, M.C. (1984). Água subterrânea no estado de São Paulo – síntese das condições de ocorrência, *Proceedings of ABAS, Third Brazilian Groundwater Congress*, Fortaleza - Ceará, October 1984.

Suguio, K. (1992). *Dicionário de Geologia marinha*. 1 ed, Editora T.A. Queiroz, ISBN 85-7182-001-5, S. Paulo, Brazil

Suguio, K., & Martin, L., (1978) Formações quaternárias marinhas do litoral paulista e sul fluminense: *Boletim IG-USP*, especial number, pp. 1-55, ISSN 0102-6275

Suguio, K. (2004). O papel das variações do nível relativo do mar durante o Quarternário tardio na origem da baixada litorânea de Juréia, SP. In: *Estação ecológica Juréia – Itatins: ambiente físico, flora e fauna*. Marques, O.A.V. & Duleba, W. (eds), pp. 34-41. Holos Editora, ISBN 85-86699-37-3, São Paulo, Brazil

Ricardi-Branco, F., Ianniruberto, M., Silva, M.A. & Branco, F. (2011). Plant debris Accumulations in the Rio Preto sub-basin, Itanhaém, SP, Brazil: insights from geotechnology. *Palaios,Vol.* 26 (May 2011), pp. 264-274. ISSN 0883-1351

Ricardi-Branco, F., Branco, F.C, Garcia, R.F., Faria, F.S, Pereira, S.Y., Portugal, R, Pessenda, L.C., Pereira, P.R.B. (2009) Features of plant accumulations along the Itanhaém river, on the southern coast of the Brazilian state of São Paulo. *Palaios*, Vol. 24, No. 7 (July 2009), pp. 416-424, ISSN 0883-1351

Zincone, S.A, Ricardi-Branco, F., Vidal, A.C. & Pessenda, L.C. (2007). Caracterização de depósitos transgressivos quaternários na bacia do rio Itanhaém, litoral sul do Estado de São Paulo. In: *Paleontologia: Cenários da Vida*, I. S. Carvalho, R.C. T. Cassab, C., Schwanke, M.A, Carvalho, A.C.S., Fernandes, M.A.C.R., Rodrigues, M.S.S., Carvalho, M., Arai & E.Q.O., Oliveira (eds.), 295-307, Editora Interciência. ISBN 978-85-7193-184-8, Rio de Janeiro, Brazil.

10

The Permo-Triassic Tetrapod Faunal Diversity in the Italian Southern Alps

Marco Avanzini[1], Massimo Bernardi[1] and Umberto Nicosia[2]

[1]*Science Museum, Trento,*
[2]*Sapienza University of Rome,*
Italy

1. Introduction

The Permian and Triassic palaeogeography of the Alpine region originated a peculiar geological situation, now well exposed in several sections, in which marine sediments, continental deposits, and volcanites interfinger. The study of the resulting mixed sections allowed to build a framework of biostratigraphic and chronological data in which tetrapod footprints play a key role (e.g., Avanzini & Mietto, 2008).

Global track record is much more abundant than the skeletal record and, although suffering from problems related to a correct attribution to the trackmakers, provides data as reliable as those obtained from skeletal remains (Carrano & Wilson, 2001 and references therein).

During the Permian and the Early-Middle Triassic, tetrapods, and especially reptiles, radiated and entire new land-dwelling groups originated (e.g., the archosaurs). Consequently, in this temporal interval the tetrapod track record shows a huge increase in variability reflecting the morphological diversity spanning from a stem-reptile to a 'mammalian' foot, from basal crocodilomorph to a dinosauromorph foot. How is this pattern documented in the Dolomites region and surrounding areas (Southern Alps, NE Italy)?

In this geographical sector the recent discovery of many new tetrapod footprint-bearing outcrops has yield to a phase of renewed interest for ichnological data. As a result, the stratigraphical, palaeoecological and palaeogeographical importance of tetrapod footprints in this geographical sector is becoming more and more widely acknowledge.

In this contribution we provide an overview of Permian and Triassic tetrapod faunal composition as deduced from the study of several ichnosites located in the Italian Southern Alps with special reference to the pattern exhibited around the PT boundary.

2. Geological setting

In the Permian succession of the Southern Alps tetrapod tracks are found within two tectonosedimentary cycles. The lower cycle is represented by the alluvial-lacustrine continental deposits of the Collio and Tregiovo Formations while the upper cycle is represented by clastic red-beds of the Verrucano Lombardo Formation and Val Gardena Sandstone (Trentino Alto Adige). Towards the top, the latter interfingers with the sulphate evaporites and shallow-marine carbonate sequences of the Bellerophon Fm. The two cycles are put in contact by a regional unconformity documented by erosional surfaces and palaeosoil horizons. Floristic and radiometric data (Cassinis & Doubinger, 1991, 1992; Cassinis et al., 1999, 2002; Schaltegger & Brack, 1999) indicate that the lower cycle is Artinskian-Kungurian in age (Early Permian). The upper cycle has been dated by means of sporomorphs and foraminifers and is considered Wuchiapingian in age (Late Permian, Cassinis et al., 2002).

Triassic units follow. The precise position of the PT boundary in the Southern Alps has been debated for long time (see Broglio Loriga & Cassinis, 1992 for a review). Although some early works put the boundary at or very close to the Bellerophon-Werfen Formation limit (Bosellini, 1964; Assereto et al., 1973; Posenato, 1988), the PT boundary is now thought to lie within (Pasini, 1985; Broglio Loriga et al., 1986; Visscher & Brugman, 1988; Farabegoli & Perri, 1998; Farabegoli et al., 2007) or at the top (Wignall et al., 1996) of the Tesero Member, the lowermost part of the Werfen Fm. The first Triassic tetrapod tracks are found within the terrigenous and terrigenous-carbonate units of the Werfen Fm. and are Olenekian in age (Early Triassic). Several terrigenous and terrigenous-carbonate units of Anisian age (Middle Triassic) follow: Gracilis Fm., Voltago Cgl., Richthofen Cgl., Morbiac dark Limestones. All were deposited in lagoonal-peritidal to continental environments and have yielded well preserved ichnoassociations. No tetrapod tracks are known from the Dolomites' Ladinian (Middle Triassic) which is represented mainly by basinal units (Buchenstein Fm. and Wengen Group) and carbonate platform deposits (Sciliar Dolomite).

3. Permian and Triassic ichnoassociations

Permian and Triassic ichnoassociations in the Southern Alps have been documented since the XIX century (e.g., Curioni, 1870). Recent reviews have been published by Conti et al. (1997), Conti et al. (1999), Conti et al. (2000), Avanzini et al. (2001), Avanzini & Mietto (2008). Here we provide an updated overview on these studies focusing on the contribution that tetrapod ichnology can give to the debates concerning the large scale patterns of extinction and subsequent recovery on lands near the PT boundary (Benton & Twitchett, 2003; Twitchett et al., 2001; Retallack et al., 2006; Sheldon & Chakrabarti, 2010).

In the Alpine area several different types of bioevents were recognized (Conti et al., 2000; Ronchi et al., 2005). According to their relative importance, they had different degrees of utility for correlation. The first-appearance datum (FAD) and last appearance datum (LAD) concern a taxon, at either genus and species level, and are important at the regional level. The first occurrence (FO) and last occurrence (LO) as well as the disappearance event (DE) concern a faunal complex and have very high correlation power.

Here, ichnoassociations have been grouped in four discrete time bins: Artinskian-Kungurian (Early Permian) (Fig. 1), Wuchiapingian (Late Permian) (Fig. 2), Olenekian (Early Triassic) and Anisian (Middle Triassic) (Fig. 3). Traditionally, these discrete associations have been considered 'Faunal Units' (FU) (Conti et al., 1997; Conti et al., 1999; Conti et al., 2000; Avanzini et al., 2001; Ronchi et al., 2005).

In the present study: i) no vertical differences within a single bin have been considered even though some are well known; and, ii) only species-level determinations have been considered except where a genus or a higher-level taxon is represented by an undetermined species only (e.g., *Procolophonichnium* isp.), or in the case of *Rhynchosauroides* isp.1 and *Rhynchosauroides* isp.2 of Valentini et al. (2007) which underwent a formal revision and, although not yet renamed, can be considered as distinct ichnospecies.

3.1 Artinskian-Kungurian (Early Permian)

The Permian first cycle ichnoassociation is documented within the Collio and Tregiovo Formations. Data have been recorded in the Collio basin (Geinitz, 1869; Curioni, 1870; Cassinis, 1966; Berruti, 1969; Ceoloni et al., 1987; Conti et al., 1991) and from sediments cropping out in other localities of the Orobic basin (Gümbel, 1880; Dozy, 1935; Casati, 1969; Casati & Forcella, 1988; Nicosia et al., 2000; Santi, 2003, 2005; Santi & Krieger, 2001) and in the Tregiovo and Luco basins (Conti et al., 1997; Avanzini et al., 2008).

The ichnoassociation includes the following ichnospecies (after Conti et al., 1997; Conti et al., 2000, updated) (Fig. 1).

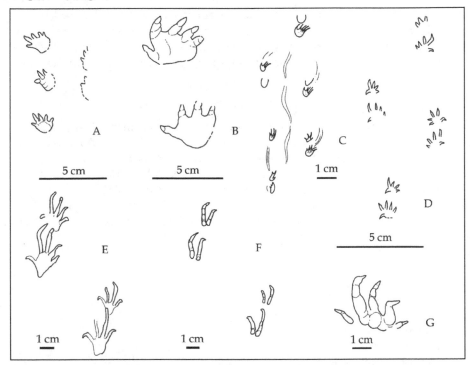

Fig. 1. Artinskian and Kungurian (Early Permian) tetrapod tracks form the Southern Alps. A) *Amphisauropus imminutus* Haubold, 1970; B) *Amphisauropus latus* Haubold, 1970; C) *Batrachichnus salamandroides* (Geinitz, 1861); D) *Camunipes cassinisi* Ceoloni et al., 1987; E) *Dromopus lacertoides* (Geinitz, 1861); F) *Dromopus didactylus* (Moodie, 1896); G) *Varanopus curvidactylus* Moodie, 1929. From Santi & Krieger (1999), Ceoloni et al. (1987), Avanzini et al. (2008).

Amphisauropus imminutus Haubold, 1970
Amphisauropus latus Haubold, 1970
Batrachichnus salamandroides (Geinitz, 1861)
Camunipes cassinisi Ceoloni et al., 1987
Dromopus lacertoides (Geinitz, 1861)
Dromopus didactylus (Moodie, 1896)
Ichniotherium cottae (Pohlig, 1885)

Varanopus curvidactylus Moodie, 1929

Santi (2007) has clearly shown that *Camunipes cassinisi* Ceoloni et al., 1987 is a junior synonym of *Erpetopus* Moodie, 1929.

The Early Permian ichnoassociation is bounded by the *Amphisauropus latus* FO and *Ichniotherium cottae* FO at the base and by the *Dromopus didactylus* LO at the top. This association is very similar to those found in the Permian of various European countries. It is named Collio FU and it can in turn be subdivided into two subunits. The lower subunit (Pulpito subunit) is characterized by the presence of *Amphisauropus latus* Haubold, 1970, *Dromopus lacertoides* (Geinitz, 1861) and *Ichniotherium cottae* (Pohlig, 1885). The upper subunit (Tregiovo subunit) is bounded by the disappearance event (DE) for all but one of the preceding forms, yielding a drastic reduction of faunal diversity; indeed the subunit is monotypic being represented by very frequent specimens of the only relict species *Dromopus didactylus* (Moodie, 1896). This pattern was interpreted as a bioevent, given the relatively unchanged sedimentation conditions.

3.2 Wuchiapingian (Late Permian)

The Permian second cycle ichnoassociation has been recorded in the Bletterbach section (Ceoloni et al., 1988; Conti et al., 1975; 1977; 1980; Leonardi & Nicosia, 1973; Leonardi et al., 1975; Nicosia et al., 1999) and other quasi-coheval outcrops (SS. 48, Col Fratton (Kittl, 1891; Abel, 1926); S. Pellegrino Pass (Conti et al., 1977), Seceda (Valentini et al., 2009), Nova Ponente (Wopfner, 1999), Recoaro (Mietto, 1975; 1981) San Genesio-Meltina Plateau in the Adige basin (Avanzini, unpublished).

The ichnoassociation includes the following ichnospecies (after Avanzini et al., 2001; Conti et al. 2000; Valentini et al., 2007, updated) (Fig. 2).

Chelichnus tazelwurmi Ceoloni et al., 1988
Dicynodontipus geinitzi (Hornstein, 1876)
Ganasauripus ladinus Valentini et al., 2007
Hyloidichnus tirolensis Ceoloni et al., 1988
Ichniotherium accordii Ceoloni et al., 1988
Ichniotherium cottae (Pohlig, 1885)
Janusichnus bifrons Ceoloni et al., 1988
Pachypes dolomiticus Leonardi et al., 1975
Paradoxichnium radeinensis Coeloni et al., 1988
Protochirotherium isp.
Rhynchosauroides pallinii Conti et al., 1977
Rhynchosauroides isp.1 of Valentini et al., 2007
Rhynchosauroides isp.2 of Valentini et al., 2007
Therapsida indet.

Note that the specimens from Val Gardena Fm. ascribed to *Chelichnus tazelwurmi* Ceoloni et al., 1988 and *Ichniotherium cottae* (Pohlig, 1885) are currently under revision and that their nomenclature is expected to change in the near future.

The Upper Permian tetrapod footprint complex is characterized by a diversified fauna almost completely new as compared with other Permian faunas (Ceoloni et al., 1986). Bounded in its lower part by the *Rynchosauroides* and *Protochirotherium* FAD, its records also the *Dycinodontipus* FAD and the *Ichniotherium* LAD. It is named the Bletterbach FU.

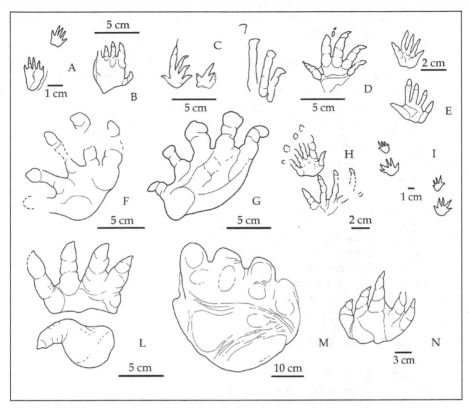

Fig. 2. Wuchiapingian (Late Permian) tetrapod tracks form Southern Alps. A) *Chelichnus tazelwurmi* Coeloni et al., 1988; B) *Dicynodontipus geinitzi* (Hornstein, 1876); C) *Ganasauripus ladinus* Valentini et al., 2007; D) *Rhynchosauroides pallinii* Conti et al., 1977; E) *Hyloidichnus tirolensis* Ceoloni et al., 1988; F) *Ichniotherium accordii* Ceoloni et al., 1988; G) *Ichniotherium cottae* (Pohlig, 1885); I) *Janusichnus bifrons* Ceoloni et al., 1988; M) *Pachypes dolomiticus* Leonardi et al., 1975; H) *Paradoxichnium radeinensis* Ceoloni et al., 1988; L) cf. *Protochirotherium* isp.; M) Therapsida indet. From Conti et al. (1977, 2000), Valentini et al. (2007).

Early Permian and Late Permian associations of the Southern Alps are separated by a stratigraphic break that correspond to a time-gap of 14-27 Ma (Italian IGCP 203 Group, 1986; Cassinis et al., 1988; Cassinis et al., 1999). A similar hiatus, equivalent to part of the Middle Permian, has been recognized in the global Permian tetrapod fossil record and has been termed Olson's gap (Lucas and Heckert, 2001). The gap is thought to be associated with a significant remodelling of the global Permian tetrapod fauna (Lucas and Heckert, 2001; Lucas, 2004). As stated before, the two Southern Alpine FUs have been considered two different 'evolutionary stages' (Conti et al., 2000).

3.3 Olenekian (Early Triassic)
In the studied sector, the Early Triassic is represented by the Werfen Fm.: a complex succession of carbonate, terrigenous and mixed sediments. The formation is divided into

nine members and tracks are present in the upper three (Campil Member, Cencenighe Member and San Lucano Member).

Tracks have been found near Recoaro, in Val Gardena (Bulla/Pufels), and Val Travignolo (Conti et al., 2000; Mietto, 1986). Olenekian tracks are generally poorly preserved and some, as those signaled by Leonardi (1967, p. 118, 119), are here discarded. Furthermore (*contra* e.g. Avanzini & Mietto, 2008) levels bearing *Capitosauroides* and *Procolophonichnium* ichnogenera (e.g., cropping out in the Monte Marzola) previously considered to be part of the Werfen Fm. are here reconsidered within the 'Gracilis Formation' which is Anisian (Middle Triassic) in age. The ichnoassociation includes the following ichnospecies (after Avanzini et al., 2001; Conti et al. 2000, updated) (Fig. 3):

Rhynchosauroides palmatus (Lull, 1942)

Rhynchosauroides schochardti (Rühle von Lilienstern, 1939)

Both *R. palmatus* (Lull, 1942) and *R. schochardti* (Rühle von Lilienstern, 1939) have not been validated by Valentini et al. (2007) in their review of the ichnogenus *Rhynchosauroides* from the Val Gardena Sandstone. No review has been conducted of this ichnogenus in the Werfen Formation. We acknowledge that both ichnospecies could follow in an undetermined *Rhynchosauroides* isp., this is however nearly irrelevant (i.e., Early Triassic ichnodiversity could be 1 instead of 2) to the data analyses presented here.

The Lowermost Triassic of the Southern Alps is characterized by the scarce presence of vertebrates, a factor that is undoubtedly linked to environmental conditions (palaeogeography) which did not favor the permanence in this region of complex and consistent faunal associations. The ichnoassociations are dominated by *Rhynchosauroides* forms that are likely mainly Permian survivor.

3.4 Anisian (Middle Triassic)

Several Anisian ichnoassociations have been found in the Dolomites and surrounding areas. These are found within several formations (Gracilis Fm., Voltago Cgl., Richthofen Cgl., Morbiac dark Limestones) which document transitional continental to marine environments. Tracksites are located in the Braies Dolomites (Northern Dolomites), in the eastern Dolomites, in the upper Val di Non and Val d'Adige, and in the Recoaro-Vallarsa area (Abel, 1926; Brandner, 1973; Mietto, 1987; Avanzini, 1999, 2000, 2002; Avanzini et al., 2001; Avanzini & Leonardi, 2002; Avanzini & Lockley, 2002; Avanzini & Neri, 1998; Avanzini & Mietto, 2008; Todesco et al., 2008; Todesco & Bernardi, 2011).

The ichnoassociation includes the following ichnospecies (after Avanzini & Mietto, 2008, updated) (Fig. 3):

Brachychirotherium paeneparvum Demathieu & Leitz, 1982

Brachychirotherium circaparvum Demathieu, 1967

Capitosauroides cf. *bernburgensis* Haubold, 1971

Chirotherium barthii Kaup, 1835

Chirotherium rex Peabody, 1948

Isochirotherium infernense Avanzini & Leonardi, 2002

Isochirotherium delicatum Courel & Demathieu, 1976

Parasynaptichnium gracilis Mietto, 1987

Procolophonichnium isp.

Rhynchosauroides tirolicus Abel, 1926

Rhynchosauroides peabodyi Faber, 1958

Rotodactylus cf. *cursorius* Peabody, 1948

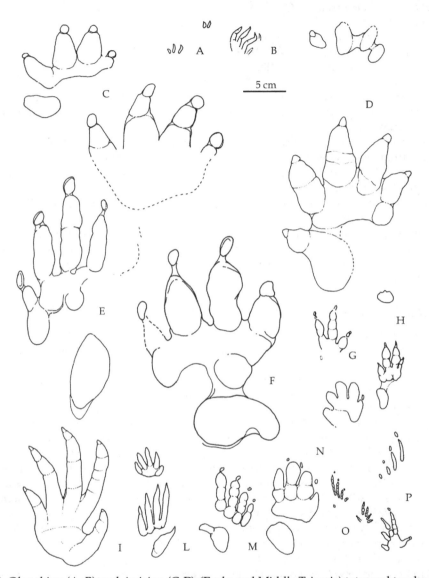

Fig. 3. Olenekian (A, B) and Anisian (C-P) (Early and Middle Triassic) tetrapod tracks from Southern Alps. A) *Rhynchosauroides palmatus* (Lull, 1942); B) *Rhynchosauroides schochardti* (Rühle von Lilienstern, 1939); C) *Chirotherium rex* Peabody, 1948; D) *Isochirotherium infernense* Avanzini and Leonardi, 2002; E) cf. *Sphingopus* isp.; F) *Chirotherium barthii* Kaup, 1835; G) tridactyl footprint, possible dinosauriformes; H) *Isochirotherium delicatum* Courel & Demathieu ,1976; I) *Parasynaptichnium gracilis* Mietto 1987; L) *Synaptichnium pseudosuchoides* Nopcsa, 1923; M) *Synaptichnium cameronense* (Peabody, 1948); N) *Brachychirotherium* aff. *Parvum*; O) *Rhynchosauroides tirolicus* Abel, 1926; P) *Rhynchosauroides peabodyi* Faber, 1958. From Avanzini (2003), Avanzini & Mietto (2008).

Rotodactylus lucasi Demathieu & Gand, 1973
Sphingopus isp.
Synaptichnium cameronense (Peabody, 1948)
Synaptichnium pseudosuchoides Nopcsa, 1923
Synaptichnium diabloense Peabody, 1948
Therapsida indet.

In the considered sector, tetrapod tracks are not recorded from Ladinian deposits, in correspondence of a great rise of the sea level. Well documented associations reappear only in the Late Triassic (Tuvalian).

In the Anisian, a progressive increase in the complexity of the ichnoassociations and the size of the taxa from the Bithynian to the Illyrian is documented in the Southern Alps.

The analysis of the stratigraphic distribution of ichnofaunas crossed with sequence stratigraphy and the ammonite biostratigraphy has led to the identification, within the Pelsonian - Illyrian interval of several taxa characterized by a narrow vertical distribution (Avanzini & Mietto, 2008).

While the Aegean-Bithynian interval is characterized by several lizard-like ichnotaxa and small archosauriform (*Parasynaptichnium gracilis* and *Synaptichnium pseudosuchoides* FAD), the middle Pelsonian and the Illyrian are mainly dominated by medium-large chirotheroids (i.e., *Chirotherium barthii, C.* cf. *rex, Isochirotherium herculis*).

4. Ichnodiversity trends trough the Permo-Triassic

The likely producer of an ichnospecies is generally uncertain; tracks and trackmakers can only be associated with certainty when found in association (e.g., Voigt et al., 2007). However, for most studies the identity of the trackmaker does not need to be (and cannot be) determined at a specific or generic level. The range of the possible producers can be constrained within the least inclusive group that bounds all the taxa sharing similar morphological characteristics that fit with the features of the studied tracks (Carrano & Wilson, 2001). In the case of the track record here analysed this implies that different ichnogenera can be attributed to trackmakers belonging to different taxonomic categories. Furthermore, acknowledging that different substrates, behaviours and even different ichnotaxonomic traditions can bias our palaeobiodiversity and abundance estimates, the here suggested taxonomic designations are based on: i) the most conservative found in literature (i.e., broad), and ii) the most widely acknowledge (i.e., some single alternative hypothesis have been discarded) (Tab. 1).

For example the ichnogenus *Procolophonichnium* has been attributed to different biological groups such as procolophonids (Anapsida), therapsids (Synapsida) and basal amniotes (see Klein & Lucas, 2010; Klein et al., in press). The least inclusive group that bounds these taxa is the Amniota and *Procolophonichnium* is here conservatively considered as produced by an indeterminate stem group amniote (i.e., basal amniote). A coarse assignment of trackmaker identity has been shown to be still useful for evolutionary studies (e.g., Wilson and Carrano, 1999; Carrano & Wilson, 2001).

Although possibly unsafe, a link between ichnofaunal and faunal record in the Dolomites area is the only possible source of information about tetrapod life on lands in Permian and Triassic since, as state above, the bony record is nearly absent.

Ichnodiversity and faunal composition through time are shown in Figs. 4 and 5. Taken at face value, available data show that diversity (shear number of ichnospecies) grows through the Permian, then suddenly drops in (or until) the Olenekian and then reaches its maximum in the Anisian. A finer stratigraphic and temporal resolution would show that lowest and highest diversity are separated by circa 6 million years only (Avanzini & Mietto, 2008).

Ichnotaxon	Attribution
Amphisauropus	Seymouriamorpha
Batrachichnus	Stem group Amphibia
Brachychirotherium	Archosauria
Capitosauroides	Stem-group Amphibia
Chelichnus	Non-therapsid Eupelycosauria
Chirotherium	Archosauriformes
Dicynodontipus	Non-mammalian Therapsida
Dromopus	Araeoscelida
Erpetopus (*Camunipes*)	Captorhinidae
Ganasauripus	Lepidosauromorpha
Hyloidichnus	Captorhinidae
Ichniotherium	Diadectomorpha
Isochirotherium	Archosauriformes
Janusichnus	Stem-group Amniota
Pachypes	Pareiasauridae
Paradoxichnium	Lepidosauromorpha
Parasynaptichnium	Archosauriformes
Procolophonichnium	Stem-group Amniota
Protochirotherium	Archosauriformes
Rhynchosauroides	Lepidosauromorpha/Eosuchia
Rotodactylus	Dinosauromorpha
Sphingopus	Dinosauromorpha
Synaptichnium	Archosauriformes
Varanopus	Captorhinomorpha

Table 1. Permian-Triassic tetrapod ichnogenera and their inferred trackmakers (mainly after Haubold, 1996, 2000; Klein et al., 2011)

The Artinskian-Kungurian (Early Permian) ichnoassociation (Fig. 1) is comparable with coeval associations around the World (the so-called *Batrachichnus* ichnofacies; Lucas & Hunt, 2006) and document the presence of 5 different groups: amphibians, seymouriamorphs, diadectomorphs, captorhinomorphs and araeoscelids (Fig. 4). These are represented, in similar proportion, by 8 different ichnospecies. The most notable difference with other Lower Permian tracksites is the absence of tracks ascribed to large non-therapsid Eupelycosaur grade trackmakers (e.g. *Dimetropus*).

The richer Wuchiapingian (Late Permian) ichnoassociation (Fig. 2) is constituted of 14 ichnospecies which document the presence of 6 groups: diadectomorphs and captorhinomorphs are still present, together with lepidosauromophs, pareiasaurs, synapsids and archosauriforms (Fig. 4). This ichnoassociations, according to Lucas and Hunt (2006, p. 152), "the most characteristic ichnofauna being that of the Val Gardena Formation" allowed defining a *Pachypes* ichnocoenosis in the *Brontopodus* ichnofacies (Lucas and Hunt, 2006). As widely acknowledge, however, Southern Alps Late Permian ichnoassociation is far richer than elsewhere and, although ichnodiversity-based estimates can be misleading for the problems acknowledge before, it is interesting to note that, on a global scale, Wuchiapingian land vertebrate life experienced a dramatic drop in diversity (the end-Guadalupian crisis; see Benton, 1995; King, 1991; Retallack et al., 2006; Sahney & Benton, 2008).

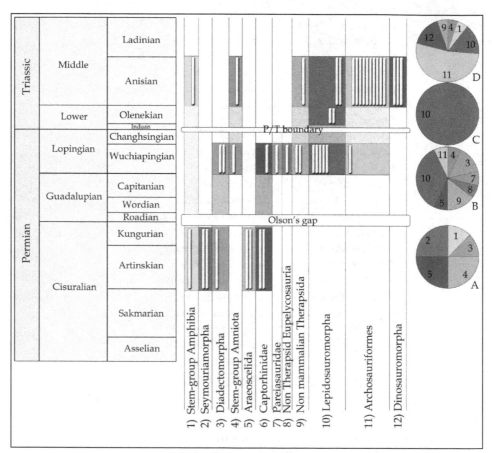

Fig. 4. Tetrapod ichnodiversity and faunal composition trough the Permian and Triassic in the Dolomites region (Southern Alps, NE Italy). White bars represent the ichnospecies listed in the text for each of the four studied ichnoassociation: A) Artinskian-Kungurian (Early Permian), B) Wuchiapingian (Late Permian), C) Olenekian (Early Triassic), and D) Anisian (Middle Triassic). Each color, corresponding to the cardinal numbers 1-12, represent any least inclusive group to which the ichnospecies are referred.

Olenekian (Early Triassic) tracks (Fig. 3) are all found isolated and poorly preserved. This surely has to do with the low preservation potential for tracks in the Southern Alps formations. However the few track-bearing levels within the three upper members of the Werfen Fm. undoubtedly record transitional continental-marine environments with high preservation potential: where tracks could be preserved a monotypic lepidosauromorph fauna is documented (2 ichnospecies). The interpretation of this datum in the considered geographical sector is however poorly supported: both an "ecological bias" (e.g., only the coast dwellers who were able to live on the ephemeral emerged lands left their footprints), or a real "evolutionary pattern" (e.g., recovery in diversity from PT mass extinction was extremely slow) would produce the observed pattern of diversity.

Anisian (Middle Triassic) ichnoassociation is very diverse being constituted by 18 ichnospecies. 5 groups are represented: lepidosauromophs, stem amphibians, "synapsids", archosauriforms and dinosauromorphs (Fig. 4). It is interesting to note that as soon as the Anisian, nearly 70% of the ichnodiversity in the considered geographical sector documents the presence of archosaurs (10 archosaur-produced + 3 dinosauromorph-produced ichnospecies). The pattern is consistent with those observed in Germany, France and Arizona (USA) (see Hunt and Lucas, 2007 and references therein).

5. Discussion and conclusions

Our analysis of the Permo-Triassic faunal composition in the Dolomites region as deduced from the track record well matches the skeletal record with the obvious exclusion of few, very rare forms, dubiously referred to high level taxonomic groups too inclusive to be meaningful (Amphibia, Amniota). In this way, incidentally, the footprint reliability for evolutionary studies was, once again, confirmed. This is in particular true when data are collected from rich and deeply studied ichnoassociations. Once stated the reliability of data, we can analyse the impact of two well known biological crises that have been documented world-wide in the Middle and Late Permian (see Hallam & Wignall, 1997; Retallak et al. 2006; Sheldon, 2006 and references therein) on mid Early Permian to late Early Triassic faunas in the Dolomites.

The effects of the first crisis, the so-called Olson's extinction, are documented in the considered geographical sector by: i) the progressive disappearance of long ranging taxa, typical of the Early Permian (and Carboniferous) as the Seymouriamorpha and the Araeoscelidia; ii) the persistence of groups as the Diadectids and the Captorhinomorphs; iii) the appearance of groups that will have their explosion during the Late Permian (Pareiasaurs and non-mammalian Therapsids). Contemporaneously it is worth of note the appearance of the first Lepidosauromorphs and of rare Archosauromorphs. On the contrary, the events recorded between Late Permian and late Early Triassic, across the end-Permian crisis, suggest a very low survival, and only of the groups that appeared during the Late Permian (Lepidosauromorph, non Mammalian Therapsid, Archosauromorphs). Noticeably, an early appearance and the abundant presence of *Rotodactylus* and *Sphingopus* in the Southern Alps also supports the hypotheses of an early origin, soon after the PT boundary, for the Dinosauromorpha (e.g., Brusatte et al., 2011).

The recognized trends, suggest different controlling factors for the two Permian crises (Fig. 5). The end-Guadalupian "crisis" (Benton, 1995, 1989; King, 1991; Retallack et al., 2006), or "extinction" (Sahney & Benton, 2008) seems to show the typical feature of the depletion of a long-lived association, mainly characterized by ecological replacement of the old taxa with new ones (famous examples, based on bone remains, are the replacement of Caseids with Pareiasaurs in the niche of large plant-eaters and of large Sphenacodontids with the non-mammalian Therapsid among carnivorous), and thus an event controlled by normal environment changes, more or less slow and gradual. The end-Permian crisis, instead, is recorded by the termination of short lived taxa, just appeared during the same Late Permian and can be better interpreted as the result of a mass-extinction, connected to a series of catastrophic events.

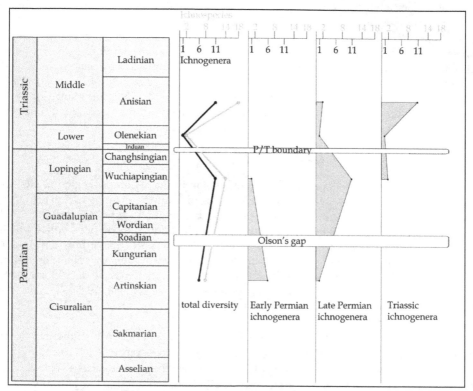

Fig. 5. Tetrapod ichnodiversity trough time in the Southern Alps. Colored polygons represent the duration of each of the studied ichnoassociations.

Thus, paradoxically, it may have been the end-Guadalupian, slow speed, crisis that, anticipating the substitution of obsolescent anatomical and ecological 'models' and forcing the appearance of entirely new taxa, allowed some land-animal groups overcoming the following and much more acute, end-Permian crisis.

6. Acknowledgments

We thank F.M. Petti and R. Todesco (Science Museum, Trento) for constructive discussions and M.C. Deflorian (Science Museum, Trento) for assistance with figures. This study is part of the project "The Permian-Triassic ecological crisis in the Dolomites: extinction and recovery dynamics in Terrestrial Ecosystems" financed by the Promotion of Educational Policies, University and Research Department of the Autonomous Province of Bolzano - South Tyrol. The research was supported also by the Science Museum, Trento and the Sapienza University of Rome.

7. References

Abel, O. (1926). Der erste Fund einer Tetrapodenfährte in den unteren alpinen Trias. *Paläontologische Zeitschrift*, Vol.7, pp. 22-24.

Assereto, R.; Bosellini, A., Fantini Sestini, N. & Sweet, W.C. (1973). The Permian–Triassic boundary in the Southern Alps (Italy). In: *The Permian and Triassic Systems and their Mutual Boundary*, A., Logan & L.V., Hills (Eds.), *Alberta Society* of *Petroleum* Geologists Memories, Vol.2, pp. 176-199.

Avanzini, M. (1999). New Anisian vertebrate tracks from the Southern Alps. In: *Third International Symposium on lithographic limestones*, S. Renesto, (Ed.), 17-21, suppl. Rivista Museo Civico Scienze Naturali "E.Caffi" Bergamo, Bergamo, Italy.

Avanzini, M. (2000). *Synaptichnium* tracks with skin impressions from the Anisian (Middle Triassic) of the Southern Alps (Val di Non – Italy). *Ichnos*, Vol.7, pp. 243-251.

Avanzini, M. (2002). Dinosauromorph tracks from the middle Anisian (middle Triassic) of the Southern Alps (Valle di Non - Italy). *Bollettino della Società Paleontologica Italiana*, Vol.41, pp. 37-40.

Avanzini, M. (2002). Tetrapod footprints from the Mesozoic carbonate platforms of the Italian Alps. *Zubía*, Vol.21, pp. 175-186.

Avanzini, M.; Ceoloni, P., Conti, M.A., Leonardi, G., Manni, R., Mariotti, N., Mietto, P., Muraro, C., Nicosia, U., Sacchi, E., Santi, G. & Spezzamonte, M. (2001). Permian and Triassic tetrapod ichnofaunal units of Northern Italy: Their potential contribution to continental biochronology. *Natura Bresciana*, Vol.25, pp. 89-107.

Avanzini M.; Ferretti, P., Seppi, R. & Tomasoni, R. (2001). Un grande esemplare di orma 'dinosauroide' *Parachirotherium* Kuhn 1958 sp. dall'Anisico superiore (Illirico) del Sudalpino. *Studi Trentini di Scienze Naturali, Acta Geologica*, Vol.76, pp. 201-204.

Avanzini, M. & Leonardi, G. (2002). *Isochirotherium inferni* ichnosp. n. in the Upper Anisian (Illyrian) of Adige Valley (Bozen, Italy). *Bollettino Società Paleontologica Italiana*, Vol.41, pp. 41-50.

Avanzini, M. & Lockley, M. (2002). Middle Triassic archosaur ontogeny and population structure: interpretation based on *Isochirotherium delicatum* fossil footprints (Southern Alps – Italy). *Palaeogeography, Palaeoclimatology, Palaeoecology*, Vol.185, pp. 391-402.

Avanzini, M. & Mietto P. (2008). Lower and Middle Triassic footprints-based biochronology in the Italian Southern Alps. *Oryctos*, Vol.8, pp. 3-13

Avanzini, M & Neri, C. (1998). Impronte di tetrapodi da sedimenti anisici della Valle di Non (Trentino occidentale - Italia): nota preliminare. *Annali del Museo Civico di Storia Naturale di Ferrara*, Vol.1, pp. 5-19.

Avanzini M.; Neri C., Nicosia, U. & Conti, M.A. (2008). A new Early Permian ichnocoenosis from the "Gruppo vulcanico atesino"(Mt. Luco, Southern Alps, Italy). *Studi Trentini di Scienze Naturali, Acta Geologica*, Vol.83, pp. 231-236.

Benton, M.J. (1989). Mass extinctions among tetrapods and the quality of the fossil record. *Proceedings of the Royal Society, Series B*, Vol.325, pp. 369-386.

Benton, M.J. (1995). Diversification and extinction in the history of life. *Science*, Vol.268, pp. 52-58.

Benton, M.J. & Twitchett, R.J. (2003). How to kill (almost) all life: the end-Permian extinction event. *Trends in Ecology and Evolution*, Vol.18, pp. 358-365.

Berruti G., (1969). Osservazioni biostratigrafiche sulle formazioni continentali pre-quaternarie delle valli Trompia e Sabbia. II. Sulla fauna fossile della Formazione di Collio (alta Val Trompia). *Natura Bresciana, Annali del Museo Civico di Scienze Naturali, Brescia*, Vol.6, pp. 3-32.

Bosellini, A. (1964). Stratigrafia, petrografia e sedimentologia delle facies carbonatiche al limite Permiano-Trias nelle Dolomiti occidentiali. Memorie del Museo di Storia Naturale della Venezia Tridentina, Vol.15, pp. 59-110.

Brandner, R. (1973). Tetrapodenfährten aus der unteren Mitteltrias der Südalpen. *Veröffentiichungen der Universiätt Innsbruck*, Vol.86, pp. 57-71.

Broglio-Loriga, C. & Cassinis, G. (1992). The Permo-Triassic boundary in the Southern Alps (Italy) and in adjacent Periadriatic regions. In: *Permo-Triassic events in the eastern Tethys*, W.C. Sweet, Y. Zunyi, J.M. Dickins & Y. Hongfu (Eds.), 78–97, Cambridge University Press, Cambridge, United Kingdom.

Broglio Loriga, C.; Neri, C. & Posenato, R. (1986). The Early macrofaunas of the Werfen Formation and the Permian Triassic boundary in the Dolomites (Southern Alps, Italy). *Studi Trentini di Scienze Naturali, Acta Geologica*, Vol.62, pp. 3-18.

Brusatte, S.L.; Niedźwiedzki, G. & Butler, R.J. (2011). Footprints pull origin and diversification of dinosaur stem-lineage deep into Early Triassic. *Proceedings of the Royal Society of London, Series B*, Vol.278, pp. 1107-1113.

Carrano, M.T. & Wilson, J.A. (2001). Taxon distributions and the tetrapod track record. *Paleobiology*, Vol.27, pp. 563-581.

Casati, P., 1969 - Strutture della Formazione di Collio (Permiano Inferiore) Nelle Alpi Orobie. *Natura*, Vol.60, pp. 301-312.

Casati, P. & Forcella, F. (1988). Alpi Orobie. *'Il Bollettino'*, CAI-Annuario del Comitato Scientifico centrale, Vol.87, pp. 81-86.

Cassinis, G. (1966). La Formazione di Collio nell'area-tipo dell'alta Val Trompia (Permiano inferiore bresciano). *Rivista Italiana di Paleontologia e Stratigrafia*, Vol.72, pp. 507-588.

Cassinis, G. & Doubinger, J. (1991). On the geological time of the typical Collio and Tregiovo continental beds in the Southalpine Permian (Italy), and some additional observations. *Atti Ticinensi di Scienze della Terra*, Vol.34, pp. 1-20.

Cassinis, G. & Doubinger, J. (1992). Artinskian to Ufimian palynomorph assemblages from the Central Southern Alps, Italy and their regional stratigraphic implications. In: *Contributions to Eurasian Geology*, A.E.M. Nairn & V. Koroteev (Eds), 9-18, Occasional Publication ESRI New Series 8b, Part I, Columbia University of South Carolina, South Carolina, USA.

Cassinis, G.; Cortesogno, L., Gaggero, L., Massari, F., Neri, C., Nicosia, U. & Pittau, P. (1999). *Stratigraphy and facies of the Permian deposits between Eastern Lombardy and the Western Dolomites*, Field Trip Guidebook, Earth Scences Department, Pavia University, Brescia, Italy.

Cassinis, G.; Massari, F., Neri, C. & Venturini, C. (1988). The continental Permian in the Southern Alps (Italy). A review. *Zeitschrift geologische Wissenshaftliche*, Vol.16, pp. 1117-1126.

Cassinis, G.; Nicosia, U., Pittau, P. & Ronchi, A. (2002). Palaeontological and radiometric data from the Permian continental deposits of the Central-Eastern Southern Alps (Italy), and their stratigraphic implications. *Mémoires de l'Association des Géologues du Permien*, Vol.2, pp. 53-74.

Ceoloni, P.; Conti, M.A., Mariotti, N. & Nicosia, U. (1986). New Late Permian tetrapod footprints from Southern Alps. *Memorie della Società Geologica Italiana*, Vol.34, pp. 45-65.

Ceoloni, P.; Conti, M.A., Mariotti, N., Mietto, P. & Nicosia, U. (1987). Tetrapod footprints from Collio Formation (Lombardy, northern Italy). *Memorie Scienze Geologiche Università di Padova*, Vol.39, pp. 213-233.

Ceoloni, P.; Conti, M.A., Mariotti, N., Mietto, P., and Nicosia, U. (1988). Tetrapod footprint faunas from Southern and Central Europe. *Zeitschrift Geologische Wissenschaften*, Vol.16, pp. 895-906.

Conti M.A.; Leonardi, G., Mariotti, N. & Nicosia, U. (1975). Tetrapod footprints, fishes and molluscs from the Middle Permian of the Dolomites (N. Italy). *Memorie Geopaleontologiche dell'Università di Ferrara*, Vol.3, pp. 139-150.

Conti M.A.; Leonardi, G., Mariotti, N. & Nicosia, U. (1977). Tetrapod footprints of the "Val Gardena Sandstone" (North Italy). Their paleontological, stratigraphic and paleoenvironmental meaning. *Palaeontographia Italica*, Vol.70, pp. 1-91.

Conti M.A.; Leonardi G., Mariotti, N. & Nicosia, U., (1980). Nuovo contributo alla stratigrafia delle "Arenarie di Val Gardena". *Memorie della Società Geologica Italiana*, Vol.20, pp. 357-363.

Conti, M.A.; Leonardi, G., Mietto, P. & Nicosia, U. (2000). Orme di tetrapodi non dinosauriani del Paleozoico e Mesozoico in Italia. In: *Dinosauri in Italia. Le orme giurassiche dei Lavini di Marco (Trentino) e gli altri resti fossili italiani*, G. Leonardi & P. Mietto (Eds.), 297-320, Accademia Editoriale, Pisa, Italy.

Conti M.A.; Mariotti, N., Manni, R., & Nicosia, U. (1999). Tetrapod footprints in the Southern Alps: an overview. In: *Stratigraphy and facies of the Permian deposits between eastern Lombardy and the western Dolomites*, G. Cassinis, L. Cortesogno, L. Gaggero, F. Massari, C. Neri, U. Nicosia & P. Pittau (Eds.), 137-138, International Field Conference on 'The continental Permian of the Southern Alps and Sardinia (Italy). Regional reports and general correlations', Field Trip Guidebook.

Conti, M.A.; Mariotti, N., Mietto, P. & Nicosia, U. (1991). Nuove ricerche sugli icnofossili della Formazione di Collio in Val Trompia. *Natura Bresciana, Annali del Museo Civico di Storia Naturale Brescia*, Vol.26, pp. 109–119.

Conti, M.A.; Mariotti N., Nicosia, U. & Pittau, P. (1997). Selected bioevents succession in the continental Permian of the Southern Alps (Italy): improvements of intrabasinal and interregional correlations. In: *Late Palaeozoic and Early Mesozoic Circum-Pacific Events and Their Global Correlation*, J.M., Dickins, Y., Zunyi, Y., Yhongfu, S.G., Lucas & S.J., Acharyya (Eds.), 51-65, Cambridge University Press, Cambridge, United Kingdom.

Curioni, G. (1870). Osservazioni geologiche sulla Val Trompia. *Rendiconti Istituto Lombardo Scienze Lettere Arti, Memorie*, Vol.3, pp. 1–60.

Dozy, J.J. (1935). Einige Tierfärhten aus dem Unteren Perm der Bergamasker Alpen. *Paläontologische Zeitschrift*, Vol.17, pp. 45-55.

Farabegoli, E. & Perri, M.C. (1998). Permian/Triassic boundary and early Triassic of the Bulla section (Southern Alps, Italy): Lithostratigraphy, facies and conodont biostratigraphy. *Giornale di geologia*, Vol.60, pp. 292-311.

Farabegoli, E.; Perri, M.C. & Posenato, R. (2007). Environmental and biotic changes across the Permian-Triassic boundary in western Tethys: The Bulla parastratotype, Italy. *Global and Planetary Change*, Vol.55, pp. 109-135.

Geinitz, H.B. (1869). Über Fossile Pflanzenreste aus der Dyas von Val Trompia. *Neue Jährbuch für Mineralogie Geologie Paläontologie*, pp. 456-461.

Gümbel, C.W. (1880). Geognostische Mitteilungen aus den Alpen. VI. Ein geognostischer Streifzug durch die Bergamasker Alpen. *Sitzungsberichte der königl Akademie der Wissenschaften Mathematisch-Naturwissenschaftliche Klasse*, Vol.10, pp. 164-240.

Hallam, A. & Wignall, P.B. (1997). *Mass extinctions and their aftermath*. Oxford University Press, Oxford.

Haubold, H. (1996). Ichnotaxonomie und Klassifikation von Tetrapodenfährten aus dem Perm. *Hallesches Jahrbuch für Geowissenschaften B*, Vol.18, pp. 23-88.

Haubold, H. (2000). The tetrapod tracks of the Permian – state of knowledge and progress 2000. *Hallesches Jahrbuch für Geowissenschaften, B*, Vol.22, pp. 1-16.

Hunt, A.P. & Lucas, S.G. (2007). Late Triassic tetrapod tracks of western North America. *New Mexico Museum of Natural History and Science Bulletin*, Vol.40, pp. 215–230.

Italian IGCP 203 Group (1986). *Permian and Permian-Triassic boundary in the South-Alpine segment of the western Tethys*. Field guide-book, SGI and IGCP Project 203, Pavia.

King, G.M. (1991). Terrestrial tetrapods and the end Permian event: a comparison of analysis. *Historical Biology*, Vol.5, pp. 239–255.

Kittl, E. (1891). Saurier Färthe von Bozen. *Mitteilungen/Österreichischer Touristen-Club Wien*, Vol.3, pp. 7.

Klein, H. & Lucas, S.G. (2010).Tetrapod footprints - their use in biostratigraphy and biochronology of the Triassic. *Geological Society, London, Special Publications*, Vol.334, pp. 419-446.

Klein, H.; Voigt, S., Saber, H., Schneider, J.W., Hminna, A., Fischer, J., Lagnaoui A. & Brosig, A. (In press). First occurrence of a Middle Triassic tetrapod ichnofauna from the Argana Basin (Western High Atlas, Morocco). *Palaeogeography, Palaeoclimatology, Palaeoecology*.

Leonardi, P. (1967). *Le Dolomiti: Geologia dei monti tra Isarco e Piave*. Edizioni Manfrini, Rovereto.

Leonardi P.; Conti, M.A., Leonardi, G., Mariotti, N. & Nicosia, U. (1975). *Pachypes dolomiticus* n. gen. n. sp.; Pareiasaur footprint from the "Val Gardena Sandstone" (Middle Permian) in the Western Dolomites (N. Italy). *Atti dell'Accademia Nazionale dei Lincei, Rendiconti, Classe di Scienze Fisiche Matematiche e Naturali*, Vol.57, pp. 221-232.

Leonardi, G. & Nicosia, U. (1973). Stegocephaloid footprint in the Middle Permian Sandstone ("Grödener Sandsteine") of the western Dolomites. *Annali dell'Università di Ferrara*, Vol.5, pp. 116-124.

Lucas, S.G. (2004). A global hiatus in the Middle Permian tetrapod fossil record. *Stratigraphy*, Vol.1, pp. 47-64.

Lucas, S.G. & Hecker, T.A.B. (2001). Olson's gap: A global hiatus in the record of Middle Permian tetrapods. *Journal of Vertebrate Paleontology*, Vol.21, pp. 75A.

Lucas, S.G. & Hunt, A.P. (2006). Permian tetrapod footprints: biostratigraphy and biochronology. In: *Non-marine Permian biostratigraphy and biochronology*, 179-200, S.G., Lucas, G., Cassinis & J.W., Schneider (Eds.), Geological Society, London, Special Publications, 265.

Mietto P. (1975). Orme di tetrapodi nelle arenarie permiche di Recoaro (Vicenza). *Studi Trentini di Scienze Naturali, Acta Geologica*, Vol.52, pp. 57-67.

Mietto P. (1981). Una grande impronta di pareiasauro nel Permiano di Recoaro (Vicenza). *Rendiconti della Società Geologica Italiana*, Vol.4, pp. 363-364.

Mietto, P. (1986). Orme di tetrapodi nella Formazione di Werfen del Recoarese. *Rivista Italiana di Paleontologia e Stratigrafia*, Vol.92, pp. 321-326.

Mietto, P. (1987). *Parasynaptichnium gracilis* nov. ichnogen., nov. isp. (Reptilia: Archosauria Pseudosuchia) nell'Anisico inferiore di Recoaro (Prealpi vicentine - Italia). *Memorie Scienze Geologiche*, Vol.39, pp. 37-47.

Nicosia, U.; Sacchi, E. & Spezzamonte, M. (1999). New Palaeontological data on the Val Gardena Sandstone. In: *Stratigraphy and facies of the Permian deposits between eastern Lombardy and the western Dolomites*, G. Cassinis, L. Cortesogno, L. Gaggero, F. Massari, C. Neri, U. Nicosia & P. Pittau (Eds.), 33, International Field Conference on 'The continental Permian of the Southern Alps and Sardinia (Italy). Regional reports and general correlations', Abstract book.

Nicosia, U.; Ronchi, A. & Santi, G. (2000). Permian tetrapod footprints from W Orobic Basin (Northern Italy). Biochronological and evolutionary remarks. *Geobios*, Vol.33, pp. 753-768.

Pasini, M. (1985). Biostratigrafia con i foraminiferi del limite Formazione a Bellerophon / Formazione di Werfen fra Recoaro e la Val Badia (Alpi Meridionali). *Rivista Italiana di Paleontologia e Stratigrafia*, Vol.90, pp. 481-510.

Posenato, R. (1988). The Permian/Triassic boundary in the western Dolomites, Italy. Review and proposal. *Annali dell'Università di Ferrara, Scienze della Terra*, Vol.1, pp. 31-45.

Retallack, G.J.; Metzger, C.A., Greaver, T., Jahren, A.H., Smith, R.M.H., Sheldon, N.D. (2006). Middle-Late Permian mass extinction on land. *Geological Society of America Bulletin*, Vol.118, pp. 1398–1411.

Ronchi, A.; Santi, G. & Confortini, F. (2005). Biostratigraphy and facies in the continental deposits of the central Orobic Basin: A key section in the Lower Permian of the southern Alps (Italy). *New Mexico Museum of Natural History & Science Bulletin*, Vol.30, pp. 273-281.

Santi, G. (2003). Early Permian tetrapod ichnology from the Orobic Basin (Southern Alps-Northern Italy). Data, problems, hypotheses. *Bollettino della Società Geologica Italiana, Volume Speciale*, Vol.2, pp. 59-66.

Santi, G. (2005). Lower Permian palaeoichnology from the Orobic Basin (Northern Alps). *GeoAlp*, Vol.2, pp. 77-90.

Santi, G. (2007). A short critique of the ichnotaxonomic dualism *Camunipes-Erpetopus*, Lower Permian ichnogenera from Europe and North America. *Ichnos*, Vol.14, pp.185-191.

Santi, G. & Krieger, C. (2001). Lower Permian tetrapod footprints from Brembana Valley–Orobic Basin (Lombardy, Northern Italy). *Revue de Paléobiologie*, Vol.20, pp. 45-68.

Schaltegger, U. & Brack, P. (1999). Radiometric age constraints on the formation of the Collio Basin (Brescian Prealps). In: *Stratigraphy and facies of the Permian deposits between eastern Lombardy and the western Dolomites*, G. Cassinis, L. Cortesogno, L. Gaggero, F. Massari, C. Neri, U. Nicosia, P. Pittau (Eds.), 15-25, International Congress on 'The continental Permian of the Southern Alps and Sardinia (Italy). Regional reports and general correlations', Field trip guide-book, 71, Brescia, Italy.

Sahney, S. & Benton, M.J. (2008). Recovery from the most profound mass extinction of all time. *Proceedings of the Royal Society, Series B*, Vol.275, pp. 759-765.

Sheldon, N.D. (2006). Abrupt chemical weathering increase across the Permian–Triassic boundary. *Palaeogeography, Palaeoclimatology, Palaeoecology,* Vol.231, pp. 315-321.

Sheldon, N.D. & Chakrabarti, R. (2010). Mass extinctions, climate change, and enhanced terrestrial weathering?: The End-Permian and End-Guadalupian events compared. *Geochimica et Cosmochimica Acta,* Vol.74S1, p. A944.

Todesco, R. & Bernardi, M. (2011). Una nuova icnoassociazione nel Triassico medio (Anisico) del Trentino meridionale (val Gerlano, Vallarsa). *Studi Trentini di Scienze Naturali,* Vol.88, pp. 203-218.

Todesco, R.; Wachtler, M., Kustatscher, E. & Avanzini, M. (2008). Preliminary report on a new vertebrate track and flora site from Piz da Peres (Anisian-Illyrian): Olang Dolomites, Northern Itlay. *Geo.Alp,* Vol.5, pp. 121-137.

Twitchett, R.J.; Looy, C.V., Morante, R., Visscher, H. & Wignall, P.B. (2001). Rapid and synchronous collapse of marine and terrestrial ecosystems during the end-Permian biotic crisis. *Geology,* Vol.29, pp.351–354.

Valentini, M.; Conti, M.A. & Mariotti, N. (2007). Lacertoid footprints of the Upper Permian Arenaria di Val Gardena formation (Northern Italy). *Ichnos,* Vol.14, pp. 193-218.

Valentini M.; Nicosia, U. & Conti, M.A. (2009). A re-evalution of *Pachypes*, a pareiasaurian track from the Late Permian. *Neues Jahrbuch für Geologie und Paläontologie Abhandlungen,* Vol.251, pp. 71-94.

Visscher, H. & Brugman, W.A. (1988). The Permian-Triassic boundary in the Southern Alps: A palynological approach. *Memorie della Società Geologica Italiana,* Vol.34, pp. 121- 128.

Voigt, S.; Berman, D.S. & Henrici, A.C. (2007). First well established track-trackmaker association of Paleozoic tetrapods based on *Ichniotherium* trackways and diadectid skeletons from the Lower Permian of Germany. *Journal of Vertebrate Paleontology,* Vol.27, pp. 553-570.

Wignall, P.B.; Kozur, H. & Hallam, A. (1996). The timing of palaeoenvironmental changes at the Permo-Triassic boundary using conodont biostratigraphy. *Historical Biology,* Vol.11, pp. 39–62.

Wilson, J.A. & Carrano, M.T. (1999). Titanosaurs and the origin of "wide-gauge" trackways: a biomechanical and systematic perspective on sauropod locomotion. *Paleobiology,* Vol.25, pp. 252–267.

Wopfner, H. (1999). Über Tetrapoden-Fährten, Kohlen und versteinerte Holzer aus dem Grödner Sandstein (Perm) dei Deutschnofen. *Der Schlern,* Vol.73, pp. 23-32.

11

Genomics of Bacteria from an Ancient Marine Origin: Clues to Survival in an Oligotrophic Environment

Luis David Alcaraz[1], Varinia López-Ramírez[2], Alejandra Moreno-Letelier[3],
Luis Herrera-Estrella[4], Valeria Souza[5] and Gabriela Olmedo-Alvarez[2]

[1]*Department of Genomics and Health,*
Center for Advanced Research in Public Health, Valencia,
[2]*Departamento de Ingeniería Genética, Cinvestav Unidad Irapuato,*
[3]*Division of Biology, Imperial College London, Silwood Park Campus, Ascot,*
[4]*Langebio, Cinvestav, Mexico*
[5]*Departamento de Ecologia Evolutiva, Instituto de Ecologia,*
Universidad Nacional Autónoma de México
[1]*Spain*
[2,4,5]*Mexico*
[3]*UK*

1. Introduction

Genomics has certainly changed the way that biology is studied, and has had a substantial impact on many other scientific disciplines as well. Life and Earth have had an interdependent history since the early establishment of the biogeochemical cycles in the Archean. Genomic sequences provide historical information that can be correlated with the geological record. Thus, it is not surprising that comparative genomics aids in understanding current findings in geological sciences. Genomics allows us to explain the evolutionary history of an organism by analyzing and comparing the set of shared genes with all respective relatives. Additionally, by examining the genes that are unique to some strains or taxonomic groups, we can make inferences about their ecology. Since molecular biology was initially developed in bacterial model organisms and we have extensive knowledge about the enzymes that participate in different biochemical pathways, inferences and functional predictions can be made about numerous sequenced genes. Moreover, due to the energetic and evolutionary costs of preserving a gene in bacteria, where specialists tend to have small size genomes, unique genes may aid in exploiting a given niche. Using comparative genomics, we have undertaken the study of the diversity, evolutionary relationships, and adaptations of bacteria to the oligotrophic (low nutrient content) conditions of the unique ecosystem of Cuatro Cienegas, Coahuila, Mexico. In this chapter, we review how genomics can be used to describe and understand microbial diversity, evolution, and ecology in different environments as well as the relationship with geological data by analyzing current studies, and our own work as a case study (Fig. 1).

2. Bacterial comparative genomics: Clues to evolution, geological history, and ecology

Since the early 1980's, science has had technology to determine the precise nature of DNA, which is the molecule found in all genomes. The era of genomics began with the sequencing of the first bacterial genome, *Haemophilus influenzae,* in 1995. The sequencing of complete genomes has since gained speed and precision, and with the costs of this technology continually decreasing, the sequencing of genomes today that contain millions of base pairs of DNA is possible in just a few days. Over 1,800 bacterial genomes have been sequenced to date, with several hundred more currently in progress. More impressively, the very large genomes of several plants and animals, including the human genome with more than 3 billion bases of DNA, have also been completed. In order to understand how particular genomes give rise to different organisms with very different traits, the ability to compare and contrast features of this wide range of genomes is needed. Moreover, an understanding of the evolutionary history and ecology of the organisms for which we have genomes is required.

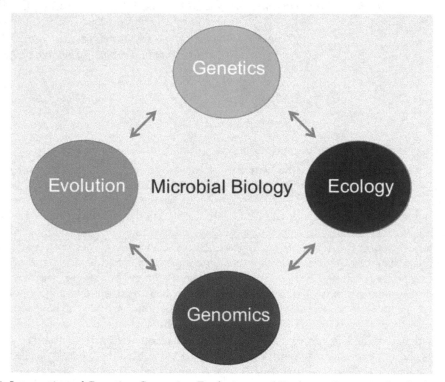

Fig. 1. Integration of Genetics, Genomics, Evolution, and Ecology. Genomics has had a great impact on Biology, particularly in the areas of Genetics, Ecology, and Evolution. Therefore, none of these areas can be studied separately. For microbial Biology, genomic studies have revealed a plethora of microbes that had eluded previous experimental strategies, which has provided the opportunity to understand their influence on the planet's history and present function in the environment.

2.1 How genomic sequences provide clues for understanding the diversity and origin of microorganisms

Classical microbiological classification systems made use of a culturable microbial diversity that only accounted for 1% of the total estimated microbial diversity (Amann et al. 1995; Khan et al. 2010), and due to the historical and clinical importance, it was a classification system that was highly biased towards clinical pathogenic strains. Although the 1% of culturable bacteria is an over-sold idea, it is clear that new culturing media should be developed and tested in order to capture more diversity within a given environment. To date, the most extensively used method for estimating the diversity of bacterial species is sequencing and comparing the 16S rRNA gene, which is an approach that has been used since the early 1970s (Woese & Fox 1977). Moreover, this method has been used in the current culture-free approaches, such as the metagenome surveys, by means of the Next Generation Sequencing (NGS) systems (Metzker 2010) in order to sequence DNA in particular environments (Fig. 2).

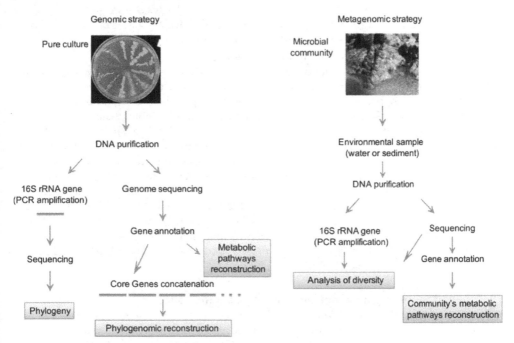

Fig. 2. Sequencing strategies for specific isolates and metagenomes.

The 16S rRNA gene sequencing approach has shed light onto the complex relationships amongst bacteria, and has even predicted the existence of new groups when using culture-free approaches. However, today we can go beyond explaining the evolution of a single gene and try to explore entire genome relationships of bacteria. With the development of NGS technologies, the associated costs are falling, and we can now envision a scenario where a small-sized lab is able to sequence the complete genome of a bacterial strain of interest. With whole genome sequencing, it is possible to perform a deep analysis to explain the relationships of every gene in genomes rather than single gene histories. Moreover, we can also explain the history and presence of bacteria by understanding the set of genes shared by all relatives, which can reveal a common evolutionary lineage as well as the genes that are exclusive to particular strains. The later genes, which are unique to certain strains or taxonomic groups, can also give us insight on the capabilities of an organism to exploit a given niche and will therefore provide information on environmental features as well.

The numerous genome sequencing projects provide clues for elucidating whole genome relatedness and functions that are exclusive to some groups. On another level of complexity, global scale efforts are conducted to attempt to determine the roles of microbes in the global biogeochemical cycles as well as the discovery and inventory of gene diversity, such as that from the Global Ocean Expedition (Rusch et al. 2007). Other projects focus on a deep understanding of the relationships between our body and the microbes with which we live. This project, known as the Human Microbiome Project (Turnbaugh et al. 2007), attempts to shed light on the close relationship between humans and our microbes. Another challenging project has to do with the sequencing of at least one representative strain of all of the major bacteria groups, known as the Genomic Encyclopedia of Bacteria and Archaea (GEBA) (Nelson et al. 2010), and the ambitious Earth Microbiome Project (http://www.earthmicrobiome.org/) will provide a reference framework for conducting comparative genomics and establishing deep relationships amongst microbes.

Some research areas, such as population genetics, are transitioning from an understanding of a very limited number of loci to the analysis of whole genome single nucleotide polymorphisms (SNPs) in order to understand fine differences between the populations of pathogens and free-living bacteria. In contrast, other research areas, such as Molecular Biology, have the power to analyze whole genome regulation by sequencing the entire RNA of an organism that is exposed to a specific condition (transcriptomics) and using the results to predict the function of hypothetical genes. The improvement in the performance of computational resources as well as decreasing costs of these technologies has created tremendous analytical power. It is now possible to rent web services, such as cloud computing, and use highly efficient software online to assemble (Chaisson et al. 2004; Miller et al. 2008; Zerbino & Birney 2008), align (Kurtz et al. 2004), annotate (Aziz et al. 2008; Van Domselaar et al. 2005), and compare the more than one thousand bacterial genomes sequenced to date with in-house datasets. In addition, we are now able to democratically access the experimental and analytical tools to dissect both the evolution and functions of microbial life.

2.2 Core and accessory genomes and the pan-genome concept

In recent years, several groups have started to make whole genome comparisons of close relative strains, which has led to new and unexpected results for what was supposed to be members of the same bacterial species, based on 16S rRNA gene analysis, DNA-DNA re-association experiments, and phenotypic traits (Medini et al. 2005; Reno et al. 2009; Tettelin et al. 2005). Comparative genomic analysis has revealed that within species with the exact same 16S rRNA gene sequence, differences in gene content can account for ± 30% of the total coding genes in one strain that are not found in their close relatives. To clearly understand this finding, we can make the comparison that if we look at this in percentages, humans are closer to chimpanzees (1% differences in coding genes) than two bacteria isolates causing the same disease.

From these observations, the genome of a species is now conceptually split into a core genome, which refers to all of the shared genes between analyzed strains, and the accessory genome, which comprises the strain-specific genes. Core genes are those that are found in all genomes from a given taxonomic group and make up the genes that provide essential functions, such as those for DNA conservation and expression (involved in translation, replication, and transcription) as well as central metabolic pathways. These genes are collectively considered housekeeping genes. Core genes are considered likely components of all isolates of a given species or genus. Since selection at these loci exerts a stabilizing rather than a diversifying influence, variations detected in these genes tend to be neutral, or nearly so, and will accumulate in a consecutive, clock-like manner (see section 9). Thus, these genes should be reliable indicators of evolutionary history. For instance, core 16S rRNA gene sequences are frequently used to reconstruct the phylogenies of microbial species, and metabolic or housekeeping genes are commonly used for genotyping.

Accessory genes are those that may or may not be present in a given strain, since they can typically be associated with mobile elements (i. e. phages or transposons), can encode genes for secondary biochemical pathways, or can code for functions that mediate interactions with the environment. However, there are examples of genes that escape detection as core genes because they have been replaced by genes from a different origin (xenologous genes) that serve the same function but cannot be identified by similarity. This is often the case for divergent alleles. In any case, the genes classified as accessory genes are very interesting from an ecological point of view, since the genes in this category would be those that reveal the function of an organism in its environment. In addition, the range of genetic diversity of a species can be discovered in these genes. Pan-genomes, on the other hand, help us to understand the gene repertoire that has yet to be discovered within the group. The pan-genome is defined as the sum of the core genome and the accessory genes. Pan-genomes are also classified as open, where in an accumulative plot there is still no plateau, and closed, where all of the expected genes are present in the group and newly discovered genes are just a product of chance regardless of how many new strains are sequenced. Pan-genomes help us determine when other genome dynamics are shaping interesting phenotypes. For example, when comparing strains of the *Bacillus cereus-anthracis* group (Anderson et al. 2005; Ehling-Schulz et al. 2005; Helgason et al. 2004), it was found that they were very clonal, and that their central pathogenic traits were the result of horizontal gene transfer in mobile elements.

A surprising finding in the many available microbial genomes was the high number of genes that had been acquired by exchange of DNA between microbes. The so-called "horizontal genetic transfer" events, confer a remarkable evolutionary potential to many species. The loss or gain of individual genes or large "genomic islands" accounts for the emergence of several specific metabolic, virulence, or drug resistance phenotypes. In fact, another characteristic of accessory genes is that these are often transferred among bacteria through the aid of phages, plasmids, or by transformation with free DNA. In a population, alleles (variant forms of a gene) may be distributed among the members, and individuals may possess either copy of the gene. These alleles (also called homeoalleles) encode enzymes with identical functions but that may have had a different evolution trajectory, and therefore exhibit differences at the gene level. Individual lineages may exhibit different homeoalleles acquired through horizontal gene transfer, but these homeoalleles may have also been lost from the lineage and replaced by other homeoalleles (reviewed in (Andam &

Gogarten 2011)). Importantly, horizontal gene transfer is responsible for the mosaicism often observed in genomes.

3. Using genes to understand the environment: A case study

3.1 The Cuatro Cienegas Basin: Water in a low phosphorous desert ecosystem

The Cuatro Cienegas Basin (CCB) is located in a valley in the central part of the State of Coahuila, Mexico (26°59′N 102°03′W). The Basin is roughly 84 km^2 in area and an average of 740 m above sea level (Fig. 3). CCB is surrounded by mountains that rise to elevations of approximately 2,500 and 3,000 m. The Geology and Physiography of the region have been thoroughly reviewed in the literature (Minckley 1969). Ancient geologic history indicates that the CCB was at the very nexus of the separation of Pangea, which created what we now know as the Northern hemisphere 220 million years ago. The CCB became isolated from the sea much later, and the subsequent uplifting of the Sierra Madre Oriental occurred approximately 35 million years. The major geological events of the Eocene epoch in northern Mexico corresponded to the genesis and development of the Sierra Madre Oriental, the fold-ranges of the Chihuahua-Coahuila and the Gulf coast plain development (Ferrusquía-Villafranca & González-Guzmán 2005).

Both climatically and geographically, the CCB belongs to the Chihuahuan Desert (Schmidt 1979), which is the second largest desert in North America. The climate is arid, with an average annual precipitation of less than 200 mm, and daytime temperatures in the summer that sometimes exceed 44°C. Despite the dry climate of the CCB, it harbors an extensive system of springs, streams, and pools. Within the valley, spring water flows on the surface and through subsurface channels in karstified alluvium. The main source for the subterranean water in these systems is old water that was deposited there in the late Pleistocene epoch (Wolaver 2008). Water in the CCB is thought to be the relict of a shallow sea that existed 35 million years ago. The oasis water contains low levels of NaCl and carbonates, but is rich in sulfates, magnesium, and calcium.

Vegetation and fauna in the CCB appear typical of an arid zone. Given the combined conditions of habitat diversity and permanence, as well as the isolation of the basin since historic times, elements of the aquatic fauna have undergone adaptive radiation and speciation that has resulted in many endemic organisms (reviewed in Holsinger & Minckley 1971; Minckley 1969).

The spring-fed ecosystems of the CCB are dominated by microbial mats and living stromatolitic features (see Fig. 3) that are supported by an aquatic sulfur cycle and a terrestrial gypsum-based ecology in large parts of the valley.

The most striking feature of the CCB ecosystem is the very low levels of phosphorus in both the water and soil, which presents an extreme elemental stoichiometry with regards to phosphorus (900:150:1- 15820:157:1 C:N:P ratio, respectively) (Elser 2005) when compared to similar environments. Phosphorus is an essential nutrient for multiple cellular processes, including energy and information. However, it is not an abundant element on the planet and can only be obtained from organic detritus or from tectonics and volcanism. Therefore, the availability of phosphorus is a limiting factor for all life forms. Nevertheless, life persevered and the CCB is characterized by a high endemism in all of the domains of life (Minckley 1969; Scanlan et al. 1993) despite the fact that phosphorus levels are below the level of detection (0.3 µM). In addition, the extremely oligotrophic waters are unable to sustain algal growth, which has caused the microbial mats to be the base of the food web (Elser 2005).

Churince system

Stromatolites in Pozas Azules

Dunes

Fig. 3. The Cuatro Cienegas Basin. (a) Location map of Cuatro Cienegas, Coahuila. (b) Photograph of the Churince sytem facing San Marcos, one of the mountains in the valley. (c) One of the ponds at Pozas Azules, where live stromatolites can be observed. (d) White gypsum dunes, which are the result of evaporating water, once covered a vast area in the basin.

3.2 Bacterial genomics of two bacterial isolates from the CCB provide clues for the mechanism of bacterial survival in a highly oligotrophic environment

We have taken different approaches to gain knowledge on the diversity and evolution of microbial communities and to understand how microbes in these communities deal with the oligotrophic conditions in the various water bodies of Cuatro Cienegas. Several microbiological surveys have been carried out, including bacterial isolation and culturing, 16S rRNA gene amplification, genome sequencing, and metagenome sequencing.

Characterization of the microbial taxonomic diversity by sequence analysis of the 16S rRNA genes directly from environmental DNA has revealed that nearly half of the phylotypes from the CCB are closely related to bacteria from marine environments (Souza et al. 2006). This makes sense, given that the geological history of Cuatro Cienegas suggests that a shallow ocean covered the region and some of the water there is most likely a part of the underground water that feeds different ponds. This has led to the hypothesis that bacteria in the CCB water systems are descendants of marine bacteria from Pre-Cambrian (Souza et al. 2006). Systematic studies have also been used to describe new species of bacteria using current criteria (Anderson et al. 2005; Cerritos et al. 2008; Ehling-Schulz et al. 2005; Escalante et al. 2009; Helgason et al. 2004) as well as by attempting to apply multivariate analysis and physical-chemical analyses in order to determine special relationships amongst the bacteria (Cerritos et al. 2011). At the genomic scale, 2 genomes (Alcaraz et al. 2010; Alcaraz et al. 2008) and 3 metagenomes (Breitbart et al. 2009; Desnues et al. 2008; Dinsdale et al. 2008) had been described, in which genes that allow for adaptation to the oligotrophic environment of Cuatro Cienegas have been identified. Two additional metagenomes from microbial mats and several more microbial genomes are currently under analysis. The entire dataset that is currently available from this area has helped us identify shared features across different microbial communities that deal with the same constraints, including phosphorous limitation, and has helped us understand the genetic strategies that are available to cope with these conditions.

However, taxonomy has only provided a single molecular marker approach. In order to gather more evidence on whether bacteria from the CCB could be recent migrants from marine environments or ancient creatures from an old sea, it required more than one gene as a tool. Therefore, we sequenced the whole genome of two isolates: *Bacillus coahuilensis* (Alcaraz et al. 2008; Cerritos et al. 2008) and *Bacillus sp.* isolate m3-13 (Alcaraz et al. 2010). On one hand, an analysis of the gene content and metabolic pathways represented by each genome revealed numerous differences (Table 1), starting with genome size, as *B. coahuilensis* turned out to have the smallest genome of all sequenced *Bacillus* spp. (3.65 Mbp) and an incomplete genome in many functions.

Another gene family overrepresented in the genomic datasets from Cuatro Cienegas includes genes involved in environment sensing mechanisms. For example, all of the two-component (histidine-kinases) as well as the unusual sensitive rhodopsins, such as in the case of *B. coahuilensis*, were found to be overrepresented, indicating that most of the responses of the bacteria that have had their genomes analyzed are environmentally-triggered when compared to generalist and cosmopolitan organisms, such as *B. subtilis*. Moreover, these comparisons also showed an underrepresentation of secondary metabolism genes in the genome of *B. coahuilensis* (Alcaraz et al. 2010; Alcaraz et al. 2008; Desnues et al. 2008)(Fig. 4). Basic capabilities, such as sulfur utilization, were found to be different between the two isolates. *B. coahuilensis* appears not to be able to use inorganic sulfur and depends on organic sources for this element, while *Bacillus* m3-13 is able to use inorganic sulfur. In addition, the two isolates seem to be exposed to different stress, as *B. coahuilensis* has an alkyl hydroperoxide reductase protein coding gene that can aid with oxidative stress. Moreover, this strain also has several sugar transporters and biosynthesis pathways for sugars that help it deal with osmotic stress, such as trehalose, choline, and betaine uptake as well as betaine biosynthesis, all of which are absent in *Bacillus* m3-13.

In contrast, isolate m3-13 of *Bacillus* seemed to have a robust metabolism, starting with a complete urea cycle, capacity for taking up inorganic sulfur, and a wide selection of sugar transporters. Theoretically, the m3-13 strain would be able to use chitin, N-acetylglucosamine, maltose, maltodextrin, and sucrose to ferment lactate. It also has genes

Genomic features (differences)	B. coahuilensis	Bacillus m3-13
Genome size	3,640 Mbp	4,294 Mbp
Nitrogen metabolism	Lacks many genes, incomplete urea cycle	Complete urea cycle (Arginine and Ornithine Degradation)
Bacteriorhodopsin	Genes for **bacteriorhodopsin** and carotene synthesis	Lacks bacteriorhodopsin gene Pigmented, *crtB y crtI* (fitoene synthase and dehydrogenase)
Phosphate utilization genes	**Sulfoquinovose and glycosyl transferase genes**, sulfolipid synthesis	Phosphonate ABC transporter Genes for phosphonate utilization (C-P)
Phosphorous recycling	Alkaline phosphatase	Predicted ATPase related to phosphate starvation-inducible protein PhoH
Membrane Transport	Dipeptide-binding ABC transporter, periplasmic substrate-binding component	
Carbohydrates	Trehalose Uptake and Utilization	Chitin and N-acetylglucosamine Maltose and Maltodextrin Sucrose Fermentations: Lactate D-gluconate and ketogluconates metabolism D-ribose utilization Formaldehyde assimilation: Ribulose monophosphate pathway Beta-galactosidase
Cofactors, Vitamins, and Prosthetic Groups		Cobalamin, heme and siroheme, and thiamin biosynthesis
Sulfur metabolism	Sulfolipid synthesis	Inorganic Sulfur Assimilation: Adenylyl-sulfate reductase Ferredoxin--sulfite reductase; Sulfate permease, Pit-type
Stress Response	Osmotic stress: Choline and Betaine Uptake and Betaine Biosynthesi; Alkyl hydroperoxide reductase	
Virulence, Disease and Defense	Arsenical-resistance protein ACR3, Bacitracin export and resistance	Multidrug Resistance Efflux Pumps

Table 1. Genome differences between *B. coahuilensis* and *Bacillus* m3-13.

involved in D-gluconate and ketogluconates metabolism as well as D-ribose utilization. In addition, most of the amino acid biosynthesis pathways are complete. As expected in a place with extreme phosphorous limitation, some notable genetic features of those genomes are related to their ability to utilize phosphonate, which is discussed in the next section.

A. Metabolic map of *Bacillus sp.* m3-13   B. Metabolic map of *Bacillus coahuilensis*

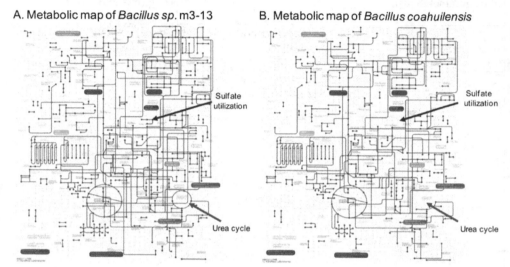

Fig. 4. Metabolic pathways in *Bacillus* m3-13 (A) and *B. coahuilensis* (B). A larger genome and more robust metabolism in m3-13 suggest that this is a generalist bacteria. In contrast, metabolism that lacks several reactions, such as those for inorganic sulfur utilization (blue arrow) and an incomplete urea cycle (red arrow), suggest that these bacteria depend on the community. The reduced genome of *B. coahuilensis* is puzzling for free-living bacteria. These maps were reconstructed from the annotated genome sequences using the KEGG (http://www.genome.ad.jp/kegg-bin/srch_orth/html) database.

4. Strategies of bacteria for dealing with limited phosphorous

Phosphorus is an essential nutrient for multiple processes, such as the synthesis of DNA, RNA, and ATP as well as many other pathways involving phosphorylation (Tetu et al. 2009). However, it is not an abundant element on the planet and can only be obtained from organic detritus or from tectonics and volcanism, and therefore the availability is a limiting factor for all life forms. Since the growth rate and primary productivity are highly dependent on phosphorus (Elser & Hamilton 2007; Elser et al. 2006; Zubkov et al. 2007), bacteria have different mechanisms for the uptake and storage of phosphates in order to cope with this limitation (Adams et al. 2008; Rusch et al. 2007; Tetu et al. 2009). For example, bacteria can use alternative phosphorus sources, use polyphosphates as storage compounds, or employ a highly effective phosphate-recycling mechanism (Fig. 5). Two major phosphate transport systems are involved: the low affinity phosphate inorganic transport (Pit) system, and the high affinity phosphate specific transport (Pst) system (van Veen, 1997). The uptake and assimilation of organic forms of phosphorus, such as phosphonates, require different transporters. Phosphonates are a class of organophosphorus compounds that are characterized by a chemically stable carbon-to-phosphorus (C-P) bond. Although phosphonates are widespread, only microorganisms are able to cleave this bond. The use of alternative phosphorus sources (phosphonates, phosphites, and hypophosphites) was determined by the presence of the high affinity transporters *pnhD* and *ptxB* as well as the C-P lyase genes *phnH* and *htxB* (White & Metcalf 2004).

Some microorganisms have been shown to accumulate relatively large amounts of polyphosphate, which has been hypothesized to have an important role in the response to changes in nutritional status or environmental conditions. Polyphosphate acts as a reservoir of intracellular phosphate, which is a strategy that seems to be particularly important for mobility and the formation of biofilms (Brown & Kornberg 2004). Amongst the genes induced under phosphorus deprivation are *ppA*, *ppK*, and *ppX* (coding for pyrophosphatase, polyphosphatase kinase, and exopolyphosphatase, respectively), which are involved in polyphosphate metabolism. Finally, extracellular phosphates are recycled by the overexpression of alkaline phosphatases *phoA* and *phoX* (Scanlan et al. 1993). Some bacteria can also utilize phytic acid (inositol hexakisphosphate (IP6)), which is the principal storage form of phosphorus in some plant tissues.

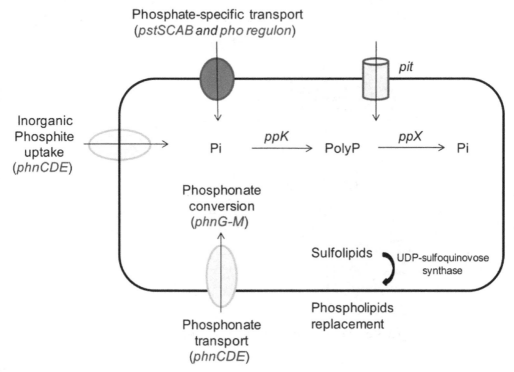

Fig. 5. Strategies for phosphorous utilization by bacteria (adapted from Hirota et al. 2010).

4.1 Phosphourous metabolism in the *Bacillus* genus

Since a low phosphorus concentration is a feature of the CCB basin, we focused on identifying the mechanisms of the *Bacillus* genus as a group, and in particular the isolates from the CCB for dealing with phosphorous. Comparative genomics can reveal the diversity of mechanisms employed by the *Bacillus* strains where the genomes have been sequenced (Fig. 6). We observed a great diversity in strategy used by different *Bacilli*. For instance, Phytases seem to be restricted to the *B. subtilis-pumilus* group. These genes are a good example of accessory genes and reveal a specific niche for this group of bacteria. However, this same group lacks genes for polyphosphate or phosphonates metabolism. Alkaline phosphatases are widespread in the *Bacillus* genus. The *B. cereus-anthracis-thuringiensis* group is very homogeneous and has quite versatile mechanisms for phosphorous usage.

Among the CCB isolates, *Bacillus* m3-13 appears to be able to use phosphonates, but *B. coahuilensis* cannot, and these isolates have different alkaline phosphatase genes. Most species harbor the high affinity transporter PstS and lack the low affinity transporter Pit, which is present in *B. cereus* and *B. subtilis*. Remarkably, only *B. coahuilensis* has genes for the replacement of cell-membrane phospholipids to sulfolipids (see below). Finally, it has been proposed that genome reduction may be a strategy for reducing phosphorous utilization (Alcaraz et al. 2008; Desnues et al. 2008), and *B. coahuilensis* has the smallest genome among all sequenced *Bacilli* to date.

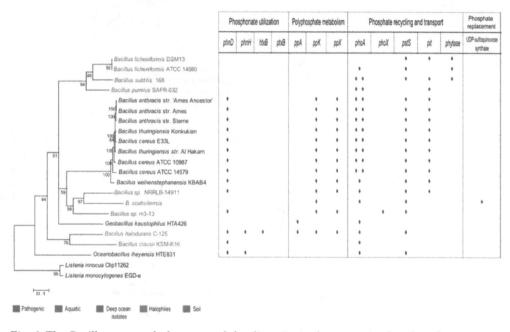

Fig. 6. The *Bacillus* genus phylogeny and the diversity in the strategies for phosphorous usage.

4.2 Sulfolipid synthesis in *B. coahuilensis*

Phospholipids constitute 30% of the total phosphate in most organisms. Interestingly, in plants and cyanobacteria subjected to phosphorous deprivation, phospholipids can be replaced by non-phosphorus lipids (such as sulfo- and galactolipids) to maintain membrane functionality and integrity, and can release phosphorus in order to sustain other cellular processes that require phosphorus (Dormann&Benning 2002). In addition, genes encoding sulfoquinovose synthase (*sqd1*) and glycosyltransferase (*sqdX*), which are the two key enzymes in the synthesis of sulfolipids, are present in *B. coahuilensis*. Thin layer chromatography and mass spectrometry analysis have confirmed the presence of sulfolipids in *B. coahuilensis*.

The remarkable acquisition of constitutively expressed genes that allow *B. coahuilensis* to replace membrane phospholipids with sulfolipids is in agreement with genomic adaptations to extreme phosphate limitation. However, not all *Bacilli* from the CCB ponds use the same strategy to cope with phosphorous limitations. *B. coahuilensis* and its relatives as well as a few other *Bacilli* species have this gene, but we did not find sulfolipids among several other *Bacillus* spp. from the CCB. Therefore, how do other bacteria strains cope with the limiting phosphorous conditions? As explained above, genomic sequencing of another strain, *Bacillus* isolate m3-13, showed that it possessed *phn* genes that code for phosphonate ABC importers, permeases, and a phosphonate lyase. We hypothesize that this strain may use these genes to take up and assimilate phosphonates. Importantly, both strategies seem to be used by bacteria from the Cuatro Cienegas as well as by marine bacteria (Fig. 6).

4.3 Mobile genes involved in phosphorous uptake

As discussed above, growth rate and primary productivity are highly dependent on phosphorus, and bacteria have different mechanisms for the uptake and storage of phosphates in order to cope with this limitation. The sulfoquinovose synthesis operon is absent in all other known *Bacillus* spp. genomes. The *B. coahuilensis* genes are closely related to cyanobacterial *sqd1* and *sqdX*, and the operon arrangement is identical to that in *Synechococcus* sp. PC7942, where these genes participate in the synthesis of sulfolipids (Benning 1998). This finding suggests that the adaptation of *B. coahuilensis* to the extremely low phosphorous concentration of the CCB may have included the acquisition of these genes through horizontal gene transfer (Alcaraz et al. 2008).

Another example of horizontal gene transfer involves genes of the *pho* regulon. The high affinity phosphate transport system (*pst*) is thought to be responsible for phosphate uptake under nutrient stress (Qi et al. 1997; Adams et al. 2008). *Pst* is a typical ABC transport system. Unlike the model bacteria *Escherichia coli* or *B. subtilis*, bacteria from the CCB as well as sequenced marine *Bacillus* lack the low affinity phosphate uptake system and must rely solely on the high affinity transport system. We found two types of operon architectures that harbor the *pst* gene and evaluated their phylogenetic congruity with housekeeping genes. We found high divergence of the two types of *pst*-operons in *Firmicutes* and incongruence with species phylogeny. In contrast to what was expected, the *pst* operon of marine *Bacillus* is not monophyletic, even though marine and the CCB *Bacilli* are resolved as a monophyletic group in the core-gene reconstruction. Therefore, the heterogeneous distribution of the different types of the *pst* operon among closely related species suggests horizontal gene transfer (Moreno-Letelier et al. 2011).

5. Carotenes and gene transfer of bacteriorhodopsin: A good combination in a high-radiation environment

B. coahuilensis has a gene encoding Bacteriorhodopsin, which is a situation similar to the abundance of BR genes in marine environmental samples, and suggests an additional adaptation of marine bacteria in the CCB. The phylogeny of *B. coahuilensis* sensory BR showed that its closest relative is the Anabaena sp. PCC7120 rhodopsin. Evidence for horizontal gene transfer of rhodopsins has recently been obtained from whole genome sequencing and metagenomic projects, and is now thought to be a frequent event in marine bacteria in the photic zone and extreme saline environments. The retinal chromophore of rhodopsin is synthesized as a cleavage product of carotenoids; thus, the combination of carotenoid synthesis and rhodopsin genes has been suggested to be sufficient for rhodopsin function (Frigaard et al. 2006). The genome of B. coahuilensis also contains genes encoding crtB (phytoene synthase) and crtICA2 (phytoene dehydrogenases) that could be involved in retinal biosynthesis. The high radiation exposure that is prevalent in shallow waters of the CCB could explain the selection pressure responsible for the maintenance and constitutive expression of the bsr gene.

6. Core and pan-genome of the *Bacillus* genus

In order to understand the cohesion of the *Bacillus* genus at the genomic level, we used the core and pan-genomes as the working units and took advantage of the large dataset available. We have analyzed the genes comprising the core genome of the *Bacillus* genus. The core genome of the *Bacillus* spp. that was analyzed (see phylogeny in Fig. 6) contained 814 genes. After annotating each gene in the genome of the two isolates, it was possible to classify and assign the different genes to specific functional categories and reconstruct and compare their metabolic pathways. Figure 7 shows the drawing of a metabolic map representation of the Bacillus pan-genome and Bacillus core genome obtained with the Kyoto Encyclopedia of Genes and Genomes. Using the pan-genome, we are able to understand the variation in functions across a cosmopolitan genus that can survive under harsh conditions, such as the bottom of the sea, hydrothermal vents, hypersaline environments, or even simply the hosts, as is the case for pathogens. The average gene content for the *Bacillus* genus was 4,973 ± 923 and the total pan-genome involved a around 75,000 genes clustered in 19,043 gene families. This is a very large number if we consider that the most recent predictions show that the human genome harbors 20,000 genes (Nelson et al. 2011). From these analyses, it is evident that a vast repertoire of functions is encoded in the *Bacillus* genus, and helps to explain the versatility of these bacteria for living and surviving in harsh conditions.

6.1 The sporulation core and accessory genes

The core genome sequence can help to identify and understand relevant, conserved genes for a trait of an entire group, and we therefore analyzed sporulation (Alcaraz et al. 2010), as the *Bacillus* is a group that is defined as endospore-forming genus. To obtain insight into the biology of this group, we described the relatedness within *Bacillus* using whole genome information to reconstruct their evolutionary history by taking advantage of the dataset available from the complete and draft genomes of 20 Bacilli isolated from a wide range of environments. Nucleotide metabolism, cell motility, and secretion showed little variability. In contrast, the features that were highly represented within the genomes and varied the most among the different genes were related to repair and transcription. Secondary metabolism, as expected, also exhibited variation among the taxonomic groups. Figure 7 shows a comparison between the functions defined by the core genome versus those defined by the pan-genome. The latter, which covers numerous metabolic pathways and reflects the potential of the whole group, explains why this is a cosmopolitan bacterial genus with the capability of colonizing diverse niches.

A. Metabolic map of a *Bacillus* Pan-genome   B. Metabolic map of a *Bacillus* core genome

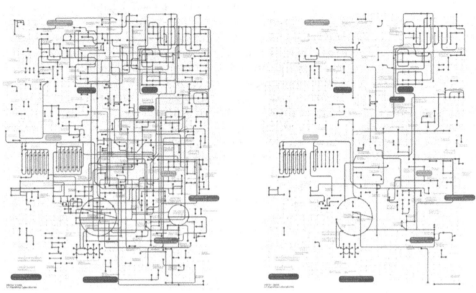

Fig. 7. Comparative metabolic reconstruction for the *Bacillus* pan-genome (A) and the *Bacillus* core genome (B). The combined pool of genes in the sequenced Bacillus genomes reflects the greater metabolic potential of the *Bacillus* genus. (B) The metabolic map of the core genome, which is represented by all of the conserved genes among the sequenced *Bacillus* genomes, reflects the basic housekeeping functions of these genera. The apparent absence of some enzymatic pathways that can be considered essential may be explained by the replacement of genes coding these capacities with xenologous genes. These maps were reconstructed from the annotated genome sequences using the KEGG (http://www.genome.ad.jp/kegg-bin/srch_orth/html) database.

When analyzing the core genome for genes involved in sporulation, we found that less than 52 of the 200 genes were conserved across 20 other *Bacillus* genomes. These genes are known to be essential for completion of the sporulation process in *B. subtilis*, but the fact that only a quarter of them are present in the other strains suggests the adaptability of genes for the same process, and particularly for the signaling circuit that responds to diverse environmental cues. Our study also identified the variable genes involved in general metabolism and allowed for the clustering of genes that made sense of their evolution and ecology.

The "accessory" sporulation genes are strain-specific and reflect the genomic flexibility of the group that allows *Bacillus* to colonize different environments that require different sensors to trigger the developmental response of the bacteria. This comparative strategy allows for the identification of the variable genes involved in the process and allows for the clustering of groups on the basis of their evolution and ecology.

7. Genomics in the aid of phylogeny

The use of the whole core genome to reconstruct the group's phylogeny helped us understand the evolutionary relationships (Alcaraz et al. 2010) (Fig. 6). As stated above, the core genome is thought to be faithful to the evolutionary history of the analyzed strains, and thus many individual genes' phylogenies are dissected to determine which can be used to define a natural group. Some other studies have analyzed what genes are necessary for a strain to be considered part of the same genus. For example, the cosmopolitan groups are either pathogenic, free-living, or extremophiles, such as *Pseudomonas* (Sarkar & Guttman 2004) and *Bacillus* (Alcaraz et al. 2010). Although the 16S rRNA gene has been traditionally used to draw a phylogeny, the resolution is often lost when working within a genus. Many genes can be concatenated and used to build a phylogeny. In fact, all genes shared by different genomes can be used to reconstruct a phylogeny; in the case of *Bacillus*, we used the 814 genes that constitute the core genome to build a *Bacillus* phylogeny. Most of these genes code for the expected housekeeping genes involved in basic cellular functions, as explained above, which fall into the traditional markers for phylogeny. However, we also included hypothetical conserved genes as well as nontraditional categories, such as genes involved in metabolism and transport mechanisms. The use of the whole core genome to reconstruct a group's phylogeny resolves the evolutionary relationships using the most available information. However, there are drawbacks of a low number of representatives of a given genus and the constant release of new genomes, which will eventually help fill in the gaps in the phylogenetic diversity and will help to understand which genes define a group.

8. Molecular clock

Traditionally, it has been difficult to set absolute dates for diversification of lineage events throughout Earth's history. Before DNA sequencing became widely available, the ages of divergence of the major groups of organisms depended solely on fossil information. Since the fossil record is incomplete, the history of lineages that lacked a rich fossil record, such as bacteria, could not be reconstructed (Kuo & Ochman 2009). When the first amino acid sequences of proteins were analyzed, they seemed to change in a rather constant fashion, which led to the hypothesis that a molecular clock existed in protein evolution. Therefore, molecular dating is based on the assumption that if most variation is neutral, then mutations will become fixed in a lineage at a constant rate that is equal to the mutation rate (Bromham & Penny 2003). One direct consequence of the evolution of a sequence at a rate that is relatively constant is that the genetic difference between any two species is proportional to the time that passed since the species last shared a common ancestor.

If the molecular clock hypothesis holds true, it serves as an extremely useful method for estimating evolutionary timescales. This is of particular value when studying organisms such as bacteria, which have left few traces of their biological history in the fossil record. Phylogenetic trees can be reconstructed that shed light onto the Earth's past. A calibration point is needed to obtain an absolute estimate of the age of a clade, and this can be set by using a known date of divergence, such as fossil information and samples from historical sources as well as paleontological, geological, atmospheric, and climatologic records. Based on that calibration point, the absolute date of a divergence can be estimated based on the number of mutations or substitutions (k) per total length of the sequence (n) per unit of time (Fig. 8), and can then be extrapolated to other parts of the phylogeny (Li 1997). This entire methodology assumes that the rates are equal in all branches of a phylogeny, which is not always the case (Ayala 1999). To deal with rate heterogeneity, several methods have been developed in the past few years that make molecular dating more accurate (Drummond & Rambaut 2007).

Geological evidence suggests that the CCB was a shallow marine environment for most of Earth's history, and when CCB became isolated due to the uplift of the Coahuila block, the bacteria that remained became relics of the diversity from the ancient sea. Thus, the high diversity of CCB would be a product of two things: the diversity already present at the time of isolation, and the new community assemblages that arose in this unique aquatic environment (Moreno-Letelier et al. 2011; Souza et al. 2006). *Firmicutes*, an abundant and widespread genera within CCB (Alcaraz et al. 2010; Alcaraz et al. 2008; Cerritos et al. 2011; Moreno-Letelier et al. 2011; Souza et al. 2006) are the focal group of our studies, in part because there are several endemic species within the site, and because the whole genome has been sequenced for both *Exiguobacterium* and *Bacillus* isolates (*Bacillus coahuilensis* m4-4 (Alcaraz et al. 2008); *Bacillus* sp. m3-13 (Alcaraz et al. 2010); and several unpublished data (from drafts *Bacillus* sp. p15.4, *Exiguobacterium* EPVM and 11-28). We have reconstructed the phylogenetic relationships and estimated the divergence times of aerobic *Firmicutes* from CCB and other similar habitats, in order to determine whether the diversity of the basin is a product of a recent adaptive radiation or the bacteria that remained became relics of the diversity we observe today. For our phylogenetic studies, calibration points were obtained from geologic events. The maximum age of the tree can not exceed the estimate of the origin of life on Earth, which is estimated to be approximately 4000 million years ago (Nisbet & Sleep 2001). The node of aerobic *Firmicutes* is set to have an age of 2300 Ma, the date of the Great Oxidation Event (Battistuzzi et al. 2004; Papineau 2010). CCB bacteria and their sister species were also constrained to have a minimum age of 35 Ma in the case of CCB bacteria and marine bacteria. This age corresponds to the final retreat of the Western Interior Seaway and the uplift of the Sierra Madre Oriental that finally isolated Cuatro Cienegas from the ocean (Ferrusquía-Villafranca & González-Guzmán 2005).

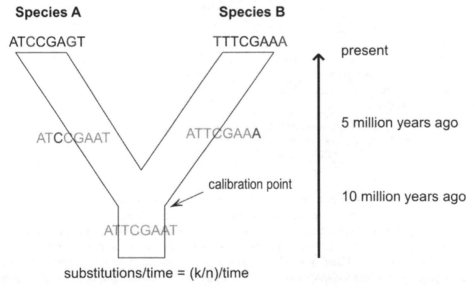

Fig. 8. The concept of the molecular clock based on DNA sequences. A representation of a set of sequences that evolves in a molecular clock fashion, where each branch contains 2 substitutions in 10 million years. For each species k = 2 and n = 8, which equals 0.025 substitutions per million years.

Our results showed that the diversity of *Firmicutes* in Cuatro Cienegas is not the product of a recent adaptive radiation. We suggest instead that ancient lineages became isolated and continued their own evolutionary path. This is especially true for the representatives of *Bacillus*, where the different taxa have diverged form their sister species at different times (Moreno-Letelier et al, under review). Overall, the Cuatro Cienegas Basin reinforces its geological importance since it is a refuge of ancient bacterial lineages, with traits that are relicts of old times, in some cases even from Precambrian times, and others that have arisen in order to survive in this endangered, extreme, and unique environment.

Questions	Experimental strategies
Have we already uncovered all of the microbial diversity in water, sediments, or mats of any pond?	16S rRNA gene libraries, *taxon*-specific libraries, metagenomes
Why is there so much microbial diversity? Does genetic diversity correlate with environmental stability or resource availability, or is it a matter of chance and history?	Explore the reasons for diversity: competition, mesocosm experiments in combination with 16S rRNA gene libraries, *taxon*-specific libraries, metagenomes
In most ponds, microbial mats are in charge of biogeochemical cycles. Therefore, what are the most important functions and are the functions similar in mats from different ponds?	Transcriptome analysis
Organisms from an ancient sea have adapted to the oligotrophic conditions of the ponds. How did they evolve and adapt? How was their genome shaped? What is the role of mobile elements?	Study transposons, plasmids, and phages in genomes and metagenomes
Do communities with extensive genetic diversity also have more functional diversity?	Comparative metagenomics and transriptomics
Do communities respond differently to environmental change?	Mesocosms, transcriptomics

Table 2. Remaining questions on the CCB microbial diversity and evolution, and the genomic approaches to address them.

9. Future directions

The current approaches that combine genomics, transcriptomics, metagenomics, and proteomics together with classical microbiology will continue to contribute to our understanding of microbial activity and strategies for cell survival and growth of bacteria in oligotrophic ecosystems. Additional explorations of the diversity of the bacteria in these ponds using taxon specific oligonucleotides will allow us to determine whether we've actually reached an understanding of diversity with more general approaches, or if some taxons were missed due to the bias of DNA isolation techniques and PCR amplification of 16S rRNA genes with universal primers. Once we confidently identify the members of a community, a future challenge will be to understand the rules of assembly for the organisms within it, including the role of history, competition (for nutrients or antagonic), and migration. In particular, transcriptomics will help us go beyond the identification of the genetic potential of the organisms afforded by the metagenomic approach and will allow us to actually look at the expression of such genes. The presence of specific mRNAs will be the actual reflection of what the microorganisms are sensing from the environment and how they are responding to them. This will be particularly fruitful once we have full genome information for many of the CCB bacterial isolates, since we will be able to assign mRNAs to specific taxons in the communities.

10. Conclusion

The application of genomic approaches to living systems uncovers the genetic bases of functional variation in nature. The revolution in high-throughput DNA sequencing and gene expression technologies has redefined the notion of a 'model' organism. The interrogation of genomes from animals, plants, microbes, or communities of organisms can identify genetic markers of processes at any scale: ecological, physiological, developmental, transcriptional, and others. Challenges lie ahead in the full interpretation of these datasets as well as understanding the connections between the gene information and the metadata of each particular environment. This will require interdisciplinary work between ecologists, microbial geneticists, biogeochemists, and computational biologists.

11. Acknowledgments

Julio Cruz for the Dune's photograph; Eugenio Reynoso for graphic art. G. Olmedo's laboratory team for their comments to the manuscript. This work was supported by a Multidisciplinario project from Cinvestav to GOA and Fondos Mixtos-Conacyt.

12. References

Adams, M. M., Gomez-Garcia, M. R., Grossman, A. R. & Bhaya, D. (2008). Phosphorus deprivation responses and phosphonate utilization in a thermophilic Synechococcus sp. from microbial mats. *J Bacteriol* 190(24): 8171-8184.

Alcaraz, L. D., Moreno-Hagelsieb, G., Eguiarte, L. E., Souza, V., Herrera-Estrella, L., et al. (2010). Understanding the evolutionary relationships and major traits of Bacillus through comparative genomics. *BMC Genomics* 11: 332.

Alcaraz, L. D., Olmedo, G., Bonilla, G., Cerritos, R., Hernandez, G., et al. (2008). The genome of Bacillus coahuilensis reveals adaptations essential for survival in the relic of an ancient marine environment. *Proc Natl Acad Sci U S A* 105(15): 5803-5808.

Amann, R. I., Ludwig, W. & Schleifer, K. H. (1995). Phylogenetic identification and in situ detection of individual microbial cells without cultivation. *Microbiol Rev* 59(1): 143-169.

Andam, C. P.& Gogarten, J. P. (2011). Biased gene transfer in microbial evolution. *Nat Rev Microbiol* 9(7): 543-555.

Anderson, I., Sorokin, A., Kapatral, V., Reznik, G., Bhattacharya, A., et al. (2005). Comparative genome analysis of Bacillus cereus group genomes with Bacillus subtilis. *FEMS Microbiol Lett* 250(2): 175-184.

Ayala, F. J. (1999). Molecular clock mirages. *Bioessays* 21(1): 71-75.

Aziz, R. K., Bartels, D., Best, A. A., DeJongh, M., Disz, T., et al. (2008). The RAST Server: rapid annotations using subsystems technology. *BMC Genomics* 9: 75.

Battistuzzi, F. U., Feijao, A. & Hedges, S. B. (2004). A genomic timescale of prokaryote evolution: insights into the origin of methanogenesis, phototrophy, and the colonization of land. *BMC Evol Biol* 4: 44.

Benning, C. (1998). Biosynthesis and Function of the Sulfolipid Sulfoquinovosyl Diacylglycerol. *Annu Rev Plant Physiol Plant Mol Biol* 49: 53-75.

Breitbart, M., Hoare, A., Nitti, A., Siefert, J., Haynes, M., et al. (2009). Metagenomic and stable isotopic analyses of modern freshwater microbialites in Cuatro Cienegas, Mexico. *Environ Microbiol* 11(1): 16-34.

Bromham, L. & Penny, D. (2003). The modern molecular clock. *Nat Rev Genet* 4(3): 216-224.

Brown, M. R .& Kornberg, A. (2004). Inorganic polyphosphate in the origin and survival of species. *Proc Natl Acad Sci U S A* 101(46): 16085-16087.

Cerritos, R., Eguiarte, L. E., Avitia, M., Siefert, J., Travisano, M., et al. (2011). Diversity of culturable thermo-resistant aquatic bacteria along an environmental gradient in Cuatro Cienegas, Coahuila, Mexico. *Antonie Van Leeuwenhoek* 99(2): 303-318.

Cerritos, R., Vinuesa, P., Eguiarte, L. E., Herrera-Estrella, L., Alcaraz-Peraza, L. D., et al. (2008). Bacillus coahuilensis sp. nov., a moderately halophilic species from a desiccation lagoon in the Cuatro Cienegas Valley in Coahuila, Mexico. *Int J Syst Evol Microbiol* 58(Pt 4): 919-923.

Chaisson, M., Pevzner, P. & Tang, H. (2004). Fragment assembly with short reads. *Bioinformatics* 20(13): 2067-2074.

Desnues, C., Rodriguez-Brito, B., Rayhawk, S., Kelley, S., Tran, T., et al. (2008). Biodiversity and biogeography of phages in modern stromatolites and thrombolites. *Nature* 452(7185): 340-343.

Dinsdale, E. A., Edwards, R. A., Hall, D., Angly, F., Breitbart, M., et al. (2008). Functional metagenomic profiling of nine biomes. *Nature* 452(7187): 629-632.

Dormann, P.& Benning, C. (2002). Galactolipids rule in seed plants. *Trends Plant Sci* 7(3): 112-118.

Drummond, A. J. & Rambaut, A. (2007). BEAST: Bayesian evolutionary analysis by sampling trees. *BMC Evol Biol* 7: 214.

Ehling-Schulz, M., Svensson, B., Guinebretiere, M. H., Lindback, T., Andersson, M., et al. (2005). Emetic toxin formation of Bacillus cereus is restricted to a single evolutionary lineage of closely related strains. *Microbiology* 151(Pt 1): 183-197.

Elser, J. J. (2005). Effects of phosphorus enrichment and grazing snails on modern stromatolitic microbial communities. *Freshwater Biol* 50: 1808–1825.

Elser, J. J. & Hamilton, A. (2007). Stoichiometry and the new biology: the future is now. *PLoS Biol* 5(7): e181.

Elser, J. J., Watts, J., Schampel, J. H. & Farmer, J. (2006). Early Cambrian food webs on a trophic knife-edge? A hypothesis and preliminary data from a modern stromatolite-based ecosystem. *Ecol Lett* 9(3): 295-303.

Escalante, A. E., Caballero-Mellado, J., Martinez-Aguilar, L., Rodriguez-Verdugo, A., Gonzalez-Gonzalez, A., et al. (2009). Pseudomonas cuatrocienegasensis sp. nov., isolated from an evaporating lagoon in the Cuatro Cienegas valley in Coahuila, Mexico. *Int J Syst Evol Microbiol* 59(Pt 6): 1416-1420.

Ferrusquía-Villafranca, I. & González-Guzmán, L. (2005). Northern Mexico's Landscape, Part 2, The biotic setting across space and time. *Biodiversity, Ecosystems and Conservation in Northern Mexico*. J. L. Cartron, G. C.-G. R. F. New York, Oxford University Press. 2: 40-51.

Frigaard, N. U., Martinez, A., Mincer, T. J. & DeLong, E. F. (2006). Proteorhodopsin lateral gene transfer between marine planktonic Bacteria and Archaea. *Nature* 439(7078): 847-850.

Helgason, E., Tourasse, N. J., Meisal, R., Caugant, D. A. & Kolsto, A. B. (2004). Multilocus sequence typing scheme for bacteria of the Bacillus cereus group. *Appl Environ Microbiol* 70(1): 191-201.

Hirota, R., Kuroda, A., Kato, J. & Ohtake, H. (2010). Bacterial phosphate metabolism and its application to phosphorus recovery and industrial bioprocesses. *J Biosci Bioeng* 109(5): 423-432.

Holsinger, J. R.& Minckley, W. L. (1971). A new genus and two new species of subterranean amphipod crustaceans (Gammaridae) from northern Mexico. *Proceedings of the Biological Society of Washington* 83: 25–44.

Khan, M. M., Pyle, B. H. & Camper, A. K. (2010). Specific and rapid enumeration of viable but nonculturable and viable-culturable gram-negative bacteria by using flow cytometry. *Appl Environ Microbiol* 76(15): 5088-5096.

Kuo, C. H. & Ochman, H. (2009). Inferring clocks when lacking rocks: the variable rates of molecular evolution in bacteria. *Biol Direct* 4: 35.

Kurtz, S., Phillippy, A., Delcher, A. L., Smoot, M., Shumway, M., et al. (2004). Versatile and open software for comparing large genomes. *Genome Biol* 5(2): R12.

Li, W. H. (1997). Molecular Evolution Sunderland, Massachusetts, Sinauer Associates.

Medini, D., Donati, C., Tettelin, H., Masignani, V. & Rappuoli, R. (2005). The microbial pan-genome. *Curr Opin Genet Dev* 15(6): 589-594.

Metzker, M. L. (2010). Sequencing technologies - the next generation. *Nat Rev Genet* 11(1): 31-46.

Miller, J. R., Delcher, A. L., Koren, S., Venter, E., Walenz, B. P., et al. (2008). Aggressive assembly of pyrosequencing reads with mates. *Bioinformatics* 24(24): 2818-2824.

Minckley, W. L. (1969). Environments of the Bolson of Cuatro Cienegas, Coahuila, Mexico, with Special Reference to the Aquatic Biota. El Paso, TX, Texas Western Press,.

Moreno-Letelier, A., Olmedo, G., Eguiarte, L. E., Martinez-Castilla, L. & Souza, V. (2011). Parallel Evolution and Horizontal Gene Transfer of the pst Operon in Firmicutes from Oligotrophic Environments. *Int J Evol Biol* 2011: 781642.

Nelson, K. E., Mullany, P., Warburton, P. & Allan, E. (2011). Metagenomics of the Human Body. K. E. Nelson, E. New York, NY: . 165-173-.

Nelson, K. E., Weinstock, G. M., Highlander, S. K., Worley, K. C., Creasy, H. H., et al. (2010). A catalog of reference genomes from the human microbiome. *Science* 328(5981): 994-999.

Nisbet, E. G. & Sleep, N. H. (2001). The habitat and nature of early life. *Nature* 409(6823): 1083-1091.

Papineau, D. (2010). Global biogeochemical changes at both ends of the proterozoic: insights from phosphorites. *Astrobiology* 10(2): 165-181.

Qi, Y., Kobayashi, Y. & Hulett, F. M. (1997). The pst operon of Bacillus subtilis has a phosphate-regulated promoter and is involved in phosphate transport but not in regulation of the pho regulon. *J Bacteriol* 179(8): 2534-2539.

Reno, M. L., Held, N. L., Fields, C. J., Burke, P. V. & Whitaker, R. J. (2009). Biogeography of the Sulfolobus islandicus pan-genome. *Proc Natl Acad Sci U S A* 106(21): 8605-8610.

Rusch, D. B., Halpern, A. L., Sutton, G., Heidelberg, K. B., Williamson, S., et al. (2007). The Sorcerer II Global Ocean Sampling expedition: northwest Atlantic through eastern tropical Pacific. *PLoS Biol* 5(3): e77.

Sarkar, S. F. & Guttman, D. S. (2004). Evolution of the core genome of Pseudomonas syringae, a highly clonal, endemic plant pathogen. *Appl Environ Microbiol* 70(4): 1999-2012.

Scanlan, D. J., Mann, N. H. & Carr, N. G. (1993). The response of the picoplanktonic marine cyanobacterium Synechococcus species WH7803 to phosphate starvation involves a protein homologous to the periplasmic phosphate-binding protein of Escherichia coli. *Mol Microbiol* 10(1): 181-191.

Schmidt, R. H. J. (1979). A climatic delineation of the 'real' Chihuahuan Desert. *J. Arid Environ* 2: 243-250.

Souza, V., Espinosa-Asuar, L., Escalante, A. E., Eguiarte, L. E., Farmer, J., et al. (2006). An endangered oasis of aquatic microbial biodiversity in the Chihuahuan desert. *Proc Natl Acad Sci U S A* 103(17): 6565-6570.

Tettelin, H., Masignani, V., Cieslewicz, M. J., Donati, C., Medini, D., et al. (2005). Genome analysis of multiple pathogenic isolates of Streptococcus agalactiae: implications for the microbial "pan-genome". *Proc Natl Acad Sci U S A* 102(39): 13950-13955.

Tetu, S. G., Brahamsha, B., Johnson, D. A., Tai, V., Phillippy, K., et al. (2009). Microarray analysis of phosphate regulation in the marine cyanobacterium Synechococcus sp. WH8102. *ISME J* 3(7): 835-849.

Turnbaugh, P. J., Ley, R. E., Hamady, M., Fraser-Liggett, C. M., Knight, R., et al. (2007). The human microbiome project. *Nature* 449(7164): 804-810.

Van Domselaar, G. H., Stothard, P., Shrivastava, S., Cruz, J. A., Guo, A., et al. (2005). BASys: a web server for automated bacterial genome annotation. *Nucleic Acids Res* 33(Web Server issue): W455-459.

White, A. K. & Metcalf, W. W. (2004). Two C-P lyase operons in Pseudomonas stutzeri and their roles in the oxidation of phosphonates, phosphite, and hypophosphite. *J Bacteriol* 186(14): 4730-4739.

Woese, C. R. & Fox, G. E. (1977). Phylogenetic structure of the prokaryotic domain: the primary kingdoms. *Proc Natl Acad Sci U S A* 74(11): 5088-5090.

Wolaver, B., Sharp, JM, Rodríguez JM. and Ibarra JC (2008). Delineation of Regional Arid Karstic Aquifers: An Integrative Data Approach. *Ground Water* 46(3): 396–413.

Zerbino, D. R. & Birney, E. (2008). Velvet: algorithms for de novo short read assembly using de Bruijn graphs. *Genome Res* 18(5): 821-829.

Zubkov, M. V., Mary, I., Woodward, E. M. S., Warwick, P. E., Fuchs, B. M., et al. (2007). Microbial control of phosphate in the nutrient depleted North Atlantic subtropical gyre. *Environmental Microbiology* 9(8): 2079-2089.

Permissions

The contributors of this book come from diverse backgrounds, making this book a truly international effort. This book will bring forth new frontiers with its revolutionizing research information and detailed analysis of the nascent developments around the world.

We would like to thank Imran Ahmad Dar and Mithas Ahmad Dar, for lending their expertise to make the book truly unique. They have played a crucial role in the development of this book. Without their invaluable contribution this book wouldn't have been possible. They have made vital efforts to compile up to date information on the varied aspects of this subject to make this book a valuable addition to the collection of many professionals and students.

This book was conceptualized with the vision of imparting up-to-date information and advanced data in this field. To ensure the same, a matchless editorial board was set up. Every individual on the board went through rigorous rounds of assessment to prove their worth. After which they invested a large part of their time researching and compiling the most relevant data for our readers. Conferences and sessions were held from time to time between the editorial board and the contributing authors to present the data in the most comprehensible form. The editorial team has worked tirelessly to provide valuable and valid information to help people across the globe.

Every chapter published in this book has been scrutinized by our experts. Their significance has been extensively debated. The topics covered herein carry significant findings which will fuel the growth of the discipline. They may even be implemented as practical applications or may be referred to as a beginning point for another development. Chapters in this book were first published by InTech; hereby published with permission under the Creative Commons Attribution License or equivalent.

The editorial board has been involved in producing this book since its inception. They have spent rigorous hours researching and exploring the diverse topics which have resulted in the successful publishing of this book. They have passed on their knowledge of decades through this book. To expedite this challenging task, the publisher supported the team at every step. A small team of assistant editors was also appointed to further simplify the editing procedure and attain best results for the readers.

Our editorial team has been hand-picked from every corner of the world. Their multi-ethnicity adds dynamic inputs to the discussions which result in innovative outcomes. These outcomes are then further discussed with the researchers and contributors who give their valuable feedback and opinion regarding the same. The feedback is then collaborated with the researches and they are edited in a comprehensive manner to aid the understanding of the subject.

Apart from the editorial board, the designing team has also invested a significant amount of their time in understanding the subject and creating the most relevant covers. They scrutinized every image to scout for the most suitable representation of the subject and create an appropriate cover for the book.

The publishing team has been involved in this book since its early stages. They were actively engaged in every process, be it collecting the data, connecting with the contributors or procuring relevant information. The team has been an ardent support to the editorial, designing and production team. Their endless efforts to recruit the best for this project, has resulted in the accomplishment of this book. They are a veteran in the field of academics and their pool of knowledge is as vast as their experience in printing. Their expertise and guidance has proved useful at every step. Their uncompromising quality standards have made this book an exceptional effort. Their encouragement from time to time has been an inspiration for everyone.

The publisher and the editorial board hope that this book will prove to be a valuable piece of knowledge for researchers, students, practitioners and scholars across the globe.

List of Contributors

Zornitza Tosheva, Harald Hofmann and Antoine Kies
University of Luxembourg, Luxembourg
Monash University, Australia

Fred Kamona
University of Namibia, Namibia

Lin-Chong Huang and Cui-Ying Zhou
Sun Yat-sen University, China

Fereshte Haghighi
Soil Conservation and Watershed Management Institute, Tehran, Iran

Mirmasoud Kheirkhah and Bahram Saghafian
Soil Conservation and Watershed Management Research Institute, Tehran, Iran

Monika Zovko and Marija Romić
University of Zagreb, Faculty of Agriculture, Croatia

Andreas Laake
WesternGeco Cairo, Egypt

Mohammad Ali Zare Chahouki
Department of Rehabilitation of Arid and Mountainous Regions, University of Tehran, Iran

Jiří Chlachula
Laboratory for Palaeoecology, Tomas Bata University in Zlín, Czech Republic

Fresia Ricardi Branco and Sueli Yoshinaga Pereira
D. Geologia e Recursos Naturais, Instituto de Geociências, Universidade Estadual de Campinas -UNICAMP, Campinas, SP, Brazil

Fabio Cardinale Branco
Environmentality, São Paulo, SP, Brazil

Paulo R. Brum Pereira
Instituto Florestal, SP, Brazil

Marco Avanzini and Massimo Bernardi
Science Museum, Trento, Italy

Umberto Nicosia
Sapienza University of Rome, Italy

Luis David Alcaraz
Department of Genomics and Health, Center for Advanced Research in Public Health, Valencia, Spain

Varinia López-Ramírez and Gabriela Olmedo-Alvarez
Departamento de Ingeniería Genética, Cinvestav Unidad Irapuato, Mexico

Alejandra Moreno-Letelier
Division of Biology, Imperial College London, Silwood Park Campus, Ascot, UK

Luis Herrera-Estrella
Langebio, Cinvestav, Mexico

Valeria Souza
Departamento de Ecologia Evolutiva, Instituto de Ecologia, Universidad Nacional Autónoma de México, Mexico

Printed in the USA
CPSIA information can be obtained
at www.ICGtesting.com
JSHW011443221024
72173JS00004B/924